ENVIRONMENTAL SCIENCES LIBRARY

Scientific Uncertainty and Environmental Problem Solving

Scientific Uncertainty and Environmental Problem Solving

edited by

John Lemons
Professor of Biology and Environmental Science
Department of Life Sciences
University of New England
Biddeford, Maine

b

Blackwell
Science

Blackwell Science

EDITORIAL OFFICES:
238 Main Street, Cambridge,
  Massachusetts 02142, USA
Osney Mead, Oxford OX2 0EL,
  England
25 John Street, London WC1N 2BL,
  England
23 Ainslie Place, Edinburgh EH3 6AJ,
  Scotland
54 University Street, Carlton, Victoria
  3053, Australia
Arnette Blackwell SA, 1 rue de Lille,
  75007 Paris, France
Blackwell Wissenschafts-Verlag GmbH
  Kurfürstendamm 57, 10707 Berlin,
  Germany
  Feldgasse 13, A-1238 Vienna,
  Austria

DISTRIBUTORS:
USA
Blackwell Science, Inc.
238 Main Street
Cambridge, Massachusetts 02142
(Telephone orders: 800-215-1000 or
  617-876-7000)

Canada
Oxford University Press
70 Wynford Drive
Don Mills, Ontario M3C 1J9
(Telephone orders: 416–441–2941)

Australia
Blackwell Science Pty, Ltd.
54 University Street
Carlton, Victoria 3053
(Telephone orders: 03-9347-0300;
  fax orders 03-9349-3016)

**Outside North America and
  Australia**
Blackwell Science, Ltd.
c/o Marston Book Services, Ltd.
P.O. Box 269
Abingdon
Oxon OX14 4YN
England
(Telephone orders: 44-01235-465500;
  fax orders 44-01235-465555)

Acquisitions: Jane Humphreys
Production: Irene Herlihy
Printed and bound by
  Edwards Brothers, Inc.

© **1996 by Blackwell Science, Inc.**

Printed in the United States of America

96 97 98 99 5 4 3 2 1

*Library of Congress Cataloging in
  Publication Data*

Lemons, John.
    Scientific uncertainty and
environmental problem solving /
John Lemons.
        p.   cm. — (Environmental
    sciences library)
    Includes bibliographical
references and index.
    ISBN 0-86542-476-4 (alk. paper)
    1. Environmental sciences—
Decision making—Case studies.
2. Problem solving. 3. Uncertainty
(Information theory)   I. Title.
II. Series.
GE105.L46   1995
363.7'05—dc20   95-36595
                                    CIP

To the people and places of the Eastern Sierra Nevada that I have known.

-J.L.

Contents

Contributors

Donald A. Brown
Director, Bureau of Hazardous Sites and Super Fund Enforcement, Commonwealth of Pennsylvania, Pennsylvania Department of Environmental Resources, Harrisburg, Pennsylvania

John Cairns, Jr.
University Distinguished Professor of Environmental Biology, Department of Biology, Director, University Center for Environmental and Hazardous Materials Studies, Virginia Polytechnic Institute and State University, Blacksburg, Virginia

Lynton K. Caldwell
Arthur F. Bentley Professor of Political Science, Emeritus, and Professor of Public and Environmental Affairs, Indiana University, Bloomington, Iowa

Larry W. Canter
George Lynn Cross Research Professor, Sun Company Professor of Ground Water Hydrology, Director, Environmental and Ground Water Institute, University of Oklahoma, Norman, Oklahoma

Richard A. Carpenter
Environmental Consultant, Charlottesville, Virginia

Charles F. Cole
Professor of Natural Resources, Emeritus, The School of Natural Resources, The Ohio State University, Integrated Ecological Management Inc.,Worthington, Ohio

Mahmoud Kh. El-Sayed
Professor of Marine Geology, Department of Oceanography, Faculty of Science, Alexandria University, Alexandria, Egypt

Carl F. Jordan
Senior Ecologist, School of Ecology, University of Georgia, Athens, Georgia

John Lemons
Professor of Biology and Environmental Science, Department of Life Sciences, University of New England, Biddeford, Maine

Christopher Miller
School of Ecology, University of Georgia, Athens, Georgia

James Perry
Professor, College of Natural Resources, University of Minnesota, St. Paul, Minnesota

Kristin Shrader-Frechette
Distinguished Research Professor, Environmental Sciences and Policy Program, Department of Philosophy, University of South Florida, Tampa, Florida

Eric P. Smith
Professor, Department of Statistics, University Center for Environmental and Hazardous Materials Studies,Virginia Polytechnic Institute and State Univeristy, Blacksburg, Virginia

Elizabeth Leigh Vanderklein
Research Associate, College of Natural Resources, University of Minnesota, St. Paul, Minnesota

Judith S. Weis
Professor, Department of Biological Sciences, Rutgers University, Newark, New Jersey

Patrick Zaepfel
Assistant Counsel, Bureau of Hazardous Sites and Super Fund Enforcement, Commonwealth of Pennsylvania, Pennsylvania Department of Environmental Resources, Harrisburg, Pennsylvania

Acknowledgments

The idea for this book arose as a result of the influence of the work of people such as Donald A. Brown, John Cairns, Jr., Charles Malone, and Kristin Shrader-Frechette. Throughout much of their careers, all of these individuals have struggled with the scientific and public policy implications of uncertainty.

I also would like to thank Brenda Smith of the Department of Life Sciences who has provided invaluable assistance in communicating with contributors and in keeping my files organized despite my efforts to the contrary.

Introduction

John Lemons

Understanding and solving environmental problems requires application of the best scientific knowledge available. However, as environmental problems have become more complex it is questionable whether or to what extent our scientific knowledge is adequate to inform environmental decision making with reasonable certainty. Although scientific and other uncertainty about environmental problems is acknowledged by most scientists and managers, there have been too few critical analyses of uncertainty and its implications for specific environmental problems. The consequence of this is threefold.

First, decision makers often base their decisions, in part, upon calculations and conclusions drawn from such fields as risk assessment and cost-benefit analysis. Scientific information often is used as the basis for such calculations and conclusions. Under conditions of scientific uncertainty, especially when scientists do not provide an explicit and rigorous treatment of scientific uncertainty in their analysis of specific problems, environmental decision makers and managers often accept scientific analyses of environmental problems as being more factual than is warranted. Such acceptance can bias the outcomes of procedures used to make decisions, such as in risk assessment and in economic cost-benefit calculations. Further, this bias often is not apparent to many environmental decision makers, resource or agency managers, or the general public unless they are made aware of the sources of the uncertainties and their implications.

Second, most environmental laws and regulations place the burden of proof for demonstrating human health or environmental harm on governmental regulatory agencies or others wishing to demonstrate harm from development or technological activities. The standard which is used to meet the burden of proof test often is the normal standard of scientific proof, such as a 95% confidence level or an equivalent criterion. This standard has been adopted by members of the scientific community in order to minimize type I error and therefore reduce speculation in the interpretation of scientific data. When this standard is adopted as a basis for environmental decisions the scientific uncertainty which pervades most environmental problems means that the burden of proof usually is not met, despite the fact that some information might indicate the existence of environmental or human health harm.

Third, the absence of a more complete understanding of scientific uncertainty and its implications for environmental decisions means that decision makers and managers will not have adequate information to guide them in terms of whether or to what extent decisions should reflect a precautionary approach. What needs to be recognized is that when decisions to protect the environment or human health are delayed in order to obtain more reasonable scientific certainty, this is tantamount to a decision to continue with status quo behaviors and activities that either are or might be causing harm. In order to avoid this situation, public-policy makers and decision makers must therefore seek ways to deal with scientific uncertainty in a more effective manner in order to avoid continued harm to the environment and human health.

The purpose of this book is to use case studies of specific environmental problems in order to examine the nature of scientific uncertainty and some of its implications for environmental problem solving. Obviously, only a relatively few examples of the many types of environmental or human health problems can be provided. Nevertheless, it is hoped that general principles and findings can be extrapolated to other situations.

In her chapter "Methodological Rules for Four Classes of Scientific Uncertainty," Shrader-Frechette analyzes four types of scientific uncertainty that are of special relevance to environmental decision making: (a) framing uncertainty, (b) modeling uncertainty, (c) statistical uncertainty, and (d) decision-theoretic uncertainty. One of the most important issues raised by the first type of uncertainty is how to frame problems of uncertainty, such as whether one should use a two-valued epistemic logic or a three-valued epistemic logic. In other words, what are the consequences of framing the solutions to particular problems in terms of two solution options or three options? Often, environmental impact assessors assume that a two-valued epistemic logic, according to which a particular solution is acceptable or unacceptable, is an adequate way to frame a particular problem. However, in other situations problem solvers frame their solution options in terms of a three-valued epistemic logic, in terms of a particular solution's being acceptable, unacceptable, or there not being adequate data to judge acceptability one way or the other.

In numerous cases of controversial environmental decision making, such as siting hazardous and nuclear waste facilities, decision makers and scientists have framed problems in terms of an inadequate two-valued logic. In ignoring the three-valued epistemic logic, they have thereby made logically fallacious appeals to ignorance. Shrader-Frechette compares and contrasts two-valued and three-valued epistemic logic and provides decision rules for when applied-problem solvers should use the respective types of epistemic logic to frame a scientific problem involving uncertainty.

Shrader-Frechette also notes that in a variety of situations involving modeling, uncertainty arises in virtually all of the environmental sciences. She addresses the frequent use of computer models in situations of hydrologic, geologic, and ecological problem solving. Often, persons who use computer models of a particular situation will assume that their models are able to validate or verify a particular conclusion. Shrader-Frechette provides some useful methodological rules for when allegedly validating and verifying models have predictive power as opposed to when they have heuristic power. These rules should be of use to modelers who often claim their models have predictive power when they do not. As a consequence of these claims, they tend to underestimate the environmental damages likely to arise from particular scientific decisions.

Virtually any scientist who uses statistical techniques faces the problem of when, in a situation of uncertainty, to minimize type I, versus type II, statistical error. Recent evidence suggests that the typical scientific practice of avoiding type I error tends to facilitate environmental damage. Shrader-Frechette also analyzes the use and implications of minimizing type I and type II statistical errors, and suggest rules which scientists might use in deciding whether and when to minimize type I, rather than type II, error in situations of scientific uncertainty. In her analysis, she notes that minimizing type II error errs on the side of protecting natural resources.

Finally, many scientists who do probabilistic risk assessment use various decision-theoretic rules for dealing with probabilistic uncertainty. One of the crucial questions faced by such risk assessors is when to use a maximin decision rule in a situation of uncertainty and when to use a rule of maximizing average utility. It is easy to show that in probabilistic risk assessments of, for example, nuclear fission, when one uses the expected-utility rule, one adopts a benevolent attitude toward technology and facilitates environmental damage, whereas when one uses a maximin rule one adopts a more conservative position regarding environmental damage and the acceptability of a particular technological risk. In her chapter Shrader-Frechette outlines various rules for using maximin, versus expected utility, decision strategies in situations of scientific uncertainty.

Because all four types of uncertainty are common to many environmental sciences, arriving at rules for dealing with them will be useful for a variety of scientists and decision makers. Moreover, because so many decisions in situations of scientific uncertainty have been questionable in precisely these four areas of uncertainty, Shrader-Frechette concludes her analysis by arguing for significant reforms.

One of the most scientifically, technologically, and politically complex and controversial issues confronting the United States during the past several decades is how and where to dispose of high-level nuclear waste. In ''Uncertainties in the Disposal of High-Level Nuclear Waste,'' John Lemons assesses the use of science in the decision to characterize Yucca Mountain, Nevada, as the United States' only high-level nuclear waste repository and the capabilities of science to yield reasonably certain information about future repository performance. His analysis includes (a) a brief description of options for storage and disposal of high-level nuclear waste, (b) a brief description of health and environmental protection criteria for high-level nuclear waste disposal, and (c) a retrospective examination of the role of science and scientific uncertainties in public-policy decisions for siting the repository at Yucca Mountain. Following this analysis, he discusses (a) performance assessment for high-level nuclear waste disposal sites, (b) scientific uncertainties inherent in the performance assessment of the planned high-level nuclear waste repository at Yucca Mountain, (c) the relationship between scientific uncertainty and protection standards, and (d) the implications of scientific knowledge to distributional problems posed by the location of the high-level nuclear waste site at Yucca Mountain.

In their chapter "Scientific Uncertainty as a Constraint to Environmental Problem Solving: Large-Scale Ecosystems,'' Jordan and Miller conclude that although ecological processes often can be quantified under controlled conditions, the extreme variability of the world outside of laboratory settings limits our ability to improve predictions of ecosystem-scale responses to human-induced environmental perturbations. Except for a few cases such as global warming and marine fisheries, where underlying mechanisms are understood poorly, additional studies likely will not increase our ability to predict the consequences of human activities upon the environment. Jordan and Miller base their conclusion on their examination of the nature of facts and probabilities in various scientific disciplines and their assessment of the predictive ability of experiments in ecology. Based upon case studies from the Florida Everglades, oil spills, national parks, and biological reserves, Jordan and Miller also conclude that politicians often take advantage of uncertainty surrounding our knowledge of ecological consequences in order to avoid supporting regulatory proposals to enhance environmental quality. Finally, they conclude that because scientific uncertainty regarding ecological phenomena likely will continue to be pervasive, decision making must by necessity be based upon limited knowledge.

In his chapter "Uncertainty in Managing Ecosystems Sustainably,'' Richard Carpenter documents the state of science for monitoring and predicting sustainability status and trends in managing important production ecosystems in agriculture, forestry, fisheries, and grazing lands. Carpenter begins his discussion by noting that ecosystems change and evolve through a highly variable set of quasi-stable conditions over time. Within larger constraints of scale, they are self-controlled insofar that they may evolve toward increasing adaptability to environmental change and that adaptability is maximized when the system has evolved to the verge of chaos.

In order to utilize ecosystems even in a sustainable manner, humans seek to impose some constancy and dependability of supply of needed products from ecosystems. Self-organizing systems evolve in a nondeterministic manner that changes their functional and structural attributes over time. The inference is that ecosystems, being far from equilibrium, and with future states highly dependent on random perturbations of initial conditions, are chaotic. If the future is inherently unknowable, then technological control of ecosystems based on objective, scientific prediction of system behavior in response to alternative management might be impossible or counterproductive in the long term. Consequently, sustainability is inherently uncertain and even its probability is uncertain. Sustainability therefore may be only relative, not absolute; a given management practice may be more or less sustainable at a certain location over a certain time with a certain mix of products. Complicating problems of understanding and maintaining sustainable ecosystems is the problem of natural variation, wherein any perturbations in natural ecosystems due to the activities of humans are often masked by the natural changes in the attributes measured.

Carpenter notes that researchers have questioned recently whether basic ecological research, with its long-term experiments, difficulties of replication, control, randomization, and evolutionary change in the systems under study, can test hypotheses and produce useful results quickly enough to serve as a basis for management. Essentially, the emerging view is that there is no way to learn how an ecosystem will respond to stress except to stress it under appropriate time and scale conditions. However, it may not be possible to achieve a 95% confidence level, which generally is regarded as an essential scientific standard. Consequently, the usual means of dealing with uncertainty through decision theory and models may not work in such a near-chaos situation.

Building upon these themes, Carpenter explores the following categories of uncertainty as applied to management of sustainable ecosystems: (a) measurement problems due to insufficient observations and natural variability, (b) problems of extrapolating from the scale in time and space of observation to the scale of management, and (c) inadequacy of models and fundamental knowledge of underlying ecosystem mechanisms. He concludes with a discussion of the reasons why even though more appropriate and better measurements, data, and extrapolation errors sometimes can be obtained, the more important and problematic environmental problems suffer from true uncertainty or indeterminacy with an unknown probability.

Using case studies focusing on the ocean dumping of sewage sludge, acid precipitation, estuarine eutrophication, and pesticides such as tributyltin, Judith Weis analyzes the successes and failures of science to influence environmental policy in her chapter "Scientific Uncertainty and Environmental Policy: Four Pollution Case Studies." She assesses the extent to which monitoring approaches and field studies (a) can detect subtle or actual effects in unequivocal terms, (b) likely will detect ecosystem effects prior to the effects being no longer subtle, (c) can link in situ effects with specific causes or pollutants, and (d) can provide a sufficient basis to explain fully the large-scale changes noted in fish or plankton populations, fish disease occurrences, or reduced oxygen conditions. Weis's findings suggest that only one of the four case studies can be considered to be a clear success, and she further concludes with recommendations on how scientific uncertainty might be used by applied-problem solvers to justify precautionary principles that might be more protective of the environment.

The developing field of ecotoxicology or environmental toxicology is the offspring of mammalian toxicology in which extrapolations were made from surrogate species to humans. As John Cairns, Jr. and Eric Smith note in their chapter "Uncertainties Associated with Extrapolating from Toxicological Responses in Laboratory Systems to the Responses of Natural Systems," this approach is laden with uncertainties because each species differs physiologically from other species. They also note that the field has developed further as a "bottom-up" approach in which extrapolations are made from single species to ecosystems. However, both ecological and statistical uncertainties are associated with this approach as illustrated by assumptions about the shape of the distribution of tolerance to a toxicant, the ability to extrapolate information on laboratory species and conditions to field species and conditions, and to the conditions of communities and ecosystems.

Additionally, assumptions are made on the appropriateness of laboratory response thresholds with regard to their correspondence to ecosystem thresholds. Although validation of hypotheses, models, and assumptions is customary in the scientific process, generally speaking this step has been disregarded in the field of ecotoxicology, at least where the extrapolations from one level of biological organization to another are concerned. Finally, Cairns and Smith analyze some of the ways in which uncertainties associated with both top-down and bottom-up approaches can be reduced. They also make recommendations on how the validation process might belatedly be integrated into the field of ecotoxicology.

Because of the variety of values associated with biodiversity, worldwide concern has arisen over its loss, which has primarily been studied in terms of species extinction. In his chapter "The Conservation of Biodiversity: Scientific Uncertainty and the Burden of Proof," John Lemons discusses the difficulties that scientists have in determining the extent to which the loss of species has been caused by human impacts. In addition to the uncertainty surrounding the rate of species extinction and the factors responsible for it, other uncertainties pose difficulties to the enhanced protection of biodiversity. Many of these have to do with knowing the spatial and temporal patterns of diversity, how to define diversity more precisely and measure it more accurately, and which theories of the mechanisms influencing biodiversity are more explanatory or robust. In addition, much uncertainty surrounds the question of how to accurately determine the minimum viable population required to minimize the risks of extinction for a threatened or endangered species.

In general, science is called on to develop methods and technologies for the conservation of biodiversity and, more recently, the sustainable use of biological resources. The recovery of endangered species and the restoration of damaged ecosystems also will demand the expertise of scientists. However, the conservation of biodiversity is not solely a scientific problem; it also involves public policy, economic, and ethical decisions whose outcome often depends on how questions regarding the scientific uncertainty pervading biodiversity issues are resolved. Lemons demonstrates this by describing some of the areas of scientific and other uncertainty pertaining to the conservation of biodiversity, including examples drawn from (a) the calculation of minimum viable populations (MVP), (b) conservation efforts focused on community, ecosystem, or landscape levels, (c) the status of ecology to provide knowledge of impacts of human activities upon ecological resources on which species depend, (d) concepts of ecological integrity used to promote protection of biodiversity, (e) value-laden methods of ecology, and (f) value-laden conservation legislation. Following a discussion of these examples, he then discusses in general terms (a) some possible roles for science in biodiversity decision making and (b) implications of scientific uncertainty in fulfilling burden of proof requirements of conservation legislation.

Charles Cole, in his chapter "Can We Resolve Uncertainty in Marine Fisheries Management?," analyzes how uncertainty in the decision-making process for marine fisheries has developed by focusing on five critical periods: (a) pre-Colonial times to 1945; (b) growing international problems from 1945 to 1965; (c) the successes of the 200-mile limit theory, 1976 to 1982; (d) 1982 to present; and (e) marine fisheries within the 3-mile limit. Drawing upon these historic periods, he discusses how concepts pertaining to natural resources, views of nature, and scientific knowledge are not concepts found in nature waiting to be discovered but instead are matters of social construction and of human creation. Likewise, modern technology did not just develop spontaneously; it evolved within specific political and economic contexts to meet human needs and wants.

Cole's historical analysis of environmental change in marine fisheries demonstrates the evolution of scientific paradigms over time and how humans have constructed knowledge to satisfy their needs and wants within certain political economic systems. These constructions of knowledge have provided us, in part, with what we know scientifically about resources as well as what we do not know. Environmental professionals who wish to deal better with problems of scientific uncertainty ignore environmental history at their own peril because it is impossible for us to understand where we are or where we are heading in terms of resource use unless we know where we have been. Building upon these themes, Cole makes recommendations on how to factor historic and social considerations into marine fisheries legislation and decision making.

In "Scientific Uncertainty and Water Resources Management," Larry Canter analyzes how water resources management encompasses both quantitative and qualitative considerations of surface and groundwater systems, including pollution prevention and control, the need for remediation, and the use of natural assimilation capacities. Because many industrial and public-works development projects can cause changes in both flow and quality parameters, Canter analyzes why and how it is important to understand the various components of water resources management in view of uncertainties related to measurement techniques, data, modeling, and decision making. He then concludes with a discussion of implications relative to effective surface and groundwater management.

In his chapter "Scientific Uncertainty and the Environmental Impact Assessment Process in the United States," Larry Canter discusses how the environmental impact assessment process can be viewed as consisting of a number of steps or activities, including (a) describing a project, (b) identifying pertinent laws and regulations, (c) identifying potential impacts, (d) describing the affected environment, (e) impact prediction, (f) impact assessment, (g) impact mitigation, (h) evaluating alternatives and selecting the proposed action, (i) preparation of written documentation, and (j) environmental monitoring.

Generally speaking, the environmental impact assessment process focuses on what is known. Often, there is little explicit recognition of what is not known or how to address the unknown in the assessment process. Consequently, there is a need to critically analyze scientific and other uncertainties related to each step of the assessment process with a view to making conclusions on how more appropriate recognition of uncertainty and the inclusion of safeguards within the overall environmental impact assessment process can be brought about to minimize the consequences of uncertainties. Canter includes specific examples of actual impact studies as one basis for critical analysis and presentation of information.

Increasingly, many developing countries are placing greater importance on the requirements of the environmental impact process to provide information on activities that may have adverse consequences to human health or the environment, either at the national, regional, or global levels. Mahmoud Kh. El-Sayed discusses in his chapter "Implications of Scientific Uncertainty for Environmental Impact Assessment: The International Environment" how the scientific and technical aspects of the environmental impact assessment process largely are based on perceived certainties regarding available data, information, and approaches. When uncertainties are inherent in proposed activities, state-of-the-art environmental impact assessment procedures need to develop methods and techniques to factor consideration of the uncertainties into the procedures to a greater extent. This probably will require methods to develop parameters or boundaries around our scientific information in order to provide levels of confidence in the information or the development of precautionary approaches for assessment of the consequences of activities in order to take into account the uncertainties. It also might require revision of legislation and procedures of the environmental impact assessment process in relation to scientific uncertainty. The need to include a greater consideration of scientific uncertainty into the environmental assessment process is especially great for activities that threaten global resources. El-Sayed focuses on these issues by analyzing a specific case study for a global environmental problem, namely, assessing the consequences of climate change for the Nile River Delta and Alexandria in Egypt.

The intended end product of scientific research applied to a major environmental problem is usually a policy decision. Environmental problems characteristically are "solved" at two levels, the scientific and political. Because major environmental problems are broadly inclusive geographically and socially, their solutions, implied or intended, usually are expressed through government. Some environmental problem solving can be expressed through strictly scientific findings. These environmental problems may, but often do not, directly imply policy decisions (e.g., problems in taxonomy, entomology, paleontology). Solutions to environmental problems involving relationships between human society and its ambient environment are seldom strictly scientific because human culture influences the way in which science is used.

James Perry and Leigh Vanderklein focus on an interesting aspect of uncertainty, which has to do with the role of communication technology and increased information in the decision-making process. In the chapter "Environmental Problem Solving in an Age of Electronic Communications: Toward an Integrated or Reductionist Model?," they note that recent developments in communication technology have increased the ease in which individuals, organizations, industries, and governmental agencies can access massive amounts of information about environmental problems. While these developments generally have been viewed as being positive, there is beginning to be a recognition that the growth of communication technology can, in some ways, serve as an impediment to sound environmental decision making.

For example, environmental problems often are controversial. Parties on different sides of an issue obtain large amounts of data quickly and inexpensively, and they often use these data with confidence to support their positions. The advantages of obtaining data quickly and inexpensively is that more parties can have access to information and hence participate in decisions that affect their lives and the environment. On the other hand, an understudied negative aspect to this situation is that it is difficult for any party to have an adequate understanding of any particular issue, because the data base is so large. Consequently, it becomes very difficult to ascertain more valid information from the less valid. In this situation, massive amounts of information might increase our uncertainty about environmental problems rather than decrease it, which in turn impedes effective decision making. Perry and Vanderklein illustrate not only that understudied problems derived from access to large amounts of information exist, but they also show ways in which information technologies can be used to promote decision making more effectively.

Environmental problems are complex, and the environmental sciences are limited in their ability to predict with reasonable certainty future impacts of actions. As Donald Brown and Patrick Zaepfel demonstrate in their chapter ''The Implications of Scientific Uncertainty for Environmental Law,'' this situation makes it difficult for the party with the burden of proof to sustain that burden if challenged by others who assert that the burden of proof has not been met. For example, scientists are very skilled in exposing technical weaknesses in an adversary's position when the adversary has the burden of proof. As a result, a governmental agency is often troubled about what decisions to make when its knowledge is likely to be viewed to be inadequate but when it is under pressure to take action, as, for example, in recent decisions allowing for harvesting of timber in the northwestern United States at the expense of the spotted owl.

Given the imprecise nature of the environmental sciences and their lack of predictive ability, scientific and technical experts within government agencies are reluctant to take strong regulatory action on controversial issues out of fear they will alienate politicians and higher-level administrators of mission-oriented agencies. Scientific experts also are trained to be very conservative when positing cause-and-effect relationships between, say, actions and environmental impacts; they do not act quickly if reasonable uncertainty exists about such relationships. In this manner, they avoid being discredited by their peers if an original position is discredited by subsequent scientific research.

The consequence of placement of the burden of proof is that those advocating environmental protection must demonstrate with reasonable certainty that human health, species, or ecosystems are threatened, the causes of the threats, and that recommended solutions have a reasonable chance of success. As Brown and Zaepfel demonstrate, placement of this burden of proof often imposes a requirement for a level of environmental knowledge that may not be possible to meet. Consequently, if precautionary principles are to be applied to environmental protection measures, the burden of proof may have to be shifted so that those proposing to undertake significant activities that potentially threaten environmental harm would have to demonstrate that adverse impacts from such activities are not likely.

Finally, in ''Science Assumptions and Misplaced Certainty in Natural Resources and Environmental Problem Solving,'' Lynton Caldwell considers the influence of scientific uncertainty or unwarranted assumptions on the process of public-policy making and demonstrates that it is important to understand the relationships between scientific and political levels of problem solving. They are seldom clearly separable because interactive influences are characteristic. An environmental problem may have a scientific answer, but the ultimate solution for human society is political and is expressed as policy.

Multidisciplinary teams usually are needed to define and propose solutions for large and complex environmental problems. Because there may be alternative assessments and answers to an environmental problem, policy solutions may require a choice involving nonscientific considerations. To the extent that the science solution is uncertain or disputed, the latitude of the policy maker to act or postpone action is broadened.

Intermediate between the scientific and political levels of problem solving is the policy analyst who may, or may not, be conversant with science and scientists. Even at the highest levels of scientific opinion and advice, as in the United States' Presidential Office of Science and Technology Policy, proposed science-based problem solutions must compete with other considerations, such as financial, economic, military, bureaucratic, and ideologic. Uncertainty in environmental problem solving may occur at all levels or stages in the problem-solving process. Because of the other than scientific considerations, an environmental policy solution can seldom be finalized at the level of scientific research only. If, as is usually the case, the problem solution must occur at the policy level, the process of policy analysis between scientific findings and policy decisions likely is to be critical for the applied solution.

Caldwell notes that at this stage, unwarranted certainty and misconception of the actual problem can lead to ineffective policy solutions. Worse, it may lead to unforeseen and unwanted consequences. The so-called paradigmatic or ideational context in which a problem is perceived influences, and may determine, the way in which the problem is defined. For example, is the problem of world hunger too little food, inadequate distribution, or too many people? Unexamined assumptions about a problem may lead to different answers based upon spurious certainty in the validity of the problem as defined. Consequently, environmental problem solving requires careful examination of underlying assumptions, and a recognition that a specious certainty about a perceived problem, especially one shaped by cultural and political biases, may be a greater threat to the problem solution than is scientific uncertainty.

Methodological Rules for Four Classes of Scientific Uncertainty

Kristin Shrader-Frechette

In the face of inadequate data or unclear theories, is a hypothesis confirmed or falsified? Is a potential scientific threat innocent till proved guilty or guilty till proved innocent? When potential harm is unknown, does one assume that an average level of danger will occur or that a worst case is possible? All of these questions raise the issue of how to deal with scientific uncertainty, particularly in cases that could have serious consequences.

Among the many classes of cases of scientific uncertainty that confront contemporary problem solvers, at least four stand out because of their special relevance to environmental decision making. The four classes involve (1) framing uncertainty, (2) modeling uncertainty, (3) statistical uncertainty, and (4) decision-theoretic uncertainty. Using examples from quantitative risk assessment, especially hazards associated with nuclear wastes, this essay argues that there are a number of epistemologic and ethical rules that scientists ought to follow in each of these cases of uncertainty and that, indeed, many of the preferred rules are contrary to established principles that members of the scientific community actually follow. Consider first the uncertainties surrounding how scientists frame their questions.

Framing Uncertainty and Using Science for Policy[1]

How scientists frame their questions often controls their answers. Newton, for example, framed his questions about mechanics in terms of the assumption that he needed to explain what caused uniform rectilinear motion *to stop*. As a consequence, he affirmed the first law of motion (1). Aristotle, however, framed his questions about mechanics in terms of the assumption that he needed to explain what caused uniform rectilinear motion *to begin*. As a result, he denied

---

[1]Much of the discussion of framing uncertainty is based on Shrader-Frechette (15, 18).

that bodies remain either at rest or in motion, unless compelled by impressed forces to change their state (2).

Like Newton and Aristotle, different scientists frequently have alternative "frames," different sets of theoretical assumptions for structuring their data and their problem solving (3, 4). Some of the most basic scientific uncertainties concern how to frame a particular question, for example, when and how to interpret data as providing grounds for accepting particular hypotheses. In framing their questions, scientists frequently use a two-value frame (falsification/provisional acceptance) specifying that, in a situation of uncertainty, when rigorous testing fails to falsify some testable hypothesis (such as "organic molecules can come in right- and left-handed mirror-image versions"), then it is reasonable to accept the hypothesis provisionally.

Why Scientists Often Use the Two-Value Frame

Scientists sometimes employ the two-value frame because, from an empirical point of view, rigorously attempting (and failing) to falsify a precise testable hypothesis provides one of the strongest criteria for accepting it (5). Scientists often attempt to devise "crucial experiments," tests for which two mutually exclusive, exhaustive hypotheses predict conflicting outcomes. Classic examples of "crucial experiments" are Millikan's (6) attempt, in the early 1900s, to show whether electric charges are integral multiples of the charge of the electron and Lenard's 1903 test of two conflicting implications concerning the light energy that a radiating point can transmit (7). A second reason that scientists sometimes use the two-value frame is pragmatic. As Duhem (8) recognized, they provisionally accept a precise hypothesis (that has survived rigorous attempts to falsify it), even one with obvious deficiencies, if there is no better (more probable) hypothesis available. Because hypotheses have an infinite number of observational consequences that can never be "verified" conclusively (9), scientists sometimes opt—in a situation of uncertainty—for provisional acceptance of the best available nonfalsified hypothesis.

A third situation in which scientists use the two-value frame is when they give provisional acceptance to null (no-effect) hypotheses that survive rigorous attempts at falsification. Because they are more interested in avoiding false positives (type I errors) rather than false negatives (type II errors) in situations of uncertainty, scientists place the greater burden of proof on the person who postulates some, rather than no, effect. For example, a geologist might postulate the effect that, because of tectonic activity, the water table will rise at a given location by at least 500 meters over the next 10,000 years. Although "no-effect" results run the risk of type II errors, scientists usually assume, as in criminal law, that null hypotheses are provisionally acceptable (innocent) until they are rigorously falsified (proved guilty). In the third section, on statistical uncertainty, we discuss type I and type II errors in more detail.

The Two-Value Frame in Quantitative Risk Assessment

Because policy makers often need immediate decisions about particular risks, scientists have used the two-value frame in many risk assessments, from studies of hazardous landfills to childhood exposure to lead (10). To illustrate potential problems with assessors using the two-value frame, consider the 1992 *Early Site Suitability Evaluation* (ESSE) completed by the U.S. Department of Energy (DOE) for the proposed Yucca Mountain (Nevada) nuclear waste repository (11). Reporting site-suitability findings for every condition specified, the 1992 ESSE used a two-value frame to assess the site: "conclusions about the site can be either that current information supports an unsuitability finding or that current information supports a suitability finding. . . . If . . . current information does not indicate that the site is unsuitable, then the consensus position was that at least a lower-level suitability finding could be supported" (11, pp. E-5, E-11). To understand why the two-value frame may be problematic here, recall that scientists typically use it for at least one of three reasons: (1) The attempted falsification is rigorous and precise. (2) The hypothesis that has survived precise, rigorous attempts at falsification is the "best" of candidate hypotheses. (3) The surviving hypothesis is a null hypothesis.

Sometimes societal risk assessments meet none of these three conditions for use of the two-value frame. At Yucca Mountain, for example, the long time period (10,000 years of site suitability) precludes the precise predictions, specified in condition (1), necessary for rigorous attempts at falsification. Indeed, the ESSE peer reviewers unanimously warned: "many aspects of site suitability are not well suited for quantitative risk assessment. . . . Any projections of the rates of tectonic activity and volcanism, as well as natural resource occurrence and value, will be fraught with substantial uncertainties that cannot be quantified" (12, p. B-2). They cautioned that although "there is . . . currently not enough defensible, site-specific information available to warrant acceptance or rejection of this site" (12, pp. 460, 257, 40–51), nevertheless they used the two-value frame (site suitable/site unsuitable) that they "were given" and agreed with the "site-suitable" conclusions of the ESSE: "The DOE General Siting Guidelines (10 CFR Part 960) do not allow a 'no decision' finding. . . . Thus the ESSE Core Team followed the intent of the guidelines" (12, p. 460). The peer reviewers' warnings suggest that, when rigorous and precise testing is impossible, using a two-value frame might beg some questions of risk evaluation and that, in such situations, a three-value frame (site suitable/site unsuitable/site suitability uncertain at present) might be preferable.

Condition (2), that the nonfalsified hypothesis be the best available, likewise appears problematic for assessments that compare neither alternative risk sites nor different risk hypotheses. Complete comparative analyses of proposed repository sites, for example, were precluded by the 1987 amendment to the Nuclear Waste Policy Act that named Yucca Mountain, Nevada, as the only candidate location for the nation's first permanent repository for commercial nuclear waste and spent fuel. Using the two-value frame to give provisional acceptance to a particular site-suitability hypothesis as "best," however, appears problematic to the degree that alternative sites are not compared and to the degree that different hypotheses have "substantial uncertainties that cannot be quantified" (12, p. B-12).

Another justification for use of the two-value frame—that (3) surviving hypotheses be provisionally accepted if they are null—likewise seems inapplicable in many societal risk assessments. Theoretical science, of course, often places the burden of proof on those arguing against the null hypothesis because science must be *epistemologically conservative* (avoid false positives). Science applied to risk assessment, however, also must be *ethically conservative*, as the National Academy of Sciences points out (13, 14), in the sense of taking account of social consequences affecting the needs, rights, and welfare of the public. To the extent that the public has limited financial resources and information or bears inequitable or involuntary risk impositions, it may need more risk protection than the proponents of a particular null hypothesis regarding risk (15). For example, because more than 80% of Nevadans say they would vote against the Yucca Mountain repository proposed for their state (16), they may need greater risk protection. Future generations, in particular, may have special needs regarding Yucca Mountain both because they cannot exercise their consent and because current regulations require no monitoring beyond the first 50 years, after which waste migration is more likely. Likewise, in cancer risk assessment, potential victims of a false null hypothesis may need special protection because many epidemiologic studies are too insensitive—owing to small samples and their dealing with rare diseases—to detect positive effects. Also, field studies of populations exposed to hazardous substances often involve more uncertainties than those based on theoretical models (10).

Framing Uncertainties: Hypotheses Versus Decisions

Other disanalogies between theoretical science and science applied to risk assessment also argue against using the two-value frame for evaluating societal risk in a situation of uncertainty. Theoretical scientists usually evaluate the *truth or falsity* of *hypotheses* (such as "convection currents have moved this geological plate"). Risk assessors, however, also evaluate the *acceptability* of risk *decisions* (such as "this site is suitable for permanent waste disposal"). As the National Academy of Sciences put it: "risk assessment must always include policy as well as science" (13, p. 76). Because the acceptability of risk decisions includes non-epistemic factors—such as decision-theoretic, social, economic, and ethical considerations—risk assessment may be more suited to a three-value frame that explicitly takes account of uncertainty. For example, classical methods of Bayesian decision making typically employ a three-value frame, in the sense of including a category for events that are "uncertain" or about which we have inadequate information to make a decision (17). Decision theorists also recognize that even a high *probability* that a site is suitable for some activity may not be "high enough" if the activity could pose serious *consequences* for public welfare.

Because of the disanalogies between theoretical science and science applied to societal risk evaluation, assessors confronting framing uncertainties may need to consider using three-value, rather than two-value, frames for scientific decisions involving both significant uncertainty and potentially serious public consequences. Regardless of the frames they choose, however, scientists and policy makers may need to recognize that uncertainty gives framers significant power. As happened with Newton, Aristotle, and risk assessors, whoever frames the questions may control the answers (18).

Modeling Uncertainty, Verification, and Validation[2]

Another type of uncertainty that occurs frequently in science arises when modelers assume that their constructs have been verified or validated because they are consistent with other computer models. This second section of the chapter argues that, in cases of modeling uncertainty created by incomplete data or failure to employ the available data, scientists ought not claim that their models have been verified or validated.

Modeling Uncertainty and Affirming the Consequent

Problems with verification and validation of models are part of a larger set of difficulties associated with the inference known as "affirming the consequent." This inference occurs whenever one postulates that a hypothesis is true or accurate merely because some test result, predicted to follow from the hypothesis, actually occurs. In fact, however, failure of predictions can only falsify theories, but success of predictions can only confirm (but not verify) theories. All that can be validly inferred from a test is that the results are consistent with the hypothesis or that the results have falsified the hypothesis. In other words, from "*h* entails *r*," one can infer "not *r* entails not *h*." To assume that one can infer "*r* entails *h*" from "*h* entails *r*" is to affirm the consequent. Of course, it is very important to test one's hypotheses in order to determine whether the data falsify them or tend to confirm them. Moreover, the greater the number of tests, and the more representative they are, the greater is the assurance that the data are consistent with the hypotheses. Indeed, one of the repeatedly acknowledged failures of the assessments of the proposed Yucca Mountain nuclear waste repository is that the models often are not tested (19). It is impor-

---

[2]Much of the discussion of modeling uncertainty is based on Shrader-Frechette (15) and Oreskes, Shrader-Frechette, and Belitz (1994).

tant to test the models, to attempt to falsify them and to determine the degree to which they are consistent with the data. If the models turn out to be consistent with the data, however, it is wrong to assume that they have been absolutely "verified" or "validated" because, short of affirming the consequent, it is impossible to verify or validate any model. It is possible merely to know—through testing—that the hypothesis or model has been confirmed to this or that degree.

At the proposed Yucca Mountain repository, risk assessors have repeatedly proposed to test some *h*, some hypothesis, such as that the number of calculated groundwater travel times is less than 10,000 years. When the calculations, data, and models are shown to be *consistent* with the hypothesis, then the assessors have erroneously assumed, in the face of modeling uncertainties, that the hypothesis has predictive power or has been "verified." For example, one group of assessors, studying groundwater travel time, concluded: "this evidence indicates that the Yucca Mountain repository site would be in compliance with regulatory requirements" (20). Many other risk assessors speak of "verifying" their models and "validating" them. For instance, one group of assessors concluded that the tools they used demonstrated "verification of engineering software used to solve thermomechanical problems" (21, p. i) at Yucca Mountain (22).

Admittedly, software and systems engineers speak of computer models being "validated" and "verified." Yet, such "validation" language obscures the fact that the alleged validation really only guarantees that certain test results are consistent with a model or hypothesis; it does not validate or verify the model or hypothesis because affirming the consequent prevents legitimate validation or verification. Hence, when computer scientists speak of "program verification" (23–25), at best they are making a problematic inference by affirming the consequent in the face of modeling uncertainty. At worst, they are trading on an equivocation between "algorithms" and "programs." As Fetzer argues (26, 27), algorithms, as logical structures, are appropriate subjects for deductive verification. As such, *algorithms* occur in pure mathematics and pure logic. They are subject to demonstration or verification because they characterize claims that are always true as a function of the meanings assigned to the specific symbols used to express them. *Programs,* however, as causal models of logical structures, are not verifiable because the premises are not true merely as a function of their meaning. As Einstein put it, insofar as the laws of mathematics refer to reality, they are not certain; insofar as they are certain, they do not refer to reality.

Modeling Uncertainties and Misleading Language

In using "verification" and "validation" language, both official U.S. Department of Energy (DOE) documents and individual risk assessments for repositories like

Yucca Mountain are systematically misleading both about the modeling uncertainties and about whether the studies are reliable. For example, explicitly affirming the consequent, the DOE affirmed (28, p. 3-11)

> Validation . . . is a demonstration that a model as embodied in a computer code is an adequate representation of the process or system for which it is intended. The most common method of validation involves a comparison of the measured response from in-situ testing, lab testing, or natural analogs with the results of computational models that embody the model assumptions that are being tested.

Authors of the same official DOE document, used to provide standards for Yucca Mountain risk assessments, also talk about the need to verify computational models of the waste site. They say (28, p. 3-7)

> Verification, according to the guidelines in NUREG-0856 . . . is the provision of assurance that a code correctly performs the operations it specifies. A common method of verification is the comparison of a code's results with solutions obtained analytically. . . . Benchmarking is a useful method that consists of using two or more codes to solve related problems and then comparing the results.

Although the term "verification," as used by DOE assessors, suggests that the computer models or codes accurately represent the phenomena they seek to predict, it is merely a misleading euphemism for "benchmarking," comparing the results of two different codes (computer models) for simulating an identical problem. On this scheme, one "verifies" a model of Yucca Mountain against another model. What is required in the real world, however, is validating a model against reality. This validation or confirmation can be accomplished only by repeated testing of the code or model against the real world, against field conditions.

Even with repeated field testing, however, modeling uncertainties remain. Compliance can never be confirmed, short of full testing of all cases throughout all time periods. Classic studies of the problem of induction show that complete testing is impossible. Therefore, the shorter the time of testing and the fewer the cases considered, the less reliable and the less confirmed are allegedly "validated" computer models or codes. The tests can only falsify or confirm a hypothesis, not validate it. To assume otherwise is to affirm the consequent. Hence, every conclusion of compliance with government regulations, or every conclusion of repository safety, on the basis of "verified" or "validated" test or simulation results, is an example of affirming the consequent. Program *verification*, in other words, "is not even a theoretical possibility" (29). One cannot *prove* safety. One can only demonstrate that one has attempted to falsify one's results and either has failed to do so or has done so.

Responses to Modeling Uncertainties

Because of the problems associated with modeling uncertainties, scientists need to be wary of claiming that they have verified or validated a model on the basis of limited data. As both the DOE risk documents and the risk assessors at Yucca Mountain illustrate, they are misleading in speaking of "validation" and "verification" of models used at Yucca Mountain. *First,* real validation and verification are impossible because of the problems of induction and affirming the consequent. Only falsification of a hypothesis, or determining that the data are consistent with it, is possible. In the latter case, when one obtains repeated results indicating that the data are consistent with the model or hypothesis, one is able merely to increase the probability that the model or hypothesis has been confirmed. *Second,* the DOE's and assessors' use of the terms "verification" and "validation" misleads the public about the reliability of models allegedly guaranteeing repository safety. *Third,* use of the term "verification" by DOE assessors is, in particular, misleading because they typically only compare different computer codes or models, with no reference to the real world, and because any model can be tuned or calibrated to fit any pattern of data, even when the model is not well confirmed. *Fourth,* it is arguable that most useful programs are not merely unverifiable but incorrect, that even programs that function correctly in isolation may not do so in combination, and that most of the important requirements of real programs are not formalizable (30).

Given these four difficulties with "verifying" programs used in real-world situations, such as repository modeling, there are both prudential and ethical problems with risk assessors' continuing to use the language of "program verification" in connection with modeling causal relationships in situations of uncertainty. The *prudential* problem is that aiming at "verification" does not tell us what we most want to know—something about complex relationships in the physical world. The more complex the system, the less likely it is to perform as desired and the less reliable is inductive testing of it. Moreover, by emphasizing verification, theorists have increased the expense of achieving "transparent software upgrades," and they have decreased software reliability because of their emphasis on "misplaced advocacy of formal analysis" (31, p. 792). The *ethical* problem is that, by encouraging confidence in the operational performance of a complex causal system, claims of "verification" oversell the reliability of software and undersell the importance of design failures in safety-critical applications like waste repositories. Such overselling and underselling not only expose the safety of the public to the dangerous consequences of risk assessors' "groupthink" (32, p. 422), but also risk misunderstanding of software in cases where the risks are greatest. To avoid affirming the consequent, invalid inferences such as that repository safety models can be "verified," scientists need to refrain from the claim that their results "indicate" or "show" or "prove" compliance with government regulations or with some standard of safety. Scientists also would do well, when they face modeling uncertainties and have not checked the models against field data, to avoid misleading claims that they have "verified" or "validated" mathematical models at Yucca Mountain or anywhere else (33). Such terms suggest a level of reliability and predictive power which, in the face of many cases of modeling uncertainty, is impossible in practice. Instead, assessors might do better to speak in terms of *probabilities* that a given model or hypothesis has been *confirmed* and to avoid misleading claims about verification.

Statistical Uncertainty, False Positives, and False Negatives[3]

Yet another class of uncertainties scientists face is statistical. In the face of such uncertainties, they often must decide whether to limit false positives or false negatives, because they cannot do both. For example, such decisions arise because scientists performing environmental risk assessments typically face uncertainties of six orders of magnitude (34, 35). These uncertainties mean that typically we do not know, for example, whether an Indian's chance of dying in a Bhopal accident is 1 in 1 million per year or 1 in 1 per year. Our ignorance of such events is astounding and potentially catastrophic.

How ought scientists make environment-related decisions when they are ignorant of basic data and probabilities? How should they behave in a situation of statistical uncertainty? Should they assume that a particular environmental condition is safe or acceptable until it is proved unsafe or unacceptable? Or should they assume that it is unsafe or unacceptable until it is proved safe or acceptable? Where ought they to place the burden of proof? Do they place the burden of proof on polluters or on potential victims of polluters? Do they place the burden of proof on developers of the rain forest or on environmentalists who protest such development? Where they decide to place the burden of proof will determine who bears enormous risks and who receives great benefits. What is fair, equitable, and ethical in a situation of statistical uncertainty?

In this section we argue that the typical scientific norm dictating behavior under uncertainty is wrong. It is wrong to be reluctant to posit effects such as serious environmental consequences in a situation of uncertainty. Therefore, it is wrong, in a situation of uncertainty in which we cannot adequately assess effects, to place the burden of proof on possible victims of pollution or development. Instead we argue that, in situations of statistical uncertainty affecting human and environmental well-being, we should be reluctant not to posit effects such as serious harm. Therefore, in a situation of statistical uncertainty in which we cannot adequately assess effects, we should place the burden of proof on the persons who create these potentially adverse effects—that is, on polluters and developers.

---

[3]This discussion of statistical uncertainty is based on Shrader-Frechette (35, 15, 18) and Shrader-Frechette and McCoy (1992).

Statistical Uncertainty and the Burden of Proof

In a situation of uncertainty, errors of type I occur when one posits some possible effect and thereby rejects a null hypothesis that is true; errors of type II occur when one decides not to posit some possible effect and thereby fails to reject a null hypothesis that is false. (One null hypothesis might be, for example, "the pesticide benzene hexachloride will cause no deaths among pesticide applicators during 5 years of using 100,000 pounds per year of the chemical, provided that the applicators follow the manufacturer's instructions.") Given a situation of uncertainty about the pesticide, which is the more serious error, type I or type II? An analogous issue arises in law. Is the more serious error to acquit a guilty person or to convict an innocent person? In a situation of uncertainty, ought one to run the risk of rejecting a true null hypothesis, of not using the benzene hexachloride technology that is really acceptable and safe? Or, in a situation of uncertainty, ought one to run the risk of not rejecting a false null hypothesis, of employing the benzene hexachloride pesticide technology that is really unacceptable and unsafe? The basic problem is that to decrease type I risk might hurt the public, especially workers in developing nations where approximately 50,000 persons per year are killed by pesticides (36, 37). Yet, to decrease type II risk might hurt those who are economically dependent on this particular pesticide industry.

Pure Science, Applied Science, and Statistical Uncertainty

In the area of pure science and statistics, most persons believe that in a situation of uncertainty one ought to minimize type-I risks so as to limit false positives, assertions of effects where there are none. Pure scientists often attach a greater loss to accepting a falsehood than to failing to acknowledge a truth (35, 38). Societal decision making under uncertainty, as in cases involving sustainable energy or agricultural technologies, however, is arguably not analogous to decision making in pure science. Societal decision making involves rights, duties, and ethical consequences that affect the welfare of persons, whereas purely scientific decision making involves largely epistemologic consequences. For this reason, it is not clear that in societal cases under uncertainty, one ought to minimize type I risks. Instead, there are a number of in-principle reasons for minimizing type II errors. For one thing, it is arguably more important to protect the public from harm (from dangerous pesticides, for example) than to provide, in some positive sense, for welfare (creating jobs as pesticide applicators, for example), because protecting from harm seems to be a necessary condition for enjoying other freedoms (39, 40). Admittedly, it is difficult to draw the line between providing benefits and protecting from harm, between positive and negative laws or duties. Nevertheless, just as there is a basic distinction between welfare rights and negative rights (41), so there is an analogous distinction between welfare policies (that provide some good) and protective policies (that prohibit some infringement). Moral philosophers continue to honor related distinctions, such as that between letting a person die and killing someone. It therefore seems more important to protect citizens from public hazards, like those created by particular pesticides, than to attempt to enhance their welfare, over the short term, by implementing a potentially dangerous technology such as use of benzene hexachloride (35).

A second reason for minimizing type-II errors under uncertainty is that the public typically needs more risk protection than do the industry or government proponents of the risky technology, like particular pesticides. The public usually has fewer financial resources and less information to deal with societal hazards that affect it, and laypersons are often faced with bureaucratic denials of public danger, as in the 1973 case of the Michigan PBB contamination of cattle feed, or the 1976 dioxin poisoning at Seveso, Italy, or the 1953 Minimata poisoning in Japan. As these and other cases illustrate, public needs for protection seem larger than those of developers or manufacturers, and the importance of minimizing type-II errors appears greater than that of minimizing type-I errors (35).

Third, it is more important to minimize type-II error, especially in cases of great uncertainty, because laypersons ought to be accorded legal rights to protection against technological decisions that could threaten their health and physical security. These legal rights arise out of the considerations that everyone has both due-process rights and rights to bodily security. In cases where those responsible or liable cannot redress the harm done to others by their faulty decisions—as they often cannot in the case of dangerous technologies—there are strong arguments for minimizing the public risk. Industrial and technological decision makers cannot adequately compensate their potential victims for the bad consequences of many pesticides, for example, because the risks involve death. Therefore, the risks are what Judith Jarvis Thomson calls "incompensable" (42, p. 158). Surely incompensable risks ought to be minimized for those who fail to give free, informed consent to them. Whenever risks are incompensable (that is, imposing a significant probability of death on another), failure to minimize the risks is typically morally unjustifiable without the free, informed consent of the victim (35). And, in cases of uncertainty, it is impossible to obtain free, informed consent of potential victims because, by definition, the risks are uncertain and we have inadequate information about them.

A final reason for minimizing type II error in cases of uncertainty is that failure to do so would result in using some persons (such as pesticide applicators) as means to the ends of other persons (such as pesticide manufacturers). It would result in their bearing a significantly higher risk from toxic chemicals than other persons, despite the fact that some of those other persons (the pesticide manufacturers) have received most of the benefits associated with benzene hexachloride, for example. Such discrimination (in this case, against pesticide applicators, particularly those in developing nations), as Frankena has pointed out, is justified only if it would work to the advantage of everyone, including those discriminated against. Any other attempt to justify discrimination fails because it would amount to sanctioning the use of some humans as means to the ends of other humans (43). Hence, in situations of uncertainty, the morally desirable position is to place the burden of proof on those who can most bear it, developers and manufacturers, rather than on the persons who are potential victims of either development or some technology.

Dealing with Statistical Uncertainty

If the arguments in this section are correct, then, in situations of statistical uncertainty, one ought not assume that potential environmental hazards are "innocent until proved guilty." Although this is the typical position adopted by most scientists and courts of law, it does not presuppose innocence. In matters of potential global harm and human catastrophe, this position places the burden of proof on those who are least able to bear it. To change this burden of proof, in cases of statistical uncertainty, environmental effects should be assumed "guilty until proved innocent." The burden of proof ought not be on the most vulnerable—potential victims of environmental hazards. Even, and especially, victims ought not bear the burden of proof.

Decision-Theoretic Uncertainty and the Maximin Rule[4]

In addition to statistical uncertainty, many scientists, especially those who work in applied areas, face decision-theoretic uncertainty. One of the crucial questions they must address is when to use a maximin decision rule in a situation of probabilistic uncertainty and when to use an expected-utility rule. This section argues that there are a number of criteria for using maximin, versus expected utility, rules in situations of scientific and decision-theoretic uncertainty. The criteria for using maximin focus on potentially catastrophic consequences and probabilistic uncertainty.

What are the consequences of using different decision-theoretic rules in situations of uncertainty? A recent U.S. government study pointed out that estimates of saccharin-caused increase in bladder cancer differed by seven orders of magnitude. Despite these uncertainties, U.S. officials have sanctioned use of saccharin, justifying their decision on the basis of a liberal risk assessment. As a result of a very conservative risk analysis, however, they banned cyclamates. Were the two risk-assessment methodologies consistent, experts have argued that cyclamates could easily have been shown to present a lower relative risk than saccharin. In Canada, for example, cyclamates are permitted and saccharin

---

[4]The discussion of decision-theoretic uncertainty, in this section of the chapter, relies heavily on Shrader-Frechette (35).

is banned, making Canadian regulations in this area exactly the reverse of those in the United States (35, 44–46). As the saccharin-cyclamates controversy illustrates, scientific conclusions may be uncertain, not only because of the wide range of predicted hazard values, but also because scientists use different decision-theoretic rules of evaluation. This section of the chapter will assess several of the prominent decision-theoretic rules for evaluating situations of uncertainty. To see how alternative rules can generate different conclusions consider the following case.

For U.S. reactors, the core-melt probability is about one in four for their lifetimes (47, 48). Risk assessments done by both the Ford Foundation-Mitre Corporation and by the Union of Concerned Scientists (UCS) *agree* on the probability and consequence estimates associated with the risk from commercial nuclear fission, but *disagree* in their recommendations regarding the advisability of using atomic energy to generate electricity. The UCS risk analysis decided against use of the technology; the Ford-Mitre study advised in favor of it (49–51). The two studies reached different conclusions because they used quite different decision-theoretic rules to evaluate the same data. The Ford-Mitre research was based on the widely accepted Bayesian decision criterion that it is rational to choose the action with the best expected value or utility, where "expected value" or "expected utility" is defined as the weighted sum of all possible consequences of the action and where the weights are given by the probability associated with each consequence. The UCS recommendation followed the maximin decision rule that it is rational to choose the action that avoids the worst possible consequence of all options (51). Ought we to be technocratic liberals and choose a Bayesian rule? Or ought we to be cautious conservatives and follow a maximin strategy (52)? The "prevailing opinion" among scholars, according to John Harsanyi, is to use the Bayesian rule (53, 54) even in conditions of uncertainty (55–57). This last section argues that scientists have compelling reasons for rejecting the Bayesian or utilitarian strategy when they face a situation of decision-theoretic uncertainty having potentially catastrophic consequences, and that it is often more rational to prefer the maximin strategy.

Utilitarians Versus Egalitarians

Perhaps the most famous contemporary debate over which decision rules ought to be followed in situations of risk and uncertainty is that between Harvard philosopher Rawls and Berkeley economist Harsanyi (53, 56, 57). Harsanyi believes that under conditions of uncertainty, we should maximize expected utility, where the expected utility of an act for a two-state problem is

$$u_1 p + u_2 (1 - p)$$

where u_1 and u_2 are outcome utilities, p is the probability of S_1, $1 - p$ is the probability of S_2, and p represents the decision maker's own subjective probability estimate (54, 56, 58). More generally, members of the dominant Bayesian school claim that expected-utility maximization is the appropriate decision rule under uncertainty (53, 59–64). They claim that we should value outcomes, or societies, in terms of the average amounts of utility (subjective determinations of welfare) realized in them (52, 54, 56, 66–67).

Proponents of using the maximin rule (like Rawls) maintain that one ought to maximize the minimum—that is, avoid the policy having the worst possible consequences (17, 53, 54), which harms the worst-off persons (68). The obvious problem is that often the maximin and the Bayesian/utilitarian principles recommend different responses to uncertainty. To illustrate these different responses, consider an easy case involving two societies. The first consists of 1000 people, with 100 being workers (workers who are exposed to numerous occupational risks) and the rest being free to do whatever they wish. We can assume that, because of technology, the workers are easily able to provide for the needs of the rest of society. Also assume that the workers are miserable and unhappy, in part because of the work and in part because of the great risks that they face. Likewise, assume that the rest of society is quite happy, in part because they are free not to work, because they face none of the great occupational risks imposed on the 100 workers and because the nonworkers' happiness is not disturbed by any feeling of responsibility for the workers. With all these (perhaps implausible) assumptions in mind, let us suppose that, using a utility scale of 1 to 100, the workers each receive 1 unit of utility, whereas the others in society each receive 90 units each. Thus the average utility in this first society is 81.1. Now consider a second society, similar to the first, but in which, under some reasonable rotation scheme, everyone takes a turn at being a worker. In this society everyone has a utility of 35 units. Bayesian utilitarians would count the first society as more just and rational, whereas proponents of maximin and the difference principle would count the second society as more just and rational (17, 54).

Although this simplistic example is meant merely to illustrate how proponents of Bayesian utilitarianism and maximin would sanction different responses to decision-theoretic uncertainty, its specific assumptions make maximin (in this case) appear the more reasonable position. Often, however, the reverse is true. In this section we attempt to determine the better decision rule for cases of scientific decision making involving both uncertainty and great consequences to welfare (69). A reasonable way to determine whether the Bayesian/utilitarian or maximin position is superior in such cases of decision-theoretic uncertainty is to examine carefully the best contemporary defenses, respectively, of these rules. The best defenses are probably provided by Harsanyi, a utilitarian, and Rawls, an egalitarian.

Utilitarian Arguments

Harsanyi's main arguments in favor of the utilitarian, and against the maximin, strategy under decision-theoretic uncertainty are as follows: (1) Those who do not follow the utilitarian strategy are irrational and ignore probabilities. (2) They cause unacceptable ethical consequences. (3) Using the utilitarian rule, with the equiprobability assumption, promotes equal treatment.

Do Nonutilitarians Ignore Probabilities?

Choosing the maximin strategy, claims Harsanyi, is wrong because "it is extremely irrational to make your behavior wholly dependent on some highly unlikely unfavorable contingencies, regardless of how little probability you are willing to assign to them" (45, 53, p. 595). To substantiate his argument, Harsanyi gives an example of maximin decision making and alleges that it leads to paradoxes. The example is this. Suppose you live in New York City and are offered two jobs, in different cities, at the same time. The New York City job is tedious and badly paid, but the Chicago job is interesting and well paid. However, to take the Chicago job, which begins immediately, you have to take a plane, and the plane travel has a small, positive, associated probability of fatality. This means, says Harsanyi, that following the maximin principle would cause you to accept the New York job. In this example, Harsanyi assumes that your chances of dying in the near future from reasons other than a plane crash are zero. Hence, he concludes that maximin, because it directs choosing so as to avoid the worst possibility, forces one to ignore both the low probability of the plane crash and the desirability of the Chicago job and to choose the New York job. However, Harsanyi claims that a rational person, using the expected-utility criterion, would choose the Chicago job for those very reasons—its desirability and the low probability of a plane crash on the way to Chicago.

How successful is Harsanyi's first argument in employing the counterexample of the New York and Chicago jobs? For one thing, the example is highly counterintuitive; even if the example were plausible, it would prove nothing about the undesirability of using maximin in situations of *societal* risk under uncertainty, such as deciding whether to open a liquefied natural gas facility. Harsanyi makes the questionable assumption in this example that the situation of uncertainty regarding *one* individual's death, caused by the same person's decision to fly to Chicago, is no different than a situation of uncertainty regarding *many* individuals' deaths, caused by a societal decision to employ a hazardous technology.

Objecting to Harsanyi's example, John Rawls claimed that the example failed because it was of a small-scale, rather than a large-scale, situation (46, 70). My claim is similar, but more specific: situations of individual risk caused by scientific or technological decisions are *voluntarily chosen*, whereas situations of societal risk are typically *involuntarily imposed;* hence they are not analogous. This means that, to convince us that societal decisions in situations of uncertainty are best made by following a utilitarian rule, Harsanyi cannot merely provide an example of an individual decision. In the individual case, one has the right to use expected utility so as to make efficient, economic decisions regarding oneself. In the societal case, scientists and policy makers do not always have the right to use expected utility so as to make efficient, economic decisions regarding others in society, since maximizing utility or even average utility might violate rights or duties. On the individual level, scientists' rules under uncertainty must be theoretically justifiable. On the societal level, because the rules have consequences for welfare, they must be democratically justifiable in terms of ethical procedure. Decision-theoretic rules under uncertainty require scientists to take account of the fairness of the allocational *process,* not merely the *outcomes* (71). Democratic process is probably more important in cases where probabilities are unknown than in cases of scientific uncertainty where they are certain, since it would be more difficult to ensure informed consent in the former cases. This, in turn, suggests that the individual case of decision making under uncertainty, involving pure science, requires merely a *substantive* concept of rationality. However, the societal case of decision making under uncertainty, involving applied science, requires a *procedural* or "process" concept of rationality (72, 73), because it must take account of conflicting points of view, possible consequences to welfare, as well as various ethical and legal obligations, such as those involving free, informed consent and due process. For example, if I use a decision-theoretic rule affecting my own risk, I can ask "how safe is rational enough?" and I can be termed "irrational" if I have a fear of flying. But if I use a decision-theoretic rule affecting risks to others in society, I do not have the right to ask, where their interests are concerned, "how safe is rational enough?" In the societal case, I must ask, because I am bound by moral obligation to others, "how safe is free enough?" or "how safe is fair enough?" or "how safe is voluntary enough?" (53, 55, 65, 66, 74).

When they discuss decision-theoretic uncertainty, many risk assessors, like Bruce Ames, assume that risk aversion ought to be a linear function of probability, and they criticize laypersons for being more averse to industrial chemicals than to natural toxins (like the mold in foods) that have a higher probability of causing injury or death. Invoking the concept of "relative risk," they fault laypersons for their "chemophobia," for greater aversion to lower-probability risks than to higher ones (76, 77). Probability, however, is neither the only, nor the most important, factor determining risk aversion. Risks that threaten consent, equity, or other values might also cause extreme aversion. Moreover, if subjective probabilities are frequently prone to error (78–81), then, contrary to Harsanyi's first argument, rational people might well avoid them in deciding how to handle uncertainty.

Harsanyi's first argument is also problematic because he assumes that it is irrational to base decisions on *consequences* and to ignore either a small or uncertain probability associated with them (45, 53). However, it is not irrational to avoid a possibly catastrophic risk (e.g., nuclear winter) even if it is small.

Does Maximin Lead to Unethical Consequences?

Harsanyi next claims that following maximin rules in situations of decision-theoretic uncertainty would lead to unacceptable *moral* consequences: benefiting the least-well-off individuals, even when they do not deserve it and even when doing so will not help society. To establish this point, Harsanyi gives two examples (53). In the first example, there are two patients, critically ill with pneumonia, but there is only enough antibiotic to treat one of them, one of whom has terminal cancer. Harsanyi says that Bayesians or utilitarians would give the antibiotic to the victim who did not have cancer, whereas maximin strategists would give it to the cancer victim, since he is the worse off. In the second example, there are only two citizens, one severely retarded and the other with superior mathematical ability. The problem is whether to use society's surplus money to help educate the mathematician or provide remedial training for the retarded person. The Bayesian utilitarian would spend the surplus money on the mathematician, says Harsanyi, whereas the maximin strategist would spend it on the retarded person, since he is the less well off.

The problem with Harsanyi's examples is that they are not cases of *societal* decision making under *uncertainty.* The risk is of fatality, in the pneumonia example, but one knows, with certainty, that the cancer victim is soon to die, since Harsanyi defines his state as "terminal." Likewise, in the second case, the risk is of improving the lot of two persons, one retarded and one gifted mathematically. Hence, one is not in a state of uncertainty about the probability of success in spending the monies for education in the two cases. But if so, then Harsanyi has not argued for using Bayesian/utilitarian rules under uncertainty.

A second difficulty with these examples is that Harsanyi defines the retarded person as "less well off" and therefore deserving of funds for remedial education under the maximin strategy. However, being "less well off" is not merely a matter of intelligence. It is also a matter of financial well-being and of having equal political and social opportunities. If society has given equal consideration to the needs and interests of both the mathematician and the retarded person, if the retarded person is happy and incapable of being made better off, regardless of what society spends on him, then it is not clear that he is less well off than the mathematician. If the mathematician could be made better off, with greater societal expenditures, then he may be less well off than the retarded person who has reached his potential, who is as happy as he is capable of being.

Does Using Expected Utility Treat People Equally?

Having given general, utilitarian justifications for his position, Harsanyi provides a final argument for using expected-utility rules in situations of decision-theoretic uncertainty. It focuses on what Harsanyi calls "the equiprobability assumption" (53, p. 598). Decision makers ought to subscribe to this assumption as part of the expected-utility rule, says Harsanyi, because doing so enables them to treat all individuals' a priori interests as equal (34, 53, 81) to give everyone an equal chance of being better off or worse off. In a situation of decision-theoretic uncertainty, Harsanyi claims that the rational person would always make the decision that assumes everyone's interests are equal and that yields the highest "average utility level" (52, 53, pp. 598, 67).

The most basic difficulty with the equiprobability assumption is that if there is no justification for assigning a set of probabilities, because one is in a situation of uncertainty, then there is no justification for assuming that the states are equally probable (82–84). Other difficulties are that to assign the states equal probabilities is to contradict the stipulation that the situation is one of uncertainty (54), that it is often impossible to specify a list of possible states that are mutually exclusive and exhaustive (17), and hence that different ways of defining states could conceivably result in different decision results, different accounts of how best to maximize average utility (17). The equiprobability assumption is also ethically questionable because using it does not assign equal a priori weight to every individual's interests, as Harsanyi claims. It merely *postulates* that in a situation of uncertainty, in different social systems or states of affairs, every individual has the same probability of being the best-off individual, or the second-best-off, and so on. Reality, however, is quite different from this postulate. Different states of affairs are rarely equally probable. To assume that they are, when one is in a situation of uncertainty, is problematic in part because equally probable states often affect different individuals' interests unequally.

Using *averages* also affects individuals unequally. This is why, even if one granted that it is rational to maximize expected utility in individual decisions, it would not necessarily be rational to choose the average of the expected utilities of different persons. Such a procedure would not maximize *my* expected utility, but only the average of the expected utilities of members of society (52, 67, 85). This means that the concepts of "average utility" and "equiprobability" could hide the very problems of discrimination and inequality that most need addressing. Moreover, even though the equiprobability assumption assigns every individual the *same* probability (in every state of affairs) of being the best off, second best off, and so forth, this does not guarantee that every individual's interests receive *equal* weight. Because Bayesian utilitarianism focuses on *average* utility, it dictates that decisions be made on the basis of highest average utility. This rule guarantees that the minority, with less-than-average utility, can receive a disproportionate risk burden. In such cases, one would not be treating the interests of each person in the minority as equal to those of each person in the majority. Thus, in at least one important sense, Harsanyi does not treat people the same, as he claims to do through his equiprobability assumption (53). Genuinely *equal* treatment requires that we treat people differently, so as to take account of different degrees of merit, need, rights to compensation or reparation, and so on. Treating people the same, in a situation in which existing relationships of economic and political power are already established, merely reinforces those relationships, apart from whether they are ethically defensible. Treating people the same, as most persons wish to do in situations of uncertainty, also ignores the fact that duties and obligations almost always require that people's interests *not* be treated the same. For example, suppose that Mr. X builds a pesticide manufacturing plant in Houston. Also suppose that Mr. Y, who lives next door, has demonstrably damaging health effects from the emissions of the pesticide facility. To say that Mr. X's and Mr. Y's interests in stopping the harmful emissions ought to be given the same weight is to skew the relevant ethical obligations. It would give license to anyone wishing to put others at risk for his own financial gain (67, 86, 87). Hence, there are rarely grounds for treating persons' interests the *same,* since they are almost always structured by preexisting obligations that determine whose interests ought to have more weight. This means that equity of treatment can only be achieved after *ethical analysis,* not after an appeal to treating everyone the same, in the name of the "equiprobability assumption."

Egalitarian Arguments

Admittedly, discovering difficulties with Harsanyi's arguments for Bayesian and utilitarian rules is not a sufficient condition for rejecting them. We also need to assess maximin, perhaps the best alternative rule for certain classes of cases under uncertainty. To assess this option, we evaluate Rawls' analysis. He has two main arguments to support the maximin strategy in situations of uncertainty: (1) It would lead to giving the interests of the least advantaged the highest priority. (2) The maximin strategy would avoid using a utility function, designed for risk taking, in the area of morals, where it does not belong.

Giving Priority to the Least Advantaged

Consider the first argument in favor of using maximin rules in situations of decision-theoretic uncertainty: it would lead to a concept of justice based on "the difference principle," which evaluates every possible societal or policy arrangement in terms of the interests of the least-advantaged or worst-off persons (52, 68). Rawls believes that this is an advantage of maximin, because he argues that the "first virtue" of social institutions is justice or fairness. We could arrive at just or fair social institutions, according to Rawls, if we were all rational individuals caring only about our own interests, and if we negotiated with each other (about the nature of these institutions) behind the "veil of ignorance" (i.e., without anyone knowing her own social or economic positions, special interests, talents, or abilities). Not knowing what our own situation would be, Rawls claims that we would arrange society so that even the least-well-off persons would not be seriously disadvantaged (68). This means choosing the risk distribution where the least well off are least disadvantaged (85, 88, 89).

The main objection to this argument is that we ought not use maximin because it might not increase the average utility of society, and average utility is more important than helping a subset of persons. Therefore, goes the argument, in the situation of scientific or technological decision making under uncertainty, one ought not try to protect those who are most at risk, since this would take away resources from society. Instead, one ought to use a Bayesian/utilitarian strategy to employ expected utility so as to maximize the average well-being of each member of the group (53, 90).

The main problem with this objection is that it could sanction using members of a minority who are most at risk so as to benefit the majority, namely, using some persons as means to the ends of other persons, something condemned by most moral philosophers. Presumably, however, every person ought to be treated as an end in her own right, not merely as a way to satisfy the desires of someone else, not merely as an object. Moreover, there are good grounds for believing that everyone ought to receive equal treatment, equal consideration of interests: (1) The comparison class is all humans, and all humans have the same capacity for a happy life (91). (2) Free, informed rational people would likely agree to principles of equal rights or equal protection (92, 93). (3) These principles provide the basic justifications for other important concepts of ethics and are presuppositions of all schemes involving consistency, justice, fairness, rights, and autonomy (93–98). (4) Equality of rights is presupposed by the idea of law; "law itself embodies an ideal of equal treatment for persons similarly situated" (99). If all members of society have an equal, prima facie right to life, and therefore to bodily security, as the most basic of human rights, then allowing one group of persons to be put at greater risk, without compensation and for no good reason, amounts to violating their rights to life and to bodily security. Indeed, if there were no obligation to equalize the burden of technological risk imposed on one segment of the population for the benefit of another segment, then there could be no authentic bodily security and no legal rights at all. The majority could simply do whatever they wished to any victimized minority. This is why John Rawls called his notion of justice "fairness" and why he spoke about maximin under the rubric of fairness (68, 100). Of course, sanctioning *equal* treatment, in the name of fairness, does not mean guaranteeing the *same* treatment (101). Establishing the prima facie duty to treat persons equally, so far as possible, does require that we use maximin in situations of societal risk under uncertainty (68, 102) unless we have relevant moral reasons for treating people differently (43, 103).

Efficiency, or increasing overall average utility, does not appear to provide relevant moral grounds for discrimination, especially discrimination against the least well off, for several reasons. First, discrimination against persons on grounds of efficiency is something that would have to be justified for each situation in which it occurs. The reason is that to argue (as we just have) that a principle of equal rights and equal treatment under the law is desirable, but that there may be morally relevant grounds for discrimination, is to argue for a principle of prima facie political equality (101). On this view, sameness of treatment of persons and communities needs no justification; it is presumed defensible, whereas only unequal (different) treatment requires defense (34, 96, 101). This means that the burden of proof is on the person who wishes to discriminate, who wishes not to give equal protection to some minority that is exposed to societal risk. But if the burden of proof is on the discriminator and if, by definition, we are dealing with a situation of decision making under uncertainty, then it is difficult to believe that the discriminator (the person who does not want to use maximin) could argue that efficiency provides *morally relevant* grounds for discrimination (43, 103). The reason is that the potential grounds justifying the discrimination (e.g., empirical factors about merit, compensation, or efficiency) would be, by definition, unknown in a situation of uncertainty.

Efficiency also does not appear to serve any higher interest (68, 104–108). Admittedly many risk assessors and policy makers claim that efficiency (i.e., disregarding maximin) serves the interests of everyone; they say that "the economy needs" particular hazardous technologies (75, 90, 109). They also claim that certain scientific or technological decisions (made in situations of decision-theoretic uncertainty) are not cost-effective and efficient and therefore beneficial to our national well-being (34, 48, 90, 100, 111, 112). However, for efficiency to serve the overall interest of everyone would mean that it was "required for the promotion of equality in the long run"; any other interpretation of "serving the overall interest" would be open to the charge that it was built upon using humans as means to the ends of other persons rather than treating them as ends in themselves (43, p. 15). But does efficiency per se (e.g., avoiding pollution controls and therefore equal distribution of risk) lead to the promotion of equality in the long run? The problem with answering this question in the affirmative, as Harsanyi would do, is that such an answer would contain a highly questionable *factual assumption,* that promoting technology, without also seeking equal risk distribution, will lead to greater equality of treatment in the long run. This is false. Historically, there is little basis for believing that efficiency will help promote a more equitable distribution of wealth and, therefore, more political equality (97, 107, 113, 114). In the United States, for example, in the past 35 years, although there has been an absolute increase in the standard of living, the relative shares of U.S. wealth held by various groups have not changed. The poorest 20% of persons still receive 5% of the wealth, while the richest 20% still hold 41%; the share of the middle three quintiles has remained just as constant (105, 107, 115, 116). These data suggest that economic and technological growth, coupled with efficiency in the form of inequity of risk abatement, have not promoted economic equality. Because of the close relationship between wealth and the ability to use equal opportunities (101, 105, 106, 117–119), it is unlikely that this efficiency and economic expansion has promoted equal political treatment (48, 120, 121). If anything, it has probably made inequities even wider (107, 115, 117, 121).

Technological expansion (achieved through economic efficiency and through failure to abate technological risks) also does not ordinarily help to create a more egalitarian society because technology generally eliminates jobs; it does not create them (122). But if so, then there are not necessarily grounds for arguing that efficiency and Bayesian/utilitarian risk strategies help to equalize opportunities (101, 120). If anything, the plight of the least advantaged, whether the poor or those who bear a heavier burden of technological risk, is exacerbated by technological progress because they must compete more frantically for scarcer jobs. Moreover, because a larger portion of the indigent are unemployable, progress makes little immediate impact on the problem of hardcore poverty (121). Scientific and technological progress, without a commitment to equal distribution of societal risks, typically fails to remove distributive inequities because the poor usually bear the brunt of technological hazards. Most environmental policies, including risk policies, "distribute the costs of controls in a regressive pattern while providing disproportionate benefits for the educated and wealthy, who can better afford to indulge an acquired taste for environmental quality [and risk mitigation]" (123, p. 274; 124, 125). This means that, for the poor, whatever risk abatement and environmental quality cannot be paid for cannot be had. For example, a number of studies have shown that "those square miles populated by nonwhites and by all low socioeconomic groups were the areas of highest pollution levels" (126–132). In fact, various adverse environmental impacts, like higher risk burdens, are visited disproportionately upon the poor, while the rich receive the bulk of the benefits (52, 120, 123, 133). This all suggests that Bayesian/utilitarian strategies, in allowing the poor (persons who are least advantaged economically and therefore most helpless politically) to be further burdened with disproportionate technological risks, are especially questionable.

Do Egalitarians Avoid Utility Functions?
What about another argument of maximin proponents, that maximin would avoid using a von Neumann–Morgenstern utility function, designed for risk

taking, in the area of morals, where it does not belong? This argument is that utility functions express the subjective importance people *do attribute* to their needs and interests, not the importance that they *ought to attribute*. Harsanyi wishes to make moral judgments on the basis of subjective utility functions rather than on the basis of unchanging moral principles, such as "grant equal justice to equal beings." For him, weighting the subjective importance attached to things is more important than guaranteeing adherence to moral principles, because people's preferences are *different*. But if people's preferences are different, then their utility functions may operate according to different psychological laws. But this conclusion contradicts two of Harsanyi's claims: (1) that "preferences and utility functions of all human individuals are governed by the same basic psychological laws" (53, p. 602); (2) that interpersonal utility comparisons are theoretically capable of being specified completely because they "have a completely specific theoretical meaning" (53, p. 602).

If the reasoning in the previous arguments is correct, then Harsanyi cannot coherently claim *both* that (A) preferences are needed as measures of welfare, because people's preferences/utility functions are *different, and* that (B) interpersonal comparisons of utility are possible because people's utility functions "are governed by the *same* basic psychological laws" (53, p. 602).

Conclusion

Because all four classes of uncertainty—framing uncertainty, modeling uncertainty, statistical uncertainty, and decision-theoretic uncertainty—are common to many environmental sciences, arriving at our four sets of rules for dealing with these cases should be useful to a variety of scientists. (1) In cases of framing uncertainty, scientists ought not make question-begging use of two-valued frames for assessing hypothesis suitability in cases of radically incomplete data. (2) In cases of modeling uncertainty, scientists ought not claim to have verified or validated their results when they have merely determined their consistency with computer models. (3) In cases of statistical uncertainty in which they are forced to choose between maximizing false positives (type I errors) or false negatives (type II errors), when scientists are faced with potentially catastrophic consequences affecting welfare, they ought to maximize false positives. (4) In cases of decision-theoretic uncertainty involving potentially catastrophic consequences, scientists ought to use maximin rather than expected-utility rules.

This chapter has argued that many decisions in situations of scientific uncertainty have been inappropriate in precisely the four ways just outlined. They have overestimated the epistemologic errors likely to result in bad science and underestimated the ethical errors likely to result in bad science policy. Doing science well thus requires us to understand the environmental contexts of its applications.

References

1. Newton I. (A Motte trans.). Philosophiae naturalis principia. New York: Greenwood Press, 1969.
2. Aristotle. (W Ross trans.). De Caelo. Oxford: Oxford University Press, 1928.
3. Minsky M. A framework for representing knowledge. In: Haugeland J, ed. Mind design. Cambridge, MA: MIT Press, 1981:95–128.
4. Fetzer JH. Artificial intelligence. Dordrecht: Kluwer, 1990.
5. Popper K. Conjectures and refutations. New York: Harper and Row, 1968.
6. Millikan R. The electron. Chicago: University of Chicago Press, 1917.
7. Frank P. Philosophy of science. Englewood Cliffs, NJ: Prentice-Hall, 1962.
8. Duhem P. (P Wiener trans.). The aim and structure of physical theory. Princeton, NJ: Princeton University Press, 1982.
9. Carnap R. Philosophical foundations of physics. New York: Basic Books, 1966.
10. Cranor C. Regulating toxic substances. New York: Oxford University Press, 1993.
11. Younker JL, et al. Report of early site-suitability evaluation of the potential repository site at Yucca Mountain, Nevada. SAIC-91/8000. Washington, DC: US Department of Energy, 1992.
12. Younker JL, et al. Report of the peer review panel on the early site suitability evaluation of the potential repository site at Yucca Mountain, Nevada. SAIC-91/8001. Washington, DC: US Department of Energy, 1992.
13. National Research Council. Risk assessment in the federal government. Washington, DC: National Academy Press, 1983.
14. National Research Council. Pharmacokinetics in risk assessment. Washington, DC: National Academy Press, 1987.
15. Shrader-Frechette K. Burying uncertainty: Geological disposal and the risks of nuclear waste. Berkeley: University of California Press, 1993.
16. Slovic P, Flynn J, Layman M. Perceived risk, trust, and the politics of nuclear waste. Science 1991;254:1603–1604.
17. Luce RD, Raiffa H. Games and decisions. New York: Wiley, 1957.
18. Shrader-Frechette, K. Environmental risk assessment and the frame problem. BioScience 1994;44:8:548–551.
19. Loux R. State of Nevada comments on the US Department of Energy site characterization plan, Yucca Mountain site, Nevada. Carson City, NV: Nevada Agency for Nuclear Projects, 1989.
20. Sinnock S, et al. Preliminary estimates of groundwater travel time and radionuclide transport at the Yucca Mountain repository site. SAND85-2701. Albuquerque, NM: Sandia National Labs, 1986.
21. Costin L, Bauer S. Thermal and mechanical codes first benchmark exercise, Part I: Thermal analysis. SAND88-1221 UC-814. Albuquerque, NM: Sandia National Labs, 1990.
22. Hayden N. Benchmarking: NNMSI flow and transport codes: Cove 1 results. SAND84-0996. Albuquerque, NM: Sandia National Labs, 1985.
23. Dijstra E. A discipline of programming. Englewood Cliffs, NJ: Prentice-Hall, 1976.
24. Hoare C. An axiomatic basis for computer programming. Comm ACM 1969;12:576–580.
25. Hoare C. Mathematics of programming. BYTE 1986 August:115–149.
26. Fetzer JH. Program verification: The very idea. Comm ACM 1988;31:9:1048–1063.
27. Fetzer JH. Mathematical proofs of computer system correctness. N Am Math Soc 1989; 36:10:1352–1353.
28. Office of Civilian Radioactive Waste Management. Performance assessment strategy plan for the geologic repository program. DOE/RW-0266P. Washington, DC: Government Printing Office, 1990.
29. Fetzer JH. Another point of view. Comm ACM 1989;32:8:921.
30. Savitzky S. Letters. Comm ACM 1989;32:3:377.
31. Nelson D. Letters. Comm ACM 1989;32:7:792.

32. Dobson J, Randell B. Program verification. Comm ACM 1989;32:4:422.

33. Hopkins P. Cone 2A Benchmarking calculations using LLUVIA. SAND88-2511-UC-814. Albuquerque, NM: Sandia National Labs, 1990.

34. Cox L, Ricci P. Legal and philosophical aspects of risk analysis. In: Paustenbach D, ed. The risk assessment of environmental and human health hazards. New York: Wiley, 1989: 1025–1038.

35. Shrader-Frechette K. Risk and rationality: philosophical arguments for populist reforms. Berkeley: University of California Press, 1991.

36. Mathews JT. World resources 1986. New York: Basic Books, 1986.

37. Repetto R. Paying the price: pesticide subsidies in developing countries. Washington, DC: World Resources Institute, 1985.

38. Axinn S. The fallacy of the single risk. Philos Sci 1966;33:1/2:154–162.

39. Shue H. Exporting hazards. In: Brown P, Shue H, eds. Boundaries: national autonomy and its limits. Totowa, NJ: Rowman and Littlefield, 1981:107–145.

40. Lichtenberg J. National boundaries and moral boundaries. In: Brown P, Shue H, eds. Boundaries: national autonomy and its limits. Totowa, NJ: Rowman and Littlefield, 1981: 79–100.

41. Becker L. Rights. In: Becker L, Kipnis K, eds. Property. Englewood Cliffs, NJ: Prentice-Hall, 1984:76–77.

42. Thomson JJ. Rights, restitution, and risk: essays, in moral theory. Cambridge, MA: Harvard University Press, 1986.

43. Frankena WK. The concept of social justice. In: Brandt R, ed. Social justice. Englewood Cliffs, NJ: Prentice-Hall, 1962:10–14.

44. Ames B, Magaw R, Gold L. Ranking possible carcinogens. In: Paustenbach DJ, ed. The risk assessment of environmental and human health hazards. New York: Wiley, 1989: 1090–1093.

45. Whipple C. Nonpessimistic risk assessment. In: Paustenbach DJ, ed. The risk assessment of environmental and human health hazards. New York: Wiley, 1989:1109–1110.

46. MacLean D. Introduction. In: MacLean D, ed. Values at risk. Totowa, NJ: Rowman and Littlefield, 1986:1–2.

47. US Nuclear Regulatory Commission. Reactor safety study: an assessment of accident risks in US commercial nuclear power plants. NUREG-75/014, WASH 1400. Washington, DC: US Government Printing Office, 1975.

48. Shrader-Frechette K. Nuclear power and public policy. Boston: Reidel, 1983.

49. Union of Concerned Scientists. The risks of nuclear power reactors: a review of the NRC reactor safety study, WASH 1400. Cambridge, MA: Union of Concerned Scientists, 1977.

50. Nuclear Energy Policy Study Group. Nuclear power: issues and choices. Cambridge, MA: Ballinger, 1977.

51. Cooke RM. Risk assessment and rational decision. Dialec 36;4:334–335. 1982.

52. Samuels S. The arrogance of intellectual power. In: Woodhead A, Bender M, Leonard R, eds. Phenotypic variation in populations. New York: Plenum, 1988:113–120.

53. Harsanyi J. Can the maximin principle serve as a basis for morality? A critique of John Rawls's theory. Am Polit Sci R 1975;69:2:596–606.

54. Resnick M. Choices. Minneapolis: University of Minnesota Press, 1987.

55. Harsanyi J. Advances in understanding rational behavior. In: Elster J, ed. Rational choice. New York: New York University Press, 1986:88–89.

56. Harsanyi J. Understanding rational behavior. In: Butts RE, Hintikka J, eds. Foundational problems in the special sciences. Vol 2. Boston: Reidel, 1977:320–323.

57. Tversky A, Kahneman D. The framing of decisions and the psychology of choice. In: Elster J, ed. Rational choice. New York: New York University Press, 1986:123–125.

58. Otway H, Peltu M. Regulating industrial risks. London: Butterworths, 1985.

59. Davis R. Introduction. In: Thrall R, Coombs C, Davis R, eds. Decision processes. London: Wiley, 1954:1–14.

60. Ellsworth L. Decision-theoretic analysis of Rawls' original position. In: Hooker CA, Leach

JJ, McClennen EF, eds. Foundations and applications of decision theory. Dordrecht: Reidel, 1978:39–40.

61. Jeffrey RC. The logic of decision. Chicago: University of Chicago Press, 1983.

62. Marschak J. Towards an economic theory of organization and information. In: Thrall R, Coombs C, Davis R, eds. Decision processes. London: Wiley, 1954:187–197.

63. McClennen EF. The minimax theory and expected utility reasoning. In: Hooker CA, Leach JJ, McClennen EF, eds. Foundations and applications of decision theory. Dordrecht: Reidel, 1978:337–338.

64. Milnor J. Games against nature. In: Thrall R, Coombs C, Davis R, eds. Decision processes. London: Wiley, 1954:49–50.

65. Harsanyi J. Understanding rational behavior. In: Butts RE, Hintikka J, eds. Foundational problems in the special sciences. Vol 2. Boston: Reidel, 1977:320–323.

66. Harsanyi J. On the rationale of the Bayesian approach. In: Butts RE, Hintikka J, eds. Foundational problems in the special sciences. Vol 2. Boston: Reidel, 1977:382–384.

67. MacLean D. Social values and the distribution of risk. In: MacLean D, ed. Values at risk. Totowa, NJ: Rowman and Littlefield, 1986.

68. Rawls J. A theory of justice. Cambridge: Harvard University Press, 1971.

69. Sen A. Rawls versus Bentham: An axiomatic examination of the pure distribution problem. In: Daniels N, ed. Reading Rawls. New York: Basic Books, 1981:283–292.

70. Rawls J. Some reasons for the maximin criterion. Am Econ Rev 1974;64:1:141–146.

71. MacLean D. Risk and consent. In: MacLean D, ed. Values at risk. Totowa: Rowman and Littlefield, NJ, 1986:17–30.

72. Andrews RN. Environmental impact assessment and risk assessment. In: Wathern P, ed. Environmental impact assessment. London: Unwin Hyman, 1988:85–97.

73. March JG. Bounded rationality. In: Elster J, ed. Rational choice. New York: New York University Press, 1986:148–149.

74. Clifford WK. Lectures and essays. London: Macmillan, 1986.

75. Michalos AC. Foundations of decisionmaking. Ottowa: Canadian Library of Philosophy, 1987.

76. Ames B, Magaw R, Gold L. Ranking possible carcinogenic hazards. Science 1987;236:271–280.

77. Shrader-Frechette K. Risk analysis and scientific method. Boston: Reidel, 1985.

78. Hacking I. Culpable ignorance of interference effects. In: MacLean D, ed. Values at risk. Totowa, NJ: Rowman and Littlefield, 1986:136–154.

79. Kahneman D, Tversky A. Subjective probability. In: Kahneman D, Tversky A, Slovic P, eds. Judgment under uncertainty: heuristics and biases. Cambridge: Cambridge University Press, 1981:32–46.

80. Kahneman D, Tversky A. On the psychology of prediction. In: Kahneman D, Tversky A, Slovic P, eds. Judgment under uncertainty: heuristics and biases. Cambridge: Cambridge University Press, 1981:68–69.

81. MacLean D. Philosophical issues for centralized decisions. In: MacLean D, ed. Values at risk. Totowa, NJ: Rowman and Littlefield, 1986:17–30.

82. Smith VK. Benefit analysis for natural hazards. Risk Anal 1986;6:3:325–326.

83. Levi I. Newcomb's many problems. In: Hooker CA, Leach JJ, McClennen EF, eds. Foundations and applications of decision theory. Vol 1. Dordrecht: Reidel, 1978:369–383.

84. Vail S. Alternative calculi of subjective probabilities. In: Thrall R, Coombs C, Davis R, eds. Decision processes. London: Wiley, 1954:87–88.

85. Gauthier D. The social contract. In: Hooker CA, Leach JJ, McClennen EF, eds. Foundations and applications of decision theory. Vol 2. Dordrecht: Reidel, 1978:47–53.

86. Borch K. Ethics, institutions, and optimality. In: Gottinger HW, Leinfellner W, eds. Decision theory and social ethics. Dordrecht: Reidel, 1978:242–243.

87. Gersuny C. Work hazards and industrial conflict. Hanover, NH: University of New England Press, 1981.

88. Gardenfors P. Fairness without interpersonal comparisons. Theoria 1978;44:part 2:57–58.

89. Zagare F. Game theory. Beverly Hills: Sage, 1984.
90. Lave L, Leonard B. Regulating coke oven emissions. In: Paustenbach DJ, ed. The risk assessment of environmental and human health hazards. New York: Wiley, 1989:1064–1081.
91. Blackstone WT. On the meaning and justification of the equality principle. In: Blackstone WT, ed. The concept of equality. Minneapolis: Burgess, 1969:121–122.
92. Baier A. Poisoning the wells. In: MacLean D, ed. Values at risk. Totowa, NJ: Rowman and Littlefield, 1986:49–74.
93. Rawls J. Justice as fairness. In: Feinberg J, Gross H, eds. Philosophy of law. Encino, CA: Dickenson, 1975:284–285.
94. Beardsley MC. Equality and obedience to law. In: Hook S, ed. Law and philosophy. New York: New York University Press, 1964:35–193.
95. Berlin I. Equality. In: Hook S, ed. Law and philosophy. New York: New York University Press, 1964:33–34.
96. Frankena WK. Some beliefs about justice. In: Feinberg J, Gross H, eds. Philosophy of law. Encino, CA: Dickenson, 1975:250–251.
97. Markovic M. The relationship between equality and local autonomy. In: Feinberg W, ed. Equality and social policy. Urbana: University of Illinois, 1978:93–94.
98. Vlastos G. Justice and equality. In: Brandt RB, ed. Social justice. Englewood Cliffs, NJ: Prentice-Hall, 1962:50–56.
99. Pennock JR. Introduction. In: Pennock JR, Chapman JW, eds. The limits of the law. Nomos 15. The Yearbook of the American Society for Political and Legal Philosophy. New York: Lieber-Atherton, 1974:2–6.
100. Kates R, et al., eds. Hazards: technology and fairness. Washington, DC: National Academy Press, 1986.
101. Shrader-Frechette K. Science policy, ethics, and economic methodology. Boston: Reidel, 1985.
102. Sen A. Welfare inequalities and Rawlsian axiomatics. In: Butts RE, Hintikka J, eds. Foundational problems in the special sciences. Boston: Reidel, 1977:271–288.
103. Taylor R. Justice and the common good. In: Blackstone WT, ed. The concept of equality. Minneapolis: Burgess, 1969:94–97.
104. Benn SI. Egalitarianism and the equal consideration of interests. In: Pennock JR, Chapman JW, eds. Equality. Nomos 9, Yearbook of the American Society for Political and Legal Philosophy. New York: Atherton Press, 1967:75–76.
105. Patterson O. Inequality, freedom, and the equal opportunity doctrine. In: Feinberg W, ed. Equality and social policy. Urbana: University of Illinois Press, 1978:31–32.
106. Pennock JR. Democratic political theory. Princeton: Princeton University Press, 1979.
107. Rees J. Equality. New York: Praeger, 1971.
108. Stewart RB. Paradoxes of liberty, integrity, and fraternity: the collective nature of environmental quality and judicial review of administrative action. Environ Law 1977;7:3:474–476.
109. Bethe H. The necessity of fission power. Sci Am 1976;234:1:26–27.
110. Drucker P. Saving the crusade. In: Shrader-Frechette K, ed. Environmental ethics. Pacific Grove, CA: Boxwood Press, 1980:200–207.
111. Maddox J. The doomsday syndrome. London: Macmillan, 1972.
112. Maxey MM. Radwastes and public ethics. Health Phys 1978;34:2:129–135.
113. Gans HJ. The costs of inequality. In: Mooney M, Stuber F, eds. Small comforts for hard times. New York: Columbia University Press, 1977:50–51.
114. Laski H. Liberty and equality. In: Blackstone WT, ed. The concept of equality. Minneapolis: Burgess, 1969:170–173.
115. Larkin A. The ethical problem of economic growth vs. environmental degradation. In: Shrader-Frechette K, ed. Environmental ethics. Pacific Grove, CA: Boxwood Press, 1980:208–220.
116. North DC, Miller RL. The economics of public issues. New York: Harper and Row, 1971.

117. Plamenatz JP. Equality of opportunity. In: Blackstone WT, ed. The concept of equality. Minneapolis: Burgess, 1969:88–89.

118. Scharr JH. Equality of opportunity and beyond. In: Pennock JR, Chapman JW, eds. Equality. Nomos 9. Yearbook of the American Society for Political and Legal Philosophy. New York: Atherton Press, 1967.

119. Williams B. The idea of equality. In: Blackstone WT, ed. The concept of equality. Minneapolis: Burgess, 1969:49–53.

120. Gibbard A. Risk and value. In: MacLean D, ed. Values at risk. Totowa, NJ: Rowman and Littlefield, 1986:97–99.

121. Mishan EJ. 21 popular economic fallacies. New York: Praeger, 1969.

122. Grossman R, Daneker G. Jobs and energy. Washington, DC: Environmentalists for Full Employment, 1977.

123. Freeman AM. Distribution of environmental quality. In: Kneese AV, Bower BT, eds. Environmental quality analysis. Baltimore, MD: Johns Hopkins, 1972:271–275.

124. Kneese AV, Schultze CL. Pollution, prices and public policy. Washington, DC: Brookings Institution, 1975.

125. Stewart RB. Pyramids of sacrifice? Problems of federalism in mandating state implementation of natural environmental policy. In: Strom FA, ed. Land use and environment law review—1978. New York: Clark Boardman, 1978:172–173.

126. Asch P, Seneca JJ. Some evidence on the distribution of air quality. Land Econ 1978;54: 3:278–297.

127. Brodine V. A special burden. Envir 1971;13:2:24–25.

128. Dane DN. Bad air for children. Envir 1976;18:9:26–34.

129. Freeman AM. Income distribution and environmental quality. In: Enthoven AC, Freeman AM, eds. Pollution, resources, and the environment. New York: WW Norton, 1973:101–102.

130. Freeman AM, Haveman RH, Kneese AV. The economics of environmental policy. New York: Wiley, 1973.

131. Kneese AV. Economics and the quality of the environment. In: Enthoven AC, Freeman AM, eds. Pollution, resources, and the environment. New York: WW Norton, 1973:74–79.

132. Ramsey DD. A note on air pollution, property values, and fiscal variables. Land Econ 1976;52:2:230–234.

133. Stein J. Water for the wealthy. Envir 1977;19:4:6–14.

Further Reading

Oreskes N, Shrader-Frechette K, Belitz K. Verification, validation, and confirmation of numerical models in the earth sciences. Science 1994;263:641–646.

Shrader-Frechette K, McCoy E. Statistics, costs, and rationality in ecological inference. Tr Ecol Evo 1992;7(3):96–99.

Uncertainties in the Disposal of High-Level Nuclear Waste

2

John Lemons

Radioactive wastes are generated in the nuclear fuel cycle as well as in nuclear applications in medicine, research, and industry. Broadly speaking, the categories of radioactive waste include high-level waste (HLW), transuranic waste, uranium mill tailings, and low-level waste. The radiologic and safety risk from radioactive waste varies from low in short-lived, low-level waste to large for HLW. The HLW contains about 99% of the radionuclides and therefore represents the largest radiologic and safety risk. High-level nuclear waste is generated primarily by nuclear power plants and militaries with nuclear capabilities and generally is defined as (a) irradiated reactor fuel, (b) liquid waste resulting from reprocessing irradiated reactor fuel, and (c) solids into which such liquid waste has been converted.

The safe and environmentally sound management of radioactive waste includes (a) policies and measures to minimize and limit, where appropriate, the generation of radioactive waste and provide for their safe processing, conditioning, transportation, and disposal; (b) efforts within the International Atomic Energy Agency (IAEA) to develop and promulgate radioactive waste safety standards or guidelines and codes of practice as an internationally accepted basis for the safe and environmentally sound management and disposal of radioactive waste; (c) safe storage, transportation, and disposal of radioactive waste, as well as spent radiation sources and spent fuel from nuclear reactors destined for final disposal in all countries, in particular developing countries, by facilitating the transfer of relevant technologies to those countries and/or the return to the supplier of radiation sources after their use, in accordance with relevant international regulations or guidelines; and (d) proper planning, including environmental impact assessment where appropriate, of safe and environmentally sound management of radioactive waste, including emergency procedures, storage, transportation, and disposal before and after activities that generate such waste.

Problems associated with the safe handling and disposal of HLW are scientifically and technically complex and controversial, and how they are resolved has significant public-policy implications to human and ecosystem health. Facilitating the understanding of the scientific issues surrounding radioactive waste management and their implications is a primary goal of this chapter. Because it is beyond the scope of this chapter to discuss all relevant issues pertaining to the safe and environmentally sound management of radioactive waste, I will focus on the most troubling aspects of nuclear waste management associated with the storage and disposal of HLW.

Understanding the role of science in HLW decisions requires an assessment of the use of science in repository siting decisions and of the capabilities of science to yield reasonably certain information about future repository performance. In order to provide this assessment, I first present (a) a brief description of options for storage and disposal of HLW, (b) a brief description of health and environmental protection criteria for HLW disposal, and (c) a retrospective examination of the role of science and scientific uncertainties in public-policy decisions for siting a prospective HLW repository in the United States. I then discuss (a) performance assessment for HLW disposal sites, (b) scientific uncertainties inherent in the performance assessment of the planned HLW repository in the United States, (c) the relationship between scientific uncertainty and protection standards, and (d) the implications of scientific knowledge to distributional problems posed by the location of the HLW site in the United States.

Options for Handling, Storage, and Disposal

International cooperation regarding the management of nuclear waste is promoted by the IAEA, the European Economic Community, and the Nuclear Energy Agency of the Organization for Economic Cooperation and Development. Presently, these agencies emphasize the development of site investigation procedures, design and feasibility studies, and design of safety assessments. The mode of disposal for HLW is being investigated in many countries, and there appears to be general agreement that it should be isolated in deep geologic burial sites, or repositories. Ideally, deep geologic disposal relies on engineered canisters to contain radionuclides for, say, up to 1000 years, and on the geo-hydrologic characteristics of the site to prevent migration of radionuclides beyond site boundaries for 10,000 years or beyond. Although there is general agreement by countries to dispose of HLW in deep geologic burial sites, they are developing or investigating various options with respect to certain aspects of handling, storage, and disposal (1). Presently, the United States is the only country which has identified a HLW repository site and is in the process of characterizing it. In addition, the United States is the only country to have established a deadline for repository development and operation. Germany has identified a prospective HLW repository site but is not yet conducting characterization activities. Other nations have not identified repository sites or rapid repository development schedules because of scientific uncertainties concerning repository development and safety.

Health and Environmental Protection Criteria

Assessment of public and environmental protection criteria is based on the radiation exposure incurred by future individuals or populations from stored or disposed wastes. The IAEA has specified the following public and environmental protection standards relevant to the period after closure of a repository: (a) radiologic detriments, as expressed by the predicted dose distributions and their probabilities of occurrence, shall be reduced to levels which are as low as is reasonably achievable, economic and social factors being taken into account; (b) the radiologic detriment to any individual in the future, as expressed by the predicted dose and its probability of occurrence, shall not be greater than is now regarded as acceptable for individuals; and (c) siting, design, construction, operation, and post-closure of a repository shall be optimized to ensure that adverse effects on the environment and natural resources now and in the future are kept as low as reasonably achievable.

The IAEA's criteria apply to the geologic environment, the repository, and the HLW package. The criteria for the geologic environment state that (a) the site should provide a geologic formation large enough to contain the disposal system as well as an adequate subsurface buffer volume and an appropriate surface exclusion area; (b) a repository should be located in a geologic medium with a lithology and depth appropriate for the categories and quantities of waste to be disposed of; (c) the geohydrologic characteristics of the geologic environment should tend to restrict groundwater flow within the repository; (d) the physicochemical and geochemical characteristics of the geologic environment should tend to limit the transport of radionuclides; (e) the repository should be located in an area of low tectonic and seismic activity, which is far enough from areas with major tectonic activity to provide assurance that the integrity of the repository will not be endangered; (f) the location of the repository should take into account the presence of human and natural features likely to generate structural instability; (g) the location of valuable geologic resources or potential future resources should be considered, and the need for the repository at a specific time and place should be balanced against the need for and value of the geologic resources now and in the future; and (h) the occurrence of geomorphologically unstable environments, extreme climatic conditions and other processes which might affect the land surface should be considered in order to ensure that they will not significantly affect repository performance.

The criteria for the repository state that (a) the repository should be designed, constructed, and operated in a manner which will avoid significant adverse effects due to repository development and the presence of emplaced wastes on the isolation capability of the geologic environment; (b) all necessary safety systems should be provided and must continue to function until closure of the repository; (c) handling, storage, and disposal systems of the repository should be designed to ensure that waste will remain subcritical during the operational phases of the repository as well as during its post-closure period; (d) where backfilling is used, methods and materials should be chosen with due consideration for the need to enhance the isolation of the waste and to limit adverse effects on the geologic medium; (e) closure of the repository should be carried out in such a way as to prevent the release of radionuclides at rates or in quantities which may result in unacceptable radiologic consequences; (f) appropriate surveillance should be carried out during the operational phase to confirm that the behavior of the repository and its contents is as predicted and to provide a basis for remedial actions if required; and (g) after closure of the repository, surveillance should not be necessary to ensure the health and safety of the public.

The HLW package criteria state that (a) the radionuclide content and composition of the waste form should be known with sufficient accuracy to ensure compliance with authorized limits; (b) the chemical composition of the waste form should be such as to avoid possibly harmful chemical or microbial interactions in the repository; (c) the chemical durability of the radioactive waste form in the disposal environment should be sufficient to restrict the release of radionuclides to levels consistent with the requirements of the overall system; (d) the waste form should be sufficiently resistant to radiation and heat to avoid changes in its physical, chemical, and mechanical properties that would unduly affect its ability to restrict the release of radionuclides; (e) the waste package should have sufficient mechanical stability that, together with the repository structure, adequate protection of the waste form is provided under disposal conditions; (f) the waste form should be packaged in such a way that deterioration during handling, transport, and disposal under normal or accident conditions is avoided; and (g) the use of packaging materials that are widely expected to become attractive resources should be weighed against the risk that future human intrusion into the repository may be encouraged by their presence. Finally, quality assurance criteria state that an overall quality assurance program for all activities from siting to closure and for components of the disposal system should be established.

The International Commission on Radiological Protection (ICRP) has promulgated recommendations on radioactive waste disposal which strongly influence the actions of governments, despite the fact that it has no formal authority (2). Specifically, it has developed several radiologic protection principles: (a) no practice shall be adopted unless its introduction produces a positive net benefit; (b) all exposures are to be kept as low as reasonably achievable, economic and social factors being taken into account; and (c) individual doses should not exceed limits recommended by the ICRP. The ICRP also noted that in dealing with the long-term aspects of radioactive waste disposal, a decision must be made on the level of protection to be afforded to future individuals, and it recommended that risks to future individuals should be limited on the same basis as are those to individuals living now.

Individual nations also have promulgated their own protection criteria. A description and analysis of the U.S. criteria will be provided later. However, a comparison of international and national criteria suggests that there are the following differences between them: (a) the IAEA and ICRP criteria provide the same protection for present and future individuals (apparently indefinitely); (b) U.S. criteria do not protect individuals outside of the repository boundaries beyond a time period of 1000 years, or populations outside of that area beyond a time period of 10,000 years; (c) with limited exceptions, U.S. criteria do not restrict radionuclide exposure within repository boundaries; and (d) IAEA and ICRP criteria specifically endorse the ALARA principle (as low as reasonably allowed), whereas U.S. regulations specifically reject it. In addition, the United States is the only country to have established quantitative probabilistic standards for radionuclide exposure; other countries have established qualitative standards or have not established standards yet. The criteria or guidelines of most other nations are more consistent with the IAEA and ICRP criteria (1). In other words, they (a) protect present and future individuals equally, (b) protect individuals outside of the repository boundaries for beyond 10,000 years, (c) restrict radionuclide exposure within repository boundaries, and (d) endorse the ALARA principle.

High-Level Nuclear Waste Disposal in the United States: A Case Study

The United States is probably further along than most countries in attempting to develop a geologic repository for HLW. In 1982, the Nuclear Waste Policy Act (NWPA) (P.L. 97-425) identified the objective of developing mined geologic repositories for disposal of commercial and defense-generated HLW, and established a site selection process to include studies of technical suitability and protection of public health, safety, and the environment. As originally envisioned by NWPA, for reasons of equity two permanent repositories were to be built ultimately, one in the western part of the country and one in the eastern. The two sites were to be selected after a screening of a number of other potential sites. The screening process was to include the use of sound science in the selection of potential sites for characterization, and cooperation and consultation with states and affected parties (3). However, it became controversial because of questions concerning whether mandates regarding the use of sound science and cooperation and consultation were being fulfilled (4).

In 1987, the Nuclear Waste Policy Amendments Act (NWPAA) (Title V of P.L. 100-203) established Yucca Mountain, Nevada, as the single site to be characterized in the United States for acceptability as a prospective HLW repository. The NWPAA was passed because Congress feared that political controversies surrounding the site selection process under NWPA would undermine efforts to build a repository. In part, the decision to conduct site characterization activities at Yucca Mountain was based on the fact that the site is underlain by a thick welded tuff in the unsaturated zone in an arid region, and is remote from population centers and adjacent to the Nevada Test Site, a major nuclear defense reservation controlled by the United States Department of Energy (DOE), the agency responsible for conducting site characterization activities.

The repository siting and development program at Yucca Mountain is expected to be accomplished through the next several decades to about 2010 or later. The performance of the potential repository depends on canisters designed to contain waste for 300–1000 years and on the integrity of the geologic, hydrologic, and geochemical setting to prevent waste released from corroded canisters from reaching the environment accessible to humans for another 9000–9700 years. Ultimately, the license for operation of the repository will be granted by the Nuclear Regulatory Commission (NRC) based on data from site characterization studies demonstrating that performance criteria would be met.

The Environmental Protection Agency (EPA) has promulgated containment standards for protecting human health and the environment from excessive releases of radionuclides from a HLW repository. The EPA regulations are promulgated at 40 Code of Federal Regulations (CFR) 191. Subpart B of these regulations establishes several different types of requirements. The first are primary requirements that limit projected releases of radioactivity to the accessible environment for 10,000 years after disposal. A second set of requirements limits exposures to individual members of the public for 1000 years after disposal. Finally, the regulations also include a set of groundwater protection requirements that limit radionuclide concentrations for 1000 years after disposal.

The regulations also specify that disposal systems for the waste shall be designed to provide reasonable expectations based on performance assessments that the cumulative releases of radionuclides to the accessible environment shall for 10,000 years after disposal (a) have a likelihood of less than 1 chance in 10 of exceeding the standards and (b) have a likelihood of less than 1 chance in 1000 of exceeding 10 times the standards. The NRC has promulgated regulations in 10 CFR 60 calling for repository licensing to be based on use of verifiable and tested predictive scientific models and data and the need to reduce uncertainties in predictions of repository performance by use of models that leave little room for probabilistic approaches. As will be discussed, because of the complex problems of building a permanent repository and the many uncertainties involved, it is questionable whether the aforementioned EPA and CFR requirements will be met (5).

Because the United States has the greatest amount of HLW (approximately 24 billion Ci at the end of 1990) and is the only country characterizing a potential repository site, an analysis of the role of science in developing the U.S. repository siting program provides insights for those of other nations attempting to manage HLW (6).

Science and Public-Policy Goals Under the National Environmental Policy Act

In advanced industrial countries, a traditional way of dealing with complex environmental problems has developed in the last two decades. The United States, like many other nations, has adopted environmental laws that empower technical experts to study problems and take regulatory action which manages the environmental problems in conformance with the goals contained in the authorizing legislation. Among other things, achieving the goals of environmental legislation requires the use of sound science to the maximum extent possible. The management of the environmental problems associated with the disposal of HLW follows this basic approach.

Congress frequently passes laws which have the effect of committing us, as a nation, to particular stances on public-policy issues (7). One of the most important environmental laws pertaining to all federal actions which potentially have environmental consequences is the National Environmental Policy Act (NEPA). As I discuss in this and the subsequent section of the chapter, certain exemptions and limitations were made to NEPA regarding the use of science in the HLW site selection process. In part, they were made because of congressional fear that scientific uncertainty would undermine the nation's ability to select a HLW site (3, 4). As I discuss, such exemptions and limitations are contrary to the public-policy goals of NEPA.

In passing NEPA, Congress enunciated an ethical goal or principle on matters environmental. Congress characterized the purpose of NEPA as follows

> To declare a national policy which will encourage productive and enjoyable harmony between man and his environment; to promote efforts which will prevent or eliminate damage to the environment and biosphere and stimulate the health and welfare of man; to enrich the understanding of ecological systems and natural resources important to the nation; and to establish a Council on Environmental Quality.

This statement of purpose suggests that Congress intended to embody in law a particular conception of the good society and thus to have enunciated and espoused a goal of the nation to have a relationship of productive and enjoyable harmony with the natural environment. However, the exact content of that goal is ambiguous because NEPA contains language requiring the nation to strike a balance between exploitation and conservation of natural resources and between the interests of present and future generations. For example, 42 USC 4331(b) states that

> . . . it is the continuing responsibility of the Federal Government to use all practicable means . . . [to] coordinate Federal Plans, functions, programs, and resources to the end that the Nation may: . . . (a) fulfill the responsibilities of each generation as trustee of the environment for succeeding generations . . . [and] (b) attain the widest range of beneficial uses of the environment without degradation, risk to health or safety, or other undesirable and unintended consequences. . . .

The policy objectives of NEPA are ethical insofar as they adopt a conception of society that involves a "productive and enjoyable harmony" with the natural environment, but the enunciated goal is ambiguous insofar as the precise nature of the balance to be struck between environmental quality and human exploitation of the environment is left ambiguous.

What is clear from the legislative history of NEPA is that it intended to force mission-oriented agencies, such as the DOE and the NRC, to consider the goals required by environmental protection objectives in conjunction with the goal of the individual agency when taking action. In particular, the NEPA requirement that agencies prepare an environmental impact statement (EIS) before undertaking any major action significantly affecting the quality of the human environment has been consistently construed by the courts to require agencies seriously and carefully to examine the environmental consequences of proposed actions and their alternatives, including that of no action (8).

Several important provisions are contained in the EIS requirements for NEPA (40 CFR 1502.14). These regulations provide that

> . . . the discussion of alternatives including the proposed action is the heart of the environmental impact statement . . . it should present the environmental impacts of the proposal and the alternatives in comparative form, thus sharply defining the issues and providing a clear basis for choice among options by the decisionmaker and the public.

These regulations go on to state that agencies shall (a) rigorously explore and objectively evaluate all reasonable alternatives, and for all alternatives that were eliminated from detailed study, briefly discuss the reasons for their having been eliminated; (b) devote substantial treatment to each alternative considered in detail including the proposed action so that reviewers may evaluate their comparative merits; (c) include reasonable alternatives not within the jurisdiction of the lead agency; (d) include the alternative of no action; (e) identify the agency's preferred alternative or alternatives, if more than one exists, in the draft statement and identify such alternative in the final statement unless the law prohibits the expression of such a preference; and (f) include appropriate mitigation measures not already included in the proposed action or alternatives. Only in this fashion is it likely that the most intelligent, optimally beneficial decision will be made. Clearly, Congress considered the analysis of alternatives and explicitly reasoning about which alternatives to undertake to be crucial means to achieving the public-policy goal it enunciated in NEPA.

Three possible approaches to interpreting the precise nature of the substantive balance (if any) which Congress intended agencies to strike between environmental and economic values pursuant to NEPA exist. These include interpretations that assert that (a) environmental values take priority over economic values, (b) environmental values should be considered and allocated efficiently with other values including economic considerations, and (c) NEPA forces technology and science to develop in appropriate ways so that there is a reduced conflict between environmental protection and a high standard of living.

Some scholars have argued that, where conflicts between environmental and economic values exist, NEPA requires that environmental values take priority over economic values. However, the Supreme Court has held that NEPA does not require an agency always to place environmental concerns over other appropriate considerations (9). Nevertheless, NEPA does place an emphasis on information acquisition during the EIS process, which is meant to be of use to federal agencies in utilizing a systematic, interdisciplinary approach in planning and in identifying and developing methods and procedures that will ensure that presently unquantified environmental amenities and values may be given appropriate consideration in decision making along with economic and technical considerations (42 USC 4332(2)(A) & (B)).

Because there is an emphasis placed on information acquisition in NEPA, and because once a federal agency collects the kind of information NEPA requires it to collect, it can still make any decision it chooses and that decision can still accurately be labeled an adequately informed decision. However, there is a distinction to be drawn between a decision being adequately informed and well reasoned. An intelligent decision is both adequately informed and well reasoned. The Council on Environmental Quality (CEQ) regulations implementing NEPA (quoted above), by requiring both that agencies consider alternatives and discuss the reasons for various alternatives that are being eliminated, make it clear that CEQ takes NEPA to require not just informed agency decisions but also intelligent agency decisions. The NEPA case law supports this interpretation. For example, in Vermont Yankee Nuclear Power Corporation *v.* Natural Resources Defense Council the court held that NEPA is not purely procedural in that NEPA does set forth significant substantive goals for the nation (10). The implications of NEPA case law are that Congress intended agencies to be able to rationalize the balance they established between environmental and economic values. A related consideration is that in a participatory democracy it is natural to require that (with certain and rare exceptions) agency decisions be publicly reviewable (11). This commitment also is evident in CEQ regulations implementing NEPA, which require not only that EISs be written but that the draft EISs be made available to public scrutiny and comment (40 CFR 1503.1(a)(4)). Where a final EIS is found wanting, citizens have the right to judicial review of the agency's decision.

Given that agencies have to rationalize their decisions in light of the environmental impact information brought to their attention by compliance with NEPA's EIS requirements, and given that agencies sometimes modify and sometimes abandon projects as a result, NEPA can be understood to have a technology-forcing role. When an important project is altered or abandoned because an agency or court judges that current technology cannot adequately mitigate adverse environmental effects, this decision forces technology to develop in ways that eventually will minimize or eliminate conflicts between environmental and economic values.

While the NEPA case law does not support interpretations that environmental values take priority over economic values, it does support interpretations that NEPA is to ensure (a) intelligent decision making insofar as environmental values are considered and then intelligently balanced against other values, including economic values, and (b) technology and science to develop in appropriate ways so that eventually there will be less conflict between environmental protection and economic values. As will be shown, in passing the NWPA and NWPAA, Congress appears to have abandoned both the intelligent decision-making and technology-forcing goals of NEPA, which meant that certain public-policy goals enunciated in NEPA also were abandoned. As a consequence, a number of public-policy issues stemming from the role and use of science in the HLW siting process were created which have not yet been resolved.

The Nuclear Waste Policy Act

Considerations of disposing of HLW in the United States date as far back as the mid-1950s. In response to decades-long political problems concerning HLW, in 1982 Congress passed the NWPA to guide the site selection process for potential repositories. In order that one area of the country would not assume all of the risks associated with storing nuclear wastes, NWPA called for the establishment of one western and one eastern repository, ultimately to be chosen after review of numerous potential sites. The intent of NWPA was to foster public confidence in the belief that state and citizen interests were reflected in both the siting and development of a repository, including considerations of issues of distributive justice, citizen participation, and decentralized decision making.

Despite the fact that NEPA serves as one of the most important pieces of environmental legislation in the United States, NWPA exempted major elements of DOE's site selection program from complying with NEPA's strongest environmental review and reporting requirements (4). The purpose of these exemptions was to avoid litigation and delay over the application of environmental requirements to the development of a repository. For instance, NWPA Sections 112(e) and 113(d) exempted DOE from major environmental protection mandates. The DOE would not have to prepare a comprehensive EIS or consider alternative courses of action that might involve conflicting uses of the same resources, as per NEPA Sections 102(2)(C) and 102(2)(E), respectively. Moreover, DOE was exempt from NEPA Section 102(2)(F), which requires recognition of the worldwide and long-range consequences of the decision. Importantly, NWPA Section 112(b)(3) prevented DOE from undertaking new investigations for the environmental assessments and restricted preparation of the documents to the use of existing information, which normally was not site-specific. Other subsections of Section 112 limited judicial review of environmental assessments to certain aspects of sufficiency. Sections 113 and 114(f) limited the discussions of alternative courses of action. These exemptions freed DOE from having to fulfill both substantive and procedural requirements of NEPA during its site selection process.

The NWPA Sections 114(d) and 114(f) limited the customary role of the NRC in evaluating environmental impacts for siting of nuclear facilities by restricting involvement to licensing once a construction site was selected and by requiring it to utilize DOE's EIS to the extent practicable. In effect, this let DOE identify potential repository sites while being exempt from the stringent environmental regulations typically enforced by NRC in siting nuclear facilities (12). Moreover, because NWPA exempted DOE from having to prepare a formal EIS during the site selection process, the substitute environmental assessment (EA) was by congressional legislation no longer subject to EIS regulations established by the CEQ.

Since the passage of NWPA, DOE has called attention to the fact that the EAs Congress wanted need not have been prepared in accordance with NEPA (13). In fact, some analysts claim DOE has interpreted NWPA to mean that Congress sanctions an abandonment of the notion of "intelligent decision making" infused throughout NEPA. For example, the EAs which DOE submitted to support its selection of potential repository sites failed to meet standards for CEQ and NRC environmental reviews (14). The DOE argued that compliance with federal environmental regulations like the Clean Water Act, Clean Air Act, and Endangered Species Act was sufficient. In other words, once freed from certain environmental protection mandates of NEPA, DOE believed a repository at its selected site will have no significant adverse impacts if it complies with existing environmental laws. This presents two problems. First, the United States does not yet have laws that cover all aspects of radioactive waste siting and its associated environmental impacts. Second, DOE's position has already been ruled legally invalid in the famous Calvert Cliffs case, where the courts judged that certified compliance with environmental standards does not relieve a federal agency from the necessity of reviewing potential impacts.

By 1987, substantial criticisms of the EAs which DOE had prepared during the site selection process under NWPA occurred; these dealt with the scientific processes used, scope of analyses, variables used for study, methodologies employed, data collection, and conclusions reached (4). Alarmed at the prospect of complete failure of the repository siting program, Congress legislatively intervened in the nuclear waste siting program a second time by adopting the NWPAA in 1987, which mandated that Yucca Mountain be selected as the only site for characterization as a HLW repository. Despite the scientific uncertainty regarding how to characterize a potential HLW site, Congress also imposed a rapid deadline for repository development.

Defendants of NWPA/NWPAA claim that despite the exemptions to NEPA, DOE must still practice "sound" environmental protection in siting a repository. For example, Sections 102(A), (B), (D), (G), (H), and (I) of NEPA still apply, requiring that DOE practice "interdisciplinary environmental decision making" through an abundant set of regulations and guidelines (15). However, critics raise two objections to NWPA/NWPAA and the way DOE interpreted the statutes. The first objection centers on general ambiguities surrounding NWPA Sections 112 and 113; the second is more specific and focuses on the fact that NWPA, NWPAA, and DOE's decisions deliberately evaded and eclipsed the importance of NEPA as a comprehensive policy act which calls for discussion of the many scientific and public-policy issues a repository decision demands.

Section 112(b)(1)(E) of NWPA required that each nomination of a site for characterization must be accompanied by an EA, which was to include a detailed statement of the basis of such nomination and of the probable impacts of site characterization activities planned for each site, and discussion of alternative activities relating to site characterization that may be undertaken to avoid such impacts. However, Congress neither indicated how the EAs were to be different from an EIS nor did it spell out guidelines for the EAs. In the absence of established guidelines, DOE had to decide what norms should govern screening and nomination of sites. Two basic approaches existed: optimization and suitability. These terms are not purely technical, but are subjective and value-laden. The optimization approach implies that DOE would identify the best sites in the country for nomination as potential repositories. Under the suitability approach, nominated sites could include those that were suitable but not necessarily the best. The DOE adopted the suitability approach, but did not provide for public discussion concerning the rationale for such a decision.

In addition, DOE had to decide the value-laden question of whether to adopt a lenient or restrictive interpretation of NWPA's requirements for scientific and technical information. While it can be argued that a restrictive approach stressing more adequate information was required to protect environmental values better, it also can be argued that this approach must be tempered with recognition that an unduly restrictive interpretation may have led to delay and increased costs that might have outweighed benefits in terms of either increased safety or decreased political opposition or reduced risk of successful litigation. Many controversies which surfaced at public hearings during the site nomination process centered on differences of opinion regarding whether DOE should have adopted a lenient or restrictive interpretation.

The NWPA also eclipsed NEPA as a comprehensive policy act. Section 114(f) of NWPA specifically bounds the analysis of alternatives which under normal provisions of NEPA would be mandatory for the first repository by stating that

> compliance with the procedures and requirements of this [Act] shall be deemed adequate consideration of the need for a repository, the time of the initial availability of a repository, and all alternatives to the isolation of high-level radioactive waste and spent nuclear fuel in a repository.

Further, Section 112(f) constitutes congressional determination that (a) a repository is needed, (b) the appropriate timing for construction of the repository is as specified in the statute, and (c) deep underground burial is the desirable means to dispose of high-level radioactive waste and spent nuclear fuel. These premises are presumably not subject to litigation.

The NWPA Sections 112(f) and 114(f) thus frustrate the purposes of NEPA because they limit the discussion of alternatives, which is of fundamental importance to fulfilling EIS requirements. Discussion of alternatives under NEPA is "intended to provide evidence that those charged with making the decision have actually considered other methods of attaining the desired goal, and to permit those removed from the decision making process to evaluate and balance the factors on their own" (16).

Section 113 of NWPA also imposed restrictions on the range of alternatives which were to be discussed in site characterization EAs by (a) prescribing procedural mechanisms to develop site characterization plans, (b) precluding site characterization activities unnecessary to supply data for the eventual EIS, and (c) limiting radioactive material and specific requirements for reclamation of the site and mitigation of significant adverse environmental effects due to site characterization activities.

Sections 112(e) and 112(b)(1)(F) together limited review of DOE guidelines and other decisions concerning nomination of sites. Section 112(e) defined each activity of the Secretary of DOE, except as otherwise provided, as a preliminary decision-making activity. In general, preliminary DOE actions are not subject to judicial review (17). Section 112(b)(1)(F) defined DOE EAs as a final agency action subject to judicial review under NEPA Section 119, but with the added provision that judicial review shall be limited to the sufficiency of such EAs.

Sections 112(e) and 112(b)(1)(F) thus seemed to limit judicial review to the question of whether DOE performed informational gathering correctly for purposes of judicial review of its final EAs.

Lastly, NWPA Sections 114(d) and 114(f) limit the customary role of the NRC in evaluating environmental impacts for siting of nuclear facilities. Section 114(d) defers NRC involvement in the HLW program until the repository development phase. Typically, the NRC must grant a license for operation of nuclear facilities, wherein the NRC role in authorizing and licensing nuclear facilities is normally considered to be an action requiring NEPA compliance and preparation of an EIS. Section 114(f) provides for the NRC to adopt the DOE EIS to the extent practicable. The effect of this provision is to remove NRC's obligation for environmental review by relying on DOE's procedures and EIS for NEPA compliance.

In passing NWPA/NWPAA, Congress seems to have acted inconsistently with many of the public-policy goals it enunciated in NEPA. It also acted inconsistently with elements of the intelligent decision-making interpretation of NEPA insofar as it lessened the requirement for publicly reviewable agency decisions to be both adequately informed and well reasoned by the use of more sound science, and by lessening the requirement of agency decisions to force technology to develop in ways that will further mitigate conflicts between environmental protection and economic values (7).

Performance Assessment

Once Congress made the decision to characterize Yucca Mountain as a potential HLW repository, DOE had to initiate a performance assessment program at the site. In the context of radioactive waste management, performance assessment is the process of quantitatively evaluating the ability of a disposal system to contain and isolate radioactive waste for a specified time period. As noted by Bernero (18), performance assessment is the bridge between regulatory decision-making and scientific uncertainties that provides a foundation to ensure that legal, societal, practical, and ethical concerns are reflected in HLW repository decisions.

Performance assessments must address at least three problems: (a) the role in determining the total safety of the different repository components and of the varying processes of interactions that can occur, (b) the predicted and long-term safety of the repository, and (c) the magnitude of the uncertainties involved in predictions of long-term repository and radionuclide behavior. Performance assessment represents a quantitative evaluation to be used to support the development of a radioactive waste repository and to determine compliance with applicable regulatory criteria (19). More specifically, performance assessment methodology is used as a basis for decisions concerning (a) definitions of principles for final HLW disposal, (b) selection of disposal concepts, (c) definitions of performance assessment procedures, (d) definition of performance requirements and safety targets, (e) selection of site and site characterization strategy, (f) selection of barrier systems and materials, (g) optimization of the total waste disposal system, (h) assessment of the acceptability of a proposed repository, and (i) demonstration of the acceptability of a repository within the scope of regulatory and licensing processes.

Campbell and Cranwell (19) and McCombie et al (20) describe the basic elements of performance assessments for HLW repositories. Structural data bases include information on radioactive and other materials in the waste and on materials and geohydrologic characteristics of the repository barriers. Mathematical models are used to describe and simulate the interactions between physical parameters of the repository such as temperature and pressure over time. Scenario analyses are conducted with different sets of models to quantify future performance of a repository under different possible geohydrologic, temperature, or other conditions, including the predictions of probabilities and consequences for the selected scenarios. Levels of confidence in performance assessment results must be evaluated by sensitivity and uncertainty analyses to provide a basis for acceptable decisions about repository performance. Finally, the safety of a repository can then be assessed and compared with relevant public health and environmental regulatory criteria. Although performance assessment methods have been studied for the past 10–15 years or so, there is no standard or accepted performance assessment methodology for a HLW repository for the simple reason that one has never been constructed.

Performance assessment requires the application of different types of models to simulate the behavior of repository systems. A conceptual model represents an assessor's understanding of the relationships between the repository features and processes needed to describe the repository system for the intended model application. Ideally, the relationships are stated in terms of testable hypotheses. The relationships of a conceptual model are represented quantitatively in a calculation model, which may be a simple closed-form analytic solution or be so complex that only a computer solution is possible. A calculation model should be verified and validated to provide assurance that it is suitable for its intended purpose. Verification is the process of obtaining assurance that a computer program implements the calculation model, and is usually performed by comparing results with known analytic solutions and with results calculated by similar calculational computer models. Validation is the process of assuring that a calculation model is a correct representation of the process or system for which it is intended. Ideally, validation requires comparison of results obtained by a calculation model with direct observations. However, this type of validation is not possible for nuclear waste repository models because the time and physical scales exceed our observational limits.

Research and assessment models also are used in performance assessment.

Research models are built to provide a relatively accurate representation of a repository system in terms of completeness and accuracy. Because of this emphasis, they usually are not applied in estimating repository performance over a wide range of scenarios because they are limited in their computational speed and efficiency. Assessment models are used to provide estimates of repository performance as a function of different scenarios because they are simplified and therefore allow more rapid calculations. However, because assessment models represent a greater degree of approximation, they are more difficult to validate.

Several countries are in the early stages of applying performance assessments to planning for nuclear waste repositories (20). All of the studies conducted for these assessments contain numerical estimates of possible consequences of a waste repository and compare different options for design, operation, and siting of the repository. In most cases, predicted doses or risks from the repository are below regulatory limits. Nevertheless, considerable debate exists about the application of performance assessment for licensing a HLW repository because of the problems associated with making quantitative and accurate predictions of future repository behavior.

Prediction of future repository behavior is made by identifying and defining phenomena or scenarios that could disrupt its performance. Such phenomena include climate change, earthquakes, volcanic activity, human activities such as exploratory drilling and mining, and failure of repository materials and geologic barriers surrounding the repository. Ideally, scenarios should focus on events and processes for which empirical data and observed frequencies are available so that probabilities of occurrence can be ascertained. Because empirical data and frequencies of occurrences for many important scenarios that might affect repository performance are not available, analysis of various scenarios and their consequences often is performed by computer simulation. This approach relies on many assumptions about factors that control events and processes that might affect repository performance. Consequently, significant reliance is placed on human judgment.

The complexities associated with designing a HLW repository have led to the identification of several types of uncertainties that confront performance assessors. These include uncertainties in (a) structural data for the repository site or geologic barriers caused either by measurement uncertainties or by natural variability, (b) models caused by simplified mathematical representation or incorrect parameters, (c) scenarios caused by omission of significant events or faulty interpretation of geohydrology and radionuclide dispersion, and (d) understanding complex geohydrologic systems over long time periods. To some extent, uncertainty can be reduced by obtaining more and better data and by improving procedures for model verification, validation, and scenario selection. However, significant uncertainty in performance assessments cannot be eliminated (21).

Quality assurance programs whose purpose is to ensure and demonstrate that the bases for decisions about repository design and implementation are not falsified by significant errors or omissions are part of performance assessments. Quality assurance programs are concerned with three areas of performance assessment: (a) quality of data, (b) quality of models, and (c) quality of treatment of scenarios. Approaches designed to ensure quality of data include the use of appropriate and documented techniques to ensure that data and associated uncertainties are identified and safeguarded, and that the origin and use of information used in the performance assessment are documented and traceable. Attempts to ensure the quality of models are based on verification and validation procedures. Ensuring the quality and treatment of scenarios is based on the ability to define, identify, quantify, and evaluate their uncertainties and consequences. Problems of verification, validation, and scenario selection will be discussed later.

Performance assessment also requires the identification of acceptance criteria to be used in evaluating long-term repository safety. Acceptance criteria vary between different countries, wherein some have provided specific guidance for local and regional targets, while others have provided criteria based on a basic global maximum dose or risk. Basically, acceptance criteria are established according to one or more of the following guidelines: (a) criteria can be set for the total repository system or also for its individual subsystems; (b) doses or risks can be used as to establish acceptable safety criteria; (c) limits can be placed on time periods for which acceptable safety criteria based on doses or risks are applicable; and (d) formal application of the ALARA principle can be required, which means that the dose or risk from radionuclides should be kept as low as reasonably achievable provided appropriate consideration of social and economic factors is made.

Geologic Disposal and Performance Assessment at Yucca Mountain, Nevada

Yucca Mountain is considered to be representative of a class of sites thought to be suited for building a HLW repository (22–24). The scientific information used to support Congress's choice of Yucca Mountain being potentially suited for a repository has been documented by DOE as a result of procedures mandated by NWPA for site selection (25–27). Several site-specific concerns have been identified about the nature of the physical environment at Yucca Mountain that are related to the adequacy of existing methods for addressing the unique requirements for siting and developing the repository with a geologic system that functions as the primary barrier for waste containment. Because a performance assessment methodology for a HLW repository does not exist, the assessment program being conducted at Yucca Mountain is in its early stages and requires considerable refinement and development. Further, because nuclear regulations do not prescribe how performance assessment is to be conducted, there is a need for agreement within the U.S. nuclear waste program on the methodology to be applied, especially with respect to the establishment of probabilities for risk assessment.

The principal focus of scenario development for Yucca Mountain has been on physical aspects of the environment that are to comprise the primary barrier to HLW migration (28). These include (a) geohydrology and groundwater hydraulics, (b) geochemistry, (c) tectonics and faulting, (d) volcanism, (e) rock properties such as thermodynamics and strength, (f) site geometry and geology (stratigraphy), (g) occurrence of mineral and energy resources, (h) dissolution of rocks, (i) formation of inorganic colloids, (j) erosion and denudation of overburden, (k) climatic change, and (l) surface hydrology and flooding. Once the components of the physical system are understood, plausible events are to be postulated that could influence specific components of the system and lead ultimately to breaching of the repository and release of radionuclides to the biosphere. The Yucca Mountain performance assessment program is based on the assumption that the most likely route for radionuclides released from the repository to take to the accessible environment is via groundwater. Consequently, emphasis is being placed on the aforementioned physical aspects of the environment and scenarios that could lead to breaching of the repository and the release of radionuclides to groundwater. Other scenarios such as extrusive magmatic activity and denudation of the natural overburden that might lead to repository breaching and release of radionuclides to the biosphere also are being considered, but to a lesser extent.

Several nonphysical processes and events also are being considered in the Yucca Mountain performance assessment (29). These include (a) natural microbial activity, (b) human intrusion in the search of natural resources, (c) future irrigation, (d) groundwater recharge or withdrawal, and (e) climate control. The paucity of biologic processes and events included in the performance assessment reflects the early stage of development of the performance assessment concept. The only biologic process considered, which is microbial growth, is based on the hypothesis that microbes naturally present in the host rock formation could enhance the mobility and accessibility of radionuclides to the accessible environment. While some emphasis has been placed on human activities that might influence repository site performance, in particular irrigation, resource exploration and mining, and climatic manipulation, other human activities such as war, sabotage, chemical waste disposal, and archeologic exhumation have been discounted as potential influences on the future performance of a repository at Yucca Mountain. For those human activities which have been considered, the emphasis has been on assessing the likelihood of natural conditions and resources at the repository site being such that future activity might occur, as opposed to the likelihood of a future society itself being such that resource exploration or use might or might not occur.

The general categories of scenarios being considered for environmental performance at Yucca Mountain include (a) hydrologic alterations, increased underground water flow, and water-table rise; (b) formation of new groundwater discharge points; (c) tectonic disturbance, faulting, and rock fracturing; (d) alteration of rock properties and geochemical changes; (e) advance of a dissolution front; (f) extrusive magmatic activity; (g) erosion and overburden denudation; (h) climatic control or change; (i) migration of inorganic colloids; (j) accelerated natural microbial activity; and (k) human intrusion by exploratory drilling. These scenarios include both physical and nonphysical factors which might potentially affect repository performance. For both the engineered barrier and the natural barrier, the combination of different events and rock types composing the natural barrier initially led to the identification of over 21,000 different scenarios (30). Subsequently, these were narrowed to 4000 scenarios, of which 400 were considered sufficiently probable to pursue in the Yucca Mountain performance assessment program. Of these, 84 different scenarios are being analyzed in the program.

The scenarios presently included in the Yucca Mountain performance assessment assume that groundwater poses the most likely route by which the environment and humans could be exposed to radionuclides in the future. In this regard, environmental standards assume that the primary risk to future individuals will be very small except for the possibility that individuals might use groundwater from the vicinity of the repository (23). The focus of the Yucca Mountain performance assessment is on groundwater itself and on the likelihood of radionuclides reaching the biosphere via that route within the next 10,000 years. Consequently, scenarios emphasize information on environmental events and processes more than the human activities that might affect repository performance. Consequently, little attempt is being made to understand food chains, ways of life, and population distributions with respect to future environmental systems and the intergenerational effects of HLW disposal.

Once events, processes, and plausible future scenarios are defined better, probabilistic and deterministic models must be developed to perform the complex computer computations that are necessary for analyzing possible interactions among climate, geohydrologic regimes, tectonic disturbances, volcanoes, geochemical alterations, and resource exploration and use. The analytical models for the Yucca Mountain site are in a rudimentary stage of development and are limited by existing knowledge of the geologic and hydrologic environment (21). Significant progress in developing definitive models to predict the behavior of the physical environment at Yucca Mountain will require new information on how the geologic and hydrologic systems function and interact. In the interim, performance assessment at Yucca Mountain will focus on developing plausible scenarios for the environmental system.

Methodological Limitations and Uncertainty at Yucca Mountain

Assessing the adequacy of a HLW repository has never been done, and because of the scientific and technical complexities inherent in such an assessment numerous scientific uncertainties and questions about the adequacy of the Yucca Mountain site remain to be resolved (31). Types of uncertainties pertain to (a) variation in the natural environmental setting and choice of parameters; (b) conceptual and probabilistic modeling, including definitions and calculations; (c) future evolution of the environmental system; and (d) measurement errors.

Repository guidelines contained in 10 CFR 960 require consideration of events and processes that can be expected to occur within 10,000 years following completion and closure of a repository and which could be expected to affect waste isolation. The instrumental seismic record available to DOE indicates that Yucca Mountain is unlikely to experience major earthquakes within the next 10,000 years, and that the region is seismically quiet (25, 26). However, Rogers et al (32) and the United States Geological Survey (33) suggest that episodic or cyclic events may occur at Yucca Mountain and that there is a potential for significant seismicity along the 32 known faults in the Yucca Mountain area such that a high probability exists for tectonic activity accompanied by surface displacement near or within the repository to occur at least once in the next 10,000 years.

Another important aspect of tectonics with respect to repository licensing is whether faults at and near Yucca Mountain are considered capable (34). The DOE (25) has assumed that all important fault scarps have been detected at Yucca Mountain and that it is reasonable to attribute the greatest potential seismic hazard to an earthquake of magnitude 6.8 or below. In an attempt to accommodate siting criteria, the concept of a 10,000 cumulative slip earthquake has been established by DOE specifically for application to Yucca Mountain (26). This is an effort to determine a 10,000-year event that would produce the average Quaternary slip rate on a fault with greater confidence than an estimate of a true maximum earthquake magnitude. Other studies have concluded that Yucca Mountain is highly faulted and fractured with active faults and could experience a magnitude 7.3 earthquake (35). In addition, DOE's method to determine a 10,000-year event is considered to be nonconservative because its estimation is completely dependent on long-term averages. Recent work has shown that fault activity in the Great Basin commonly exhibits spatial and temporal clustering of events (36). Averages and recurrence intervals over short-term periods can be greatly different from those over the long term.

With respect to volcanism, DOE states that the probability of disruption of a repository at Yucca Mountain by basaltic intrusion is from 3.3×10^{-6} to 4.7×10^{-4} (25). This estimate was based on data from Crowe et al and led DOE to conclude, without stating the actual value or how it was determined, that the mean probability for an intrusion occurring at the site is less than 1 chance in 10,000 over the next 10,000 years (37). Without knowing how DOE arrived at this conclusion and considering that the range of stated probabilities provided by DOE is as high as 4.7 chances in 10,000 of volcanic eruptions, it is not possible to determine that the siting requirements relative to volcanic disturbances will be met (26). Further, the data of Crowe et al (37) used by DOE subsequently were questioned by Crowe et al (38) as incomplete. Other data indicate that the low relative hazard estimated by DOE to occur due to radiologic releases from disruption of the repository by volcanism was based on non-conservative calculations of basalt magma cycles (39).

Highly significant among the geohydrologic issues relative to Yucca Mountain is the movement of gas and water through the unsaturated (vadose) zone and the construction of a validated model of geohydrologic processes at the site. The DOE has postulated that an unsaturated zone in an arid climate such as Yucca Mountain may prove acceptable for a HLW repository (25, 26). The DOE assumes that matrix flow predominates over fracture flow in the unsaturated zone and that the matrix must be saturated for fracture flow to occur. However, there is a paucity of detailed information of the geohydrology of such settings (22–24). The NWTRB strongly recommended that DOE study the influence of fractures and faults on the flow of water and their potential effects on waste isolation (31). Other studies suggest that the possibility exists that air and water vapor migration in the unsaturated zone could lead to the discharge of radionuclides to the immediate surface environment of the repository (28). Further, there is the question of whether water moving through the vadose zone at Yucca Mountain will remain bound within the rock when it encounters the repository, as DOE assumes will be the case, rather than breaking through the air gap (25, 26). However, the possibility is being investigated that capillary movement can result in sufficient pressure for a seepage surface to form on the roof of a subterranean hole in the unsaturated zone. Depending on the hydraulic properties of the surrounding medium, the downward seepage rate and the size and shape of the underground structure, the unsaturated flow buildup may cause water to enter the repository cavity (40, 41).

One option for repository design is to construct a "hot" repository, which would use the radioactive decay heat to create a hydrothermal umbrella over the repository that would result from groundwater boiling away from HLW canisters (42). One effect of this so-called thermal loading would be to enhance the capacity of the unsaturated zone to retard fracture flow of groundwater into the repository for 1000 years or longer. In effect, the construction of a hot repository would partially compensate for risks posed to the repository by the flow of groundwater in the unsaturated zone. This would be accomplished by allowing the HLW canisters to heat the repository's rock walls beneath Yucca Mountain to 206°C 20 years after emplacement and to over 100°C for over 1000 years. Under this scenario, the dry-out zone could extend to 200 m above the repository.

The potential consequences of thermal loading on the capability of the repository to confine HLW for at least 10,000 years is being studied in DOE's performance assessment. However, only the physical or below-ground (beyond a depth of 7 m) setting of the repository site is being assessed. Studies are not being conducted to assess the potential for thermally induced impacts to occur at the surface of the repository which might lead to altered environmental conditions which potentially could affect repository performance. There is a potential for significant ecosystem effects resulting from a hot repository. Preliminary data from Buscheck and Nitao indicate that a hot repository would elevate temperatures within the top 7 m at Yucca Mountain by an average of 13°C (42). Where rock fractures connected with the repository vent at the surface temperatures may increase an average of 43°C above the mean annual temperature. These temperature increases potentially can cause long-term ecosystem changes, including the possibility of significant or complete denudation of the vegetative cover, which would increase the rate of soil erosion and the likelihood of increased ground fractures and subsequent infiltration of water into the geologic setting of the repository. It is difficult to assess this problem because there are no good methods to assess long-term ecologic changes on the order of several thousands of years into the future.

The siting guidelines for a HLW geologic repository require that it be located where future climatic conditions will not lead to loss of waste isolation properties. The DOE has concluded that climatic conditions resulting from another period of glaciation will be cold and dry (25). These conclusions were based on paleoclimatic studies of the late Quaternary climate at Yucca Mountain that estimated the woodland and shrub boundary from evidence of fossilized plants and the extent of displacement of the boundary during the Pleistocene. The conclusions were based on the assumption that the distribution of plants was dependent on temperature, but they failed to acknowledge that precipitation exerts more control on these species than does temperature (43). If the importance of precipitation on the distribution of the woodland and shrub boundary is considered in the reconstruction of past climatic conditions, then a case can be made that past climates were cool and wet as opposed to cold and dry. These problems notwithstanding, assessing future climatic conditions is problematic for two additional reasons. One, the extent to which paleoclimatic data can be used to predict future climatic conditions is highly questionable. Two, the uncertainties of global warming pose significant problems for predicting future climatic conditions, especially with respect to temperature and precipitation patterns for local or regional scales.

Another important but little-known issue that has not been addressed by DOE is the possible interactions of thermal, mechanical, hydraulic, and chemical processes within the repository setting (25, 26). These interactions, or coupled processes, relate to considerations of how tectonicism and volcanism will affect geohydrologic conditions that might influence the migration of contaminants from the repository. Little is known about the uncertainties associated with engineering design arising from the geochemical, hydrologic, and thermal interactions such as those that could occur at Yucca Mountain. The study of coupled processes is in the theoretical state, and because of the interdisciplinary nature of the subject and the complex occurrences of coupled processes, studies are difficult to perform.

Other unresolved issues relative to the suitability of Yucca Mountain as a HLW repository include the consideration of whether valuable extractable resources exist at or near the site. The DOE concludes that there is no knowledge of such resources at or near the site, and on this basis concludes that future human intrusion into the repository is unlikely (25, 26). Although there is a dearth of information with respect to mineral deposits at the Yucca Mountain site, there are deposits of commercially valuable metals and minerals that occur both to the east and west of the site (44).

The adequacy of available technology for investigating the Yucca Mountain site also is addressed by repository siting guidelines (45). The guidelines call for ''reasonably available technology,'' which is defined as existing technology that has been demonstrated or for which any necessary development or validation will be completed within the period during which the methods or techniques are needed.

Several reviews have assessed the ability of existing models and methods to provide data and information needed to assess the geology and hydrology of the Yucca Mountain site and to resolve some of the aforementioned uncertainties. For example, Jones et al suggest that seismic reflection profiling and electromagnetic methods are of little use in determining the deep internal geologic structure due to complexity of the Yucca Mountain site (46). Other studies have concluded that there is an absence of commonly agreed upon techniques and theory for characterizing and modeling groundwater movement in the unsaturated fractured zone of Yucca Mountain (47, 48).

Difficulties also exist with respect to understanding the subsurface geochemistry as it might influence the movement of radionuclides. Only recently have preliminary efforts been made to model geochemical phenomena with respect to radionuclide transport and repository performance assessment, and many difficulties have arisen. For example, it has recently been discovered that colloids in the subsurface environment play an important role in the migration of radionuclides (49). Failure to account for colloidal movement can result in significant underestimates of the distances that radionuclides will migrate in groundwater systems. The occurrence and properties of below-ground colloids are understood poorly, so that insights necessary for predictive modeling of this mode of radionuclide transport are not well developed. With respect to Yucca Mountain, there are concerns that colloids may be important to mobilizing radionuclides in both the vadose and the saturated groundwater zones. In addition, it is possible that in combination with naturally occurring microbes, biocolloids could be formed which would further complicate understanding groundwater transport of radionuclides from the repository to the accessible environment. This possibility complicates the performance assessment at Yucca Mountain and emphasizes the uncertainties that exist with respect to biological components of the environmental system.

Methods and procedures to predict the probabilities and uncertainties of the nonphysical environment at Yucca Mountain appear to have limited applicability. Climate change has been identified as a possible influence on repository performance, but unfortunately there are no definitive methods for predicting climate change over thousands of years. Because the Yucca Mountain site is located in a region where extractable mineral resources exist, the possibility of future human intrusion into the repository also is being assessed. Hunter and Mann conclude that objective estimates of the future mineral resource potential of the Yucca Mountain area cannot be made with available techniques (50). The possibility of unintentional human intrusion into the repository also has been evaluated. The DOE concluded that the probability of human intrusion could be reduced by using comprehensive communication systems at the repository site in the form of permanent markers (51). However, this conclusion was reached without the use of scenario, probability, or uncertainty analyses.

Knepp and Dahlem discuss the uncertainties inherent in the use of geophysics and geohydrologic methods and models in attempting to evaluate the performance of a HLW over a period of thousands of years (52). Their thesis is that new methods must be developed to deal with the appreciable uncertainty that is inevitable in attempting to evaluate the suitability of the prospective repository site. This conclusion was reached because of the difficulty faced in validating deterministic geoscience models, and because of the difficulty of knowing what is meant by validation, what should be validated, and how validation can be achieved. Uncertainties regarding geoscience methods and models complicate siting decisions, because existing guidelines mandate that repository siting decisions and repository licensing must be based on the use of deterministic models (45).

Uncertainty also is introduced in performance assessment of the repository by errors associated with measuring environmental parameters (53). This can result from inaccurate instruments, inferences made from erroneous data, and from bias and arbitrariness introduced into assumptions made in data collection, analysis, and interpretation (47). Crowe has recognized intentional bias built into the Yucca Mountain studies as a result of increasing pressure to demonstrate the adequacy of the site in the face of funding and political realities (54). This concern appears increasingly valid in light of the fact that Yucca Mountain was selected by Congress as the sole site to be considered as a possible HLW repository in the United States prior to the initiation or completion of site-specific scientific studies for it or any other potential site. With no alternatives to Yucca Mountain, the success of the DOE repository program rests on the correct assumption having been made by Congress that Yucca Mountain is an acceptable, licensable site.

It is important to note that performance assessment for the Yucca Mountain site omits consideration of the biosphere and related phenomena from scenario analysis (55, 56). Consequently, a number of important objectives of environmental assessment are not taken into account. These include predicting the fate and effects of radionuclides, evaluating nonradiologic environmental impacts of HLW disposal, and calculating associated risks, probabilities, and confidence levels (29). In contrast to the U.S. program, European performance assessment programs are using comprehensive environmental simulation to address interactions between abiotic and biotic factors (57). Further, the fact that performance assessment for the Yucca Mountain site is limited to the 10,000-year containment period means that the environmental and human health effects of the changing nature of HLW over sufficiently long time periods are not included (58). For example, significant doses from the radionuclide Pb^{210} could peak shortly before 10,000 years, but continue long afterward. The possible doses from Cs^{135} peak at almost one million years. Based on environmental simulation modeling that included isotopic composition of HLW, cycling in the biosphere, probability of transport of radionuclides to humans, radiotoxicity to humans, and the changing risk potential, HLW poses risks about two orders of magnitude greater than that on which the U.S. program is based. The containment period for the U.S. HLW program was chosen because it was believed that the techniques for predicting events by performance assessment were insufficient beyond 10,000 years (59).

Value-Laden Judgments

Risk estimates conducted in support of the performance assessment program at Yucca Mountain typically are sensitive not only to a lack of scientific data but to numerous methodological value judgments, the outcomes of which have legal, public-policy, and ethical implications (5, 21, 60). Several performance assessment models have been compared, and they differ significantly with respect to the amounts of calculations performed, the use of more simplified abstract models versus use of complex models, and with respect to numerous assumptions relative to radionuclide source terms, HLW canister failure rates, assumptions about the downward flow of water through the unsaturated zone to the saturated zone, geohydrologic data sets used for calculational purposes, characteristics of both unsaturated and saturated water flow, gaseous radio-nuclide releases, volcanism, tectonically induced rise in the water table, and human intrusion into the repository (6). The use of different assumptions leads to significantly different conclusions about repository risks even when the same data are used for calculational purposes. For example, most DOE site assessors conclude that no radionuclides are likely to reach the water table at Yucca Mountain in less than 1000 years, as required by the U.S. regulations. Other assessors have concluded that the travel time from the disturbed environment to the accessible environment could be less than this time period (31). Conclu-sions about groundwater travel time and radionuclide migration are highly sen-sitive to a number of value judgments about the reliability of model simulations, such as the validity of extrapolations and interpolations, and the need (or lack of) for site-specific data.

Shrader-Frechette has provided an extensive analysis of the value-laden components of performance assessment methodologies being utilized at Yucca Mountain (21). Drawing from her work, examples of value-laden judgments, assumptions, inferences, and evaluations include (a) short-term tests provide an adequate basis for more precise predictions of long-term repository behavior; (b) simplified geologic models used to assess a heterogeneous site with nonlin-ear flow characteristics are reliable; (c) a relatively small number of geologic samples are adequate to make reliable inductive conclusions about site char-acteristics; (d) DOE's conclusion that human intrusion into the HLW repository is not a significant contributor to risk; (e) the use of subjective judgments about predicting earthquakes is an appropriate basis for seismic design of the reposi-tory; (f) a decision by DOE that scientists choose risk estimate values that are within one standard deviation of the mean value for purposes of defining the sufficiency of risk reductions; (g) the use of average risk estimates in assessing the adequacy of the site; (h) DOE statements that the risk of a repository is low because after 1000 years it would be projected to be equivalent to that from natural uranium ore bodies; (i) DOE's position that the Yucca Mountain site should be suitable if current information does not indicate that it is unsuitable; (j) relying on experts to separate determination of risks from risk perceptions; (k) acceptance of the de minimis inference wherein a particular risk is judged acceptable if it is below a certain threshold; (l) the use of geohydrologic models as a means to predict future repository performance as opposed to heuristic understanding; and (m) the use of expert judgment, which has taken the place of comprehensive inductive studies that need to be performed.

All of these types of value-laden judgments, assumptions, inferences, and evaluations exacerbate problems associated with resolving problems of scientific uncertainty. For example, with respect to the use of short-term tests, although the HLW canisters are required to isolate waste for up to 1000 years, all have failed one-year stress-corrosion tests when exposed in groundwater and tuff at the expected repository temperatures of 200°C. Despite such failings, DOE assumes that the canisters will last for the containment period specified in EPA regulations (61). Short-term data also are used by risk assessors to predict future weather and climate in order to predict long-term repository performance by relying on 30-year precipitation data for the Yucca Mountain area. However, such data may not be adequate for long-term predictions given the variability of climate over periods of thousands of years. The use of short-term data is problematic because if the long-term risk is greater than estimated there would be no opportunity to retrieve the waste or attempt to make the repository more secure because regulations mandate that the HLW would be retrievable only for 50 years after the repository is opened.

Risk assessors at Yucca Mountain also use simplified assumptions. For example, they assume that water flow is one-dimensional, that percolation of water is downward only through the unsaturated zone, that there will be a normal distribution of cumulative releases of radionuclides to the water table, and that many geohydrologic variables are independent rather than linked synergistically (28). Many of these assumptions are scientifically questionable because they are not derived from any general scientific laws about fracture flow in a heterogeneous environment. Obviously, many simplified assumptions have to be used in complex projects like Yucca Mountain because they allow the use of efficient models and because data do not exist that enable many variables to be correlated or models' reliability to be assessed. Consequently, judgments about the adequacy of models also are questionable because models cannot be validated through field studies over long time periods. Although simplified assumptions often have to be made, their use increases the probability that weaker scientific theories with respect to explanatory powers will be accepted by risk assessors. External reviewers for the Yucca Mountain project have warned that many conclusions about the Yucca Mountain environment and the models used to assess it may be inappropriate (31).

The DOE is using a small number of boreholes or tuff samples to make conclusions about fracture transmissivity, permeability, and hydraulic conductivity. It is important that the samples obtained are representative because of the fact that different samples react differently to the same environmental variables. However, the use of a small number of samples does not enable a determination of their representativeness to be made (62).

The DOE also has concluded that inadvertent human activities as well as organized sabotage and terrorism into the HLW repository are not significant contributors to risk (28). By necessity, this conclusion involves human judgment. However, it conflicts with the EPA's claim that human error and human intrusions into the repository are the most likely causes of radionuclide release (63).

Various risk assessors have indicated that predicting earthquakes that had a 1-in-10 chance of occurring was an appropriate basis for seismic design of the repository (21). Decisions about the acceptability of risks are value laden because they rely on subjective judgment insofar as there are no or few empirical data that can be used as a basis for the estimates with any reasonable certainty. With respect to the prediction of earthquakes some assessors have estimated that there has been no significant fault-related ground movement on the Yucca Mountain site in the last 500,000 years, while others have estimated that there has been no moderate earthquake activity for 10,000 years. The use of different risk estimates can have significant implications. For example, the annual probability of various seismic accidents involving radionuclide releases at Yucca Mountain is estimated to be between 10^{-6} and 10^{-9}. These risk estimates are small on an annual basis but over many years the probability of risk is significantly higher, say, 10^{-2} for 10,000 years. Consequently, inferences about annual probability of seismic events can affect the probability of risks for future people who have no say in contemporary decisions about Yucca Mountain.

The DOE's decisions about the sufficiency of risk reductions also are value laden. The agency has recommended that risk estimate values that are within one standard deviation of the mean value be accepted as a conservative estimate of a particular risk (64). However, accepting this recommendation means that only 68% of all the cases in the frequency distribution for risks will be taken into account. Accepting a risk estimate for an accident probability that is within one standard deviation of the mean value can lead to errors of several orders of magnitude when problems are likely to arise that stem from events associated with catastrophic failures of the repository that are outside of one standard deviation of the mean value for a particular risk estimate. Clearly, the criteria regarding what constitutes an acceptable risk is a public-policy and ethical decision rather than a scientific one.

Risk assessments at Yucca Mountain often are based on the use of average risk estimates. Such estimates can underestimate risks that are significantly higher than mean or average values. For example, some estimates of mean groundwater travel time to the water table from the repository are on the order of 43,000 years (65). Because such estimates are based on mean groundwater travel times, they do not disclose the fact that approximately 1% of the calculated groundwater travel times are less than 10,000 years and that a range of travel times for some rocks is between 3900 and 15,000 years. Shorter groundwater travel times could present unacceptably high adverse risks unless the estimated groundwater travel times are clustered around the mean value.

One of the decisions DOE must make is at what point in time can the Yucca Mountain repository be considered to pose negligible risks? The DOE risk assessors have stated that after 1000 years the risks are acceptably low. This statement is based on the fact that after this time radiologic risks are approximately equivalent to those from natural uranium ore bodies (21). Such a view is value laden because it presupposes that the risk of ore is acceptable only because it is "natural." However, because a risk is normal or natural does not make it morally acceptable because people can be said to have a right to avoid or prevent natural risks if they are preventable or if they are imposed by technology, unless they give free and informed consent to be subjected to the risks.

Another value-laden decision regarding Yucca Mountain that has to be made concerns the placement of the burden of proof in determining the adequacy of the site. The DOE has noted significant uncertainties regarding geohydrologic conditions at the site. However, it also states that the site should be considered suitable if current information does not indicate that it is unsuitable (5). Such a decision places the burden of proof on those who might claim that the site is unsuitable. This position is contrary to standard civil procedures that normally use a decision rule of placing the burden of proof on those who have the greater weight of evidence on their side. The DOE's position also means that less robust inductive reasoning can be used in support of a conclusion that the site is suitable because the position implicitly assumes that the absence of sound scientific evidence against site suitability provides (in part) sufficient justification for a conclusion that the site is suitable. Given the scientific uncertainties surrounding Yucca Mountain, rather than making a decision that the site is either suitable or unsuitable the DOE could consider a third option, which is that suitability cannot be reasonably determined at this time because of too much scientific uncertainty. Such an option would be more consistent with scientific peer reviewers' judgments about the level of scientific uncertainty which exists at Yucca Mountain.

Another example of a value-laden inference used at Yucca Mountain is the de minimis inference, which is the acceptance of a particular risk when it is below a certain threshold value. This inference has been accepted by DOE by virtue of its decision to exclude scenarios from performance assessment in which processes and events have less than a 10^{-4} chance of occurring during the period of interest (64). This level of risk is about two orders of magnitude lower than that commonly accepted for other nuclear facilities. The de minimis inference has value-laden implications for people exposed to cumulative risks such as radiation because it represents a claim that each of the subthreshold risks to which people might be exposed is acceptable. From a public-policy and ethical standpoint such a claim is problematic because de minimis standards are not able to provide equal protection to all people due to individual differences in sensitivities.

The DOE risk assessors often use theoretical calculations, data, and models that they demonstrate to be consistent with a hypothesis. When there is some consistency between these elements with a hypothesis, risk assessors often assume that the hypothesis has predictive power or has been verified. However, true verification or validation is not performed because it requires repeated testing of computer codes and models under field conditions. Oreskes et al conclude that verification and validation of numerical models of natural geohydrologic systems is not possible because such systems are open systems, and because access to information about natural systems always is partial (66). Consequently, model results always are nonunique, which means that their primary value is heuristic as opposed to predictive.

A final problematic inference which confronts the Yucca Mountain repository program is relying on expert judgment to support claims of risk estimates or evaluations. While the use of expert judgment often is required in decision making about complex technological projects, it should not take the place of comprehensive inductive studies that need to be performed because it is not possible to separate fully the determination of actual risks from risk perceptions. The NWTRB has recommended that DOE rely less on expert judgment and include more scientific studies in its repository program (31). Generally speaking, the DOE has admitted that subjective expert judgment has played a critical role in determinations about site suitability. Reliance on expert judgment is problematic because, on the one hand, DOE states that such judgment is adequate for determination of site suitability, yet, on the other hand, other peer reviewers of the repository program have questioned whether the conclusions are balanced or adequate to serve as a basis about site suitability (53).

The Relationship Between Scientific Uncertainty and Protection Standards

Under NWPA, Congress granted responsibility to the DOE to design and operate a deep geologic repository for HLW. The design and operation of the repository must conform to environmental and health standards pertaining to the release of radionuclides promulgated by the EPA and must be licensed by the NRC. Under NWPAA, Yucca Mountain, Nevada, was identified as the site selected for characterization and is scheduled to open in the year 2010 or later if it proves acceptable to the NRC for licensing.

The EPA standards rely on approaches which utilize numerical probabilities to establish requirements for containing radioactivity within the repository. Compliance with one of the standards depends on quantitative projections of the amount of radioactivity that will be released to the accessible environment surrounding the repository for 10,000 years. These types of standards are generally referred to as "probabilistic" because compliance hinges on the calculation of probabilities that important events will occur and the multiplication of the probabilities by predictions of the consequences (67). Although the EPA standards recognize inherent uncertainty and limitations in required analysis of environmental and health impacts, they nevertheless assume that scientific theory or methodology will, in a rulelike manner, predict future impacts of radionuclide releases from a repository.

In contrast to the approach taken by the EPA, the NRC established qualitative goals that define acceptable risks. The goals state that there should be no significant additional risk to either individuals or to society from normal operations and accidents. To achieve the goals, the NRC established safety objectives in which risks are to be calculated and expressed as a percentage of other nonnuclear risks to individuals and society. However, the NRC recognized significant uncertainties in analytical methods and data used to calculate risks. Consequently, it concluded that safety objectives must be subordinate to the qualitative safety goals, and that analyses of compliance with safety goals may not be used as the sole basis for licensing decisions. The NRC's regulations are generally qualitative in that the determination of compliance with regulations is based on engineering judgments that take into account the inherent uncertainties involved in making long-term projections of repository performance.

The contrasting approaches taken by EPA and NRC imply a formidable task for DOE in demonstrating whether a repository at Yucca Mountain can safely isolate HLW consistent with existing standards and regulations. In fact, recent studies questioned the ability of the DOE to successfully construct and operate a HLW repository site at Yucca Mountain consistent with current EPA standards and NRC regulations (31, 60, 68). Questions focus on whether demonstration of compliance with EPA's standards relies excessively on numerical calculations and analyses that DOE must conduct to project the long-term performance of the repository. The questions are based on the fact that scientific and technical information on the Yucca Mountain site may be inadequate to determine whether meaningful numerical probability estimates can be ascertained. For Yucca Mountain to be judged an acceptable site, data must be obtained to allow prediction of the performance of the geologic waste isolation barrier that will enable engineers to develop a containment system to prevent the release of radionuclides into the environment.

There is a compelling need to resolve the contrasting approaches taken by EPA and NRC as soon as possible so that site characterization can proceed in a manner that protects the environment and human health, but that also avoids lengthy and unsuccessful NRC licensing proceedings. In fact, the NRC is concerned whether EPA's probabilistic-based standards can be implemented without paralyzing the repository licensing proceedings with litigation over details of DOE's analysis supporting compliance with EPA's standards (67).

The contrasting approaches taken by EPA and NRC raise fundamental questions that administrators must deal with. Do the laws that regulate the siting of a HLW facility give guidance to administrators about how to resolve scientific uncertainty? Do these laws create a need to describe future environmental "facts" that outstrips current scientific capability? Do these laws command a probabilistic prediction of future environmental events caused by the siting of a HLW repository that rely too heavily on numerical calculations and untested or unverified models? Are the facts that must be decided in siting the HLW repository capable of being predicted through scientifically verifiable methodologies? Are EPA's standards written in such a way that it may be difficult or impossible for DOE to demonstrate compliance in a NRC licensing proceeding?

General Legal Procedures for Approval of a High-Level Nuclear Waste Repository

A brief summary of the major steps in the administrative process that may lead to an approval of the first HLW repository as provided by the NWPA includes a description of the role played by the EPA, the DOE, and the NRC. According to NWPA, DOE is the federal agency that will design, build, and operate the facility and must prepare all applications and documents necessary to obtain the approvals required by law. The NRC is the agency that must review and approve DOE's applications for licenses and construction authorization and otherwise regulate the facility. The NRC's regulatory responsibilities are conducted pursuant to its authority under the Energy Reorganization Act of 1974 (42 USC Sections 5841 et seq.) and Section 114(d) of NWPA (42 USC Section 10134(d)). Pursuant to Section 121 of NWPA, EPA has the responsibility to set standards for overall protection of the public health and the environment. These regulations have been promulgated as 40 CFR 191. The NRC and DOE are responsible for implementing these regulations and have promulgated regulations designed to satisfy their respective responsibilities on siting under the NWPA. The NRC promulgated regulations on procedural and technical requirements related to siting at 10 CFR 60. The DOE issued regulations on site selection guidance under 10 CFR 960 pursuant to its responsibility under Sections 112 and 114 of NWPA. Moreover, Section 114 of the NWPA requires DOE to develop a final EIS with any recommendation for approval that is presented to the President. This EIS must be prepared in accordance with the requirements of NEPA (42 USC Sections 4321–4370), except that under the NWPA certain elements of an EIS that are usually required under NEPA are waived.

The EPA Environmental Standards

Regulations promulgated by EPA at 40 CFR 191 are entitled "Environmental Standards for the Management and Disposal of Spent Nuclear Fuel, High-Level and Transuranic Radioactive Wastes." Subpart B of these regulations establishes several different types of requirements for disposal of these materials. The first are primary standards for disposal that are long-term containment requirements that limit projected releases of radioactivity to the accessible environment for 10,000 years after disposal. In addition to these numerical limitations on releases, a set of six quality assurance requirements is set out for facilities not regulated by NRC, such as DOE facilities. These include institutional controls, monitoring requirements, marker requirements, barrier specifications, limitations on acceptable land uses, and a requirement that wastes may be removed for a reasonable period of time after disposal (40 CFR 191.14). A third set of requirements are limitations on exposures to individual members of the public for 1000 years after disposal (40 CFR 191.15). Finally, the regulations include a set of groundwater protection requirements that limit radionuclide concentrations for 1000 years after disposal (40 CFR 191.16).

The regulations also specify that disposal systems for the applicable wastes shall be designed to provide reasonable expectations based on performance assessments that the cumulative releases of radionuclides to the accessible environment from all significant processes and events that may affect the disposal system shall for 10,000 years after disposal (a) have a likelihood of less than 1 chance in 10 of exceeding the standards and (b) have a likelihood of less than 1 chance in 1000 of exceeding 10 times the standards (40 CFR 191.13(a)).

The regulations under 40 CFR 191.13(b) also state:

> Performance assessments need not provide complete assurance that [the probability of meeting the standards stated above] will be met. Because of the long time period involved and the nature of the events and processes of interest, there will inevitably be substantial uncertainties in projecting disposal system performance. Proof of the future performance of a disposal system is not to be had in the ordinary sense of the word in situations that deal with much shorter time frames. Instead, what is required is a reasonable expectation, on the basis of the record before the implementing agency, that compliance with 191.13(a) will be achieved.

Assessment of How Regulations Mandate Consideration of Scientific Uncertainty

The aforementioned regulations, in establishing a probability standard that must be calculated in the performance assessment for meeting the containment emission limitations, at first, seem to require a "probabilistic" approach to dealing with the scientific uncertainties associated with designing and assessing the disposal facility's ability to meet the standards. However, 40 CFR 191.13(b) recognizes that scientific proof is not a criteria for ensuring that these probabilities of meeting the standards will be met. The regulations simply require that DOE and NRC establish by a reasonable expectation that the probabilities will be met based on the administrative record before the agency. Where similar provisions of environmental law allow administrative agencies to rely on their own records, courts have upheld the administrative decision unless the agency has made a decision that is "arbitrary and capricious." The arbitrary and capricious test governs review of agency actions in instances where no hearing or formal findings on a record are required and there is no statutory requirement of a different test (69). This standard amounts to a presumption that the agency's determination is supportable by a court if there is any reasonable basis for the decision. In determining whether a decision is arbitrary and capricious the reviewing court must consider whether said decision was based on a consideration of relevant factors and whether there has been a clear error of judgment (70). Consequently, mere error is insufficient for a finding of arbitrary and capriciousness: an agency's actions will not be overturned if it is supportable on any rational basis (71). In other words, a court will most likely construe the reasonable expectation provision of 40 CFR 191.13(b) to support any agency calculation of compliance with the probability requirements of the regulations if the agency can find any reason to support its calculation of a probability that is not clearly contradicted by a fact or calculation that is also in the record supplied by someone contesting the decision.

If there is a dispute about a calculation of a probability between an agency expert and an outside expert, according to this scope of judicial review, the agency will be entitled to a presumption that its view of the scientific matter is correct. It is generally not the duty of a court to pass judgment on the scientific merits of agency decisions. To prevail in argument that a scientific study is "fatally flawed," plaintiffs must show that the agency acted arbitrarily and capriciously when it relied on the scientific study (72). Unlike some other legal proceedings wherein scientific evidence is not admissible unless the proponent of the evidence can show that the scientific procedure is generally accepted by the relevant scientific community, under the reasonable expectation standard of 40 CFR 191.13(b) in situations where there is no formal adjudicatory hearing, the courts will uphold agency calculations notwithstanding that the agency methodology is not generally accepted by the scientific community. Therefore, if current scientific knowledge does not provide an empirical basis for assigning a probability to a facility's ability to meet the containment limitations, the agency's calculation of the probability of facility compliance is likely to be upheld if there is any reasonable basis for its assignment of a probability.

For example, if, in attempting to predict the probability of earthquakes that might destabilize the facility's natural barriers, DOE relied on the frequency of known earthquakes in the area during times of historic measurement, this methodology is likely to be sustained by the courts notwithstanding that historic earthquake activity may not be compelling for predicting future activity. Where a scientific basis does not exist, the agency is free to provide a speculative basis, provided the speculation is reasonable. Therefore, under the reasonable expectation standard of these regulations, any reasonable speculation is likely to be adequate as a matter of law. However, if a person who wishes to challenge the government decision can show that an agency calculation is without any rational scientific basis, they are likely to be successful under an arbitrary and capricious standard.

Therefore, the answer to the question as to whether the EPA regulations will ultimately create a legal barrier that will prevent the siting of the HLW repository at Yucca Mountain depends on whether DOE will be able to develop reasonable assumptions about site performance. Given the magnitude of the scientific uncertainties that are endemic to this project, DOE may not be able to satisfy the arbitrary and capricious scope of judicial review (67, 68). The NRC staff, in commenting on the EPA's standards, strongly questioned the ability of anyone to satisfy the quantitative probabilistic standards of the EPA regulations when it said (73):

> Numerical estimates of the probabilities or frequencies of some future events may not be meaningful. The [US]NRC considers that identification and evaluation of such events and processes will require considerable judgement and therefore will not be amenable to quantification by statistical analysis without the inclusion of very broad ranges of uncertainty.

Consequently, the scientific uncertainties associated with predicting the Yucca Mountain waste repository's impact on the biosphere are likely to be so great that DOE will be unable to withstand a legal challenge to its ability to comply with the EPA regulations under 40 CFR 191.13(a).

The individual protection requirements of 40 CFR 191.15, and the groundwater protection requirements of 40 CFR 191.16, unlike the containment standards of 40 CFR 191.13, do not have probability standards associated with the limitations but simply require that the disposal systems be designed to ensure a reasonable expectation that the standards will be met. Therefore, for groundwater standards and the individual protection standards courts are more likely to sustain any assessment of the facility's compliance with these standards provided that the agency's conclusions about a facility's conformance with the standards are based on reasonable speculations. However, because great scientific uncertainty exists in the use of the models needed to predict compliance with both the groundwater standards and the individual protection standards, DOE may not be able to sustain challenges to their conclusions about compliance with these standards (48, 52, 60, 67).

The EPA regulations at 40 CFR 191 were challenged by the Natural Resources Defense Council and by several states and environmental groups (74). In deciding whether these regulations were valid, United States Court of Appeals for the First Circuit applied the "arbitrary and capricious" test to several issues where the scientific judgment of EPA was questioned. In this 1987 decision, the Court sustained EPA's judgment on the 10,000-year time period for assessing performance of HLW facilities, concluding that EPA had a reasonable basis for this period. However, the Court remanded the 1000-year period for protection of individuals on the basis that there was no reasonable basis for this period in the record. This decision demonstrates that courts may not sustain any technical decision if there is not an adequate basis in the administrative record to support the decision. More recently, the Energy Policy Act of 1992 directed the EPA and the NRC to rewrite their HLW standards following a study by the National Academy of Sciences and DOE allegations that existing federal standards were too costly and might not be attainable.

Recommendation of the Secretary of DOE for Site Suitability Under NWPA Section 114(a)(1)

Section 114(a)(1) of NWPA specifies the conditions under which the Secretary of DOE may make a recommendation to the President that a site be approved as a disposal site. This section states in relevant part:

> (1) The Secretary shall hold public hearings in the vicinity of the Yucca Mountain site, for the purposes of informing the residents of the area of such consideration and receiving their comments regarding the possible recommendation of such site. If, upon completion of such hearings and completion of site characterization activities at the Yucca Mountain site under section 10133 of this title, the Secretary decides to recommend approval of such site to the President, the Secretary shall notify the Governor and legislature of the State of Nevada of such decision. No sooner than the expiration of the 30-day period following such notification, the Secretary shall submit to the President a recommendation that the President approve such site for the development of a repository. Any such recommendation by the Secretary shall be based on the record of information developed by the Secretary under section 10133 of this title and this section, including the information described in subparagraph (A) through subparagraph (G). Together with any recommendation of a site under this paragraph, the Secretary shall make available to the public, and submit to the President, a comprehensive statement of the basis of such recommendation, including the following:
>
> (C) A discussion of data, obtained in site characterization activities, relating to the safety of such site;
>
> (D) A final environmental impact statement prepared for the Yucca Mountain site pursuant to subsection (f) of this section, and the National Environmental Policy Act of 1969 (42 U.S.C. 4321 et seq.), together with comments made concerning such environmental impact statement by the Secretary of the Interior, the Council on Environmental Quality, the Administrator, and the Commission, except that the Secretary shall not be required in any such environmental impact statement to consider the need

for a repository, the alternatives to geological disposal, or alternative sites to the Yucca Mountain site. . . .

The regulations promulgated by EPA at 40 CFR 191 discussed above are meant to apply to any recommendation of the Secretary of DOE concerning the site's suitability. Therefore, the recommendation of the Secretary of DOE authorized by NWPA Section 114(a)(1) will be subject to the same "arbitrary and capricious" criteria for judicial review already discussed. Courts therefore are likely to sustain any recommendation unless a plaintiff can show that there was no reasonable scientific basis in the record for the secretary's recommendation. In accordance with the above analysis on the EPA regulations, the recommendation on the site's suitability need not be supported by a rigorous scientific basis or even a scientifically based "probabilistic" rationale. Strict scientific certainty about a site's suitability is therefore not a requirement for any recommendation by the Secretary of DOE. A plaintiff that desires to challenge the Secretary of DOE's recommendation must show, therefore, that there is no rational basis in the administrative record for such recommendation. In making such an argument a plaintiff may prevail if he or she can demonstrate that there is no rational basis for the secretary's conclusion that the EPA regulations at 40 CFR 191 have been complied with in DOE's facility performance analysis. However, because of the magnitude of scientific uncertainty about the Yucca Mountain site, DOE may not be able to defend itself successfully against legal challenges to any positive recommendation it may make to the president about site suitability under 40 CFR 191.

Final Environmental Impact Statement Analysis

Although 40 CFR 191 requires that a performance assessment be prepared by EPA that analyzes the facility's impact on the environment for up to 10,000 years, only the EIS requirements under NWPA require any analysis beyond this 10,000-year period. Section 114(f) of NWPA states an EIS as required by NEPA must be prepared to accompany any recommendation of the Secretary of DOE to the president to approve the site (42 USC Section 10134(f)). However, this section makes the following exceptions to the normal EIS requirements:

> With respect to the requirements imposed by the National Environmental Policy Act of 1969, compliance with the procedures and requirements of this chapter shall be deemed adequate consideration of the need for a repository, and alternatives to the isolation of high-level radioactive waste and spent fuel in repository.
>
> For the purpose of complying with the requirements of the National Environmental Policy Act of 1969 and this section, the Secretary need not consider alternate sites to the Yucca Mountain site for the repository to be developed under this part.

Any environmental impact statement prepared in connection with a repository proposed to be prepared by the Secretary under this part shall, to the extent practical, be adopted by the Commission in connection with the issuance by the Commission of a construction authorization and license for the repository. To the extent that such statement is adopted by the Commission, such adoption shall be deemed to satisfy the responsibilities of the Commission under the National Environmental Policy Act of 1969 and no further consideration shall be required, except nothing in this section shall effect any independent responsibilities of the Commission to protect the public health and safety under the Atomic Energy Act of 1954.

Therefore, although an EIS is required by the NWPA for any final recommendation of the Secretary of DOE, that EIS need not consider alternatives to Yucca Mountain or any alternatives to geologic disposal.

Regulations on NEPA include requirements on incomplete and unavailable information (51 CFR 15618, 15621). When information on reasonably foreseeable adverse impacts evaluated in an EIS is essential to making a reasoned choice, and costs of obtaining it are not exorbitant, the agency must secure it. However, if this information is incomplete or unavailable, that is the costs of obtaining it are exorbitant or the means of obtaining it are beyond the state of the art, the agency must make clear that such information is lacking. The agency must follow four prescribed steps. First, it must state that the information is incomplete or unavailable. Second, it must state the relevance of the missing information. Third, it must summarize the existing credible scientific evidence relevant to its evaluation of reasonably foreseeable impacts. Fourth, it must analyze those impacts based on theoretical approaches or scientific methods generally accepted in the scientific community. The regulation clearly states that agencies must consider impacts with low probability but catastrophic consequences as long as the analysis is "supported by credible evidence that is not based on conjecture, and is within the rule of reason." In one case (75), the D.C. Circuit Court said:

[W]e note that NEPA does unquestionably impose on agencies an affirmative obligation to seek out information concerning the environmental consequences of proposed federal actions. . . . As this court has held, the basic thrust of NEPA is to predict the environmental effects of proposed action before the action is taken and those effects fully known. . . . Predictions, however, by their very nature, can never be perfect, and the information to the agency could always be augmented. The question in each case is, "How much information is enough?" And that is not a question to which NEPA provides a clear, firm answer . . . Some element of speculation is implicit in NEPA. . . .

The above regulation creates a legal duty for DOE to rigorously describe potential impacts but allows DOE to speculate on impacts that cannot be determined with scientific precision provided that the scientific uncertainties are disclosed in the EIS. The EIS requirements, therefore, should be understood as creating a legal duty for DOE to take a hard look at environmental impacts. These requirements have been consistently construed by the courts to create judicially enforceable requirements that agencies seriously and carefully examine the environmental consequences of proposed actions (76). In theory, NEPA forces federal agencies to think about environmental impacts before an action is undertaken—to "look before one leaps." Although it is thus normally understood that NEPA requires certain serious analytical procedural efforts that examine environmental consequences, it is generally accepted that it does not require a particular substantive environmental outcome. The United States Supreme Court has stated that NEPA is to be treated as essentially a procedural statute (10). Furthermore, the court held that once an agency has made a decision subject to NEPA's procedural requirements, the only role for the court is to ensure that the agency has considered the environmental consequences (9). In other words, NEPA does not force an agency to choose the most environmentally responsible course of action. Although NEPA has forced some agencies to improve their thinking about environmental impacts, once an agency learns to write legally sufficient EISs, it is free to pursue an environmentally destructive course of action (77).

Because of the significant amount of scientific uncertainty in predicting the environmental impacts of the Yucca Mountain waste repository, opponents of the repository may successfully challenge the EIS if they can demonstrate that DOE did not rigorously consider certain impacts. However, such challenge is likely only to delay the project because DOE can always cure any defects in the EIS by amending it to consider impacts previously unexamined.

NRC Action on Repository Licensing

Section 114 of NWPA requires that the NRC consider DOE's application for a license to construct the waste repository. Regulations at 10 CFR 60 establish the rules of NRC that relate to licensing procedures for the facility. Regulations contained in 10 CFR 60.101 provide, in relevant part, that as a condition of issuing the construction license NRC must find that the issuance of a license will not constitute an unreasonable risk to the health and safety of the public. This section further states:

> It is the purpose of this subpart to set out performance objectives and site and design criteria which, if satisfied, will support such a finding of no unreasonable risk.
>
> While these performance objectives and criteria are generally stated in

unqualified terms, it is not expected that complete assurance that they will be met can be presented. A reasonable assurance, on the basis of the record before the Commission, that the objectives and criteria will be met is the general standard that is required. For §60.112 [which incorporate and reference the EPA criteria] and other portions of this subpart that impose objectives and criteria for repository performance over long times into the future, there will inevitably be greater uncertainties. Proof of the future performance of engineered barrier systems and the geologic setting over time periods of many hundreds or many thousands of years is not to be had in the ordinary sense of the word. For such long-term objectives and criteria, what is required is reasonable assurance, making allowance for the time period, hazards, and uncertainties involved, that the outcome will be in conformance with those objectives and criteria. Demonstration of compliance with such objectives and criteria will involve the use of data from accelerated tests and predictive models that are supported by such measures as field and laboratory tests, monitoring data and natural analog studies.

Subpart B of this part also lists findings that must be made in support of an authorization to construct a geologic repository operations area. In particular, §60.31(a) requires a finding that there is reasonable assurance that the types and amounts of radioactive materials described in the application can be received, possessed, and disposed of in a geologic repository operations area of the design proposed without unreasonable risk to the health and safety of the public. As stated in that paragraph, in arriving at this determination, the Commission will consider whether the site and design comply with the criteria contained in this subpart. Once again, while the criteria may be written in unqualified terms, the demonstration of compliance may take uncertainties and gaps in knowledge into account, provided that the Commission can make the specified finding of reasonable assurance as specified in paragraph (a) of this section.

The NRC regulations quoted contrast with EPA's probabilistic standards in that the NRC recognizes that performance criteria likely will not or cannot be demonstrated in accordance with scientific proof in the ordinary sense of the word. In other words, NRC recognizes that DOE may not be able to use quantitative and probabilistic data to prove that the Yucca Mountain site is safe. The NRC recognizes that DOE may only be able to give reasonable assurances that performance criteria can be achieved.

An opponent to NRC licensing procedures will have to demonstrate that DOE has not demonstrated with "reasonable assurances" that all performance criteria, including the referenced EPA criteria, have been complied with. Whether the opponent of the project will prevail will depend on his or her ability to demonstrate that the NRC application and conclusions contained therein are not based on reasonable assurances.

The NRC decision to issue the license will become final after an adjudicatory-type hearing in which evidence must be adduced by DOE that shows that performance criteria will be complied with. The NRC procedural regulations at 10 CFR 2.743 specify that in proceedings before the NRC only evidence that is reliable may be admitted. Unlike some judicial bodies that may exclude scientific evidence unless the proponent of the evidence can testify that the evidence is generally accepted as valid within the relevant scientific community, NRC procedures would allow DOE to submit any evidence into the record that is reliable. If the type of evidence adduced by DOE in support of its application for a license is less than the type of evidence generally accepted in the relevant scientific community, NRC will still consider such evidence, but in making any final decision it will consider the strength of the scientific evidence in determining the weight of the evidence before it. According to regulations at 10 CFR 2.732, DOE would have the burden of proof to show that conclusions contained in its application should be supported, while an opponent to the licensing issuance would have the burden of proof to establish any contentions raised by the opponent. According to this analysis, DOE will prevail in any challenge to a NRC licensing procedure if it can establish that all performance criteria have been complied with as established by reliable scientific evidence.

The Requirement for Scientific Certainty

Scientific proof of environmental safety is not required as a matter of law as a condition of siting a high-level nuclear repository. In fact, anticipating that such scientific proof would not be available because of the long time frames associated with repository performance and the difficulty of predicting environmental impacts over these time periods, both EPA and NRC wrote into relevant provisions of their respective regulations that such scientific proof is not expected and that DOE need only show performance compliance with siting criteria based on reasonable expectations.

However, EPA regulations that established numerical probabilistic performance standards may be impossible for DOE to comply with notwithstanding DOE's legal ability to support its conclusions on repository performance on some assumptions rather than totally on empirical evidence or scientifically accepted theories. The DOE may fail to satisfy the legal standards of proof because of the significant scientific uncertainty entailed in making the predictions about repository performance and the concomitant inability to make assumptions based on reasonable expectations derived from tested and verifiable scientific models. If EPA's standards cannot be complied with, repository licensing likely will be delayed or denied by litigation.

Consequently, the NRC is exploring options on how EPA's standards can best be implemented (67). The NRC has proposed four alternative courses of action: (a) maintain the probabilistic format of the EPA standard in conjunction with NRC's current licensing regulations, with minimal changes to resolve implementation problems and ensure consistency between the regulations; (b) make the EPA standard more qualitative and implement it through NRC's current licensing regulations; (c) maintain a probabilistic format for the EPA standard, but have EPA expand its interpretation of the standard and NRC appropriately amend its regulations; and (d) assume that revised EPA standards will not be in place before a repository licensing proceeding and that NRC would use a qualitative criterion of "no unreasonable risk to public health and safety" from its existing regulations. After evaluating these alternatives, the NRC recommended adoption of the third alternative. The NRC staff would attempt to identify and resolve potential implementation problems with EPA's containment standard and request that EPA clarify the standard. This would include a review of (a) whether changes in CFR 191 groundwater classification should occur, (b) the time frame for individual and groundwater standards starting with 1000 years and extending up to and including 10,000 years, (c) radiation risk factors for the containment requirements, and (d) the practicality of implementing numerical or probability-based standards.

The prediction of environmental impacts for a HLW at Yucca Mountain also should be understood to be problematic not only with respect to EPA standards and NRC regulations, but also in relation to EIS requirements. The EIS which will have to be completed as a basis for repository licensing will be deficient if it does not adequately reflect attempts to assess or resolve scientific uncertainties regarding impacts to the biosphere or humans for 10,000 years and beyond. However, there appears to be no readily available means to assess or resolve such uncertainties at the present time.

Distributional Problems in Nuclear Waste Repository Siting

Nuclear waste disposal involves unique variations upon classic problems of political philosophy and public policy. Disposal of nuclear wastes is a public good that inevitably imposes a disproportionate burden upon one sector of the population, either in space or time. Accordingly, public policy ideally should be based on criteria of fair decision making and distribution of burdens.

Philosophers are divided on the question of whether and to what extent present generations have responsibilities to future generations (78). In terms of intergenerational distributive justice, some philosophers argue that future people should be counted as being equal to present people. If this argument is accepted, then enough scientific knowledge should exist about Yucca Mountain to enable a reasonable conclusion that long-term risks and uncertainties about the adequacy of the site will not lead to a greater placement of risks on future people. Reasonable conclusions that disproportionate risks will be imposed on future generations are likely under conditions of scientific uncertainty about the adequacy of Yucca Mountain.

Some philosophers argue that although obligations to future generations exist, greater weight should be given to present generations. This implies that it is justifiable to choose a HLW such as Yucca Mountain under conditions of some scientific uncertainty. Practically speaking, even a small discount rate will effectively ignore future generations. For example, a 5% discount rate means that one death today counts as much as 3 billion deaths in 450 years. Yet a third view of philosophers is that present generations do not have obligations to the future. On this view, shifting risks due to scientific uncertainty about the Yucca Mountain site to future generations in a disproportionate manner would be justifiable. Although government representatives may express concern for future generations, the issue is given little more than a rhetorical role in the politics of nuclear energy. Accordingly, decisions inevitably reflect this practical attitude.

Controversies also exist with respect to intragenerational distributive justice issues. These stem from the fact that although NWPA originally contained provisions for equity by virtue of calling for site characterizations and eventual HLW repository locations in both the western and eastern parts of the country, in passing NWPAA Congress disregarded such provisions by choosing Yucca Mountain as the only site for characterization. Accordingly, risks were placed on Nevada and the western United States, even though the entire country derives benefits of nuclear power, especially eastern regions. The perception of the risks has increased as more information about the scientific uncertainties of the Yucca Mountain site have become known.

A major problem with the types of distributional burdens created by locating a HLW repository is that it is not adequately described as a classic "free-rider" problem, which is the most common way of treating philosophical problems of public goods. This is due to the fact that HLW disposal cannot be supplied without imposing disproportionate costs on certain people or sectors of the environment, in both space and time. Although several problems of distributional burdens due to the Yucca Mountain decision have been identified, further analysis of how to resolve them are required. Such analysis might include examining the distributional burdens created by establishing a HLW repository under conditions of scientific uncertainty according to (a) utilitarianism, which focuses on the consequences of policies and actions by maximizing aggregate utility for relevant populations; (b) Pareto optimalizations, which emphasize distributional changes that increase utility for some members of relevant populations without making anyone worse off; (c) equality, which considers the premises that those who receive the benefits should bear the burdens or that burdens should be distributed in proportion to ability to bear them; (d) contractual obligations, by which affected parties define allocation of benefits and burdens; (e) Rawlsian theories of justice, which stem from the concept that each person is to have equal rights to basic liberties compatible with similar liberties for others, and from the principle by which a policy is judged fair if it allocates equal benefits to affected parties; and (f) deep ecology, which considers that nature should be let alone and that humans should not interfere with its natural structure, function, or beauty.

Politically speaking, the question of the state role in nuclear waste decision making has been central from the earliest considerations of NWPA/NWPAA. Although the federal government may have a legal right to impose a repository on a state, fundamental public-policy questions nevertheless have been raised about centralized decision making by the DOE versus state-level decisions, and about states' rights to refuse to accept a repository within their borders. Clary has concluded that by enacting NWPAA and substituting its own judgment and decision to select Yucca Mountain as the only potential HLW site for characterization, Congress in effect offered a controversial solution to the problem of HLW because it was based on political considerations as opposed to the use of sound science and fundamental principles of participatory decision making as required by NWPA originally (79).

Public-policy issues relating to different notions of the balance of federal and state decision-making authority derive from the NWPAA requirements that Nevada relinquish basic and existing rights and duties in order to receive compensatory benefits for being the repository host. The NWPAA increased the authority of the federal government to locate a repository at Yucca Mountain over alternative sites and to decide which regulations it will comply with. Such authority is said to be necessary by proponents of nuclear power and by many political representatives because of the technological complexity of siting and building a repository and in order to avoid the NIMBY syndrome. On the other hand, the ability of Nevada and its citizens to express their will and actions is reduced. Although NWPAA provides monetary compensation to Nevada during the site characterization process and in the event that a repository is developed at Yucca Mountain ultimately, it reduces the freedom of Nevada in several ways.

In order to receive compensation, Nevada would have to relinquish several existing freedoms and legal rights. First, Nevada would have to relinquish the freedom to veto the selection of Yucca Mountain as the repository site, such veto subject to being overridden by a majority vote of Congress. Second, Nevada would have to participate in DOE's effort to have the site licensed by NRC. Nevada would thus become a cosponsor with the DOE by virtue of having consented to accept the HLW and agreeing to the site's suitability, prior to it having been judged suitable and safe by the DOE. Third, Nevada would give up its court-mandated right to independent oversight of the DOE program and would have to accept DOE's assertions that the site is suitable and safe despite the pervasive scientific uncertainty pertaining to site adequacy. Further, Nevada would not have the right to contest the site's suitability before the NRC in license proceedings. Fourth, Nevada would receive a fixed sum of $10 million per year during site studies and $20 million per year during site operation. Nevada views these sums of money as arbitrary and likely inadequate because it is not possible to determine at this date what actual financial assistance is required to mitigate both short-term and long-term impacts of the repository on the state's natural resources and economy. Further, by accepting this compensation Nevada would have to give up its right to additional financial compensation for impact mitigation. Fifth, the agreements between Nevada and DOE could only be terminated by the DOE. In the event DOE does not adhere to the terms of the agreement, Nevada would not have the right to take the government to court. Thus, issues of federal and state authority and freedom clearly are raised by NWPA/NWPAA and are exacerbated by the pervasive scientific uncertainty about the adequacy of Yucca Mountain as a HLW site.

Carter justifies the Yucca Mountain decision, in part, by arguing that economic and employment benefits from a HLW repository in Nevada will offset any loss in such benefits due to reductions in the work force at the Nevada Test Site in the event of a slowdown or ban on nuclear weapons testing (3). He also argues that economic and employment benefits due to a repository will lessen the dependence of Nevadans on the gambling industry. Lastly, Carter recommends that Congress should reach an understanding with Nevada in order to obtain the state's acquiescence to the repository in return for substantial benefits such as cash bonuses, generous payments in lieu of taxes, and assurances that the state will be allowed a strong voice in matters of public concern, such as how spent fuel will be shipped into the state. On the other hand, Bryan maintains that the economic benefits of a repository will be minimal and may not even offset the costs to local and state governments in terms of necessary services and facilities (79). He concludes that the public's fear of a nuclear waste repository would have devastating consequences to Nevada's tourist industry. Further, Bryan argues that the majority of land in Nevada is federally owned and constitutes a vital resource for future growth, recreation, defense, and other activities. Such lands require proper management and planning for future utilization, but to date there have been few attempts to study how these resources can be best used for present and future generations. Bryan's overriding concern is that significant scientific uncertainties exist with respect to the geologic suitability of the site, and therefore it is not prudent to risk contamination of groundwater and potential loss of economic and natural resources until the uncertainties are better known.

Conclusion

There are several aspects of the HLW program in the United States in relation to its use of science that have important implications. The NEPA is one of the nation's most important environmental laws by virtue of calling for the use of scientific information to assess the environmental and human health impacts of proposed actions and their alternatives. Congress eclipsed NEPA as a comprehensive environmental policy act by exempting the HLW site selection process from several of NEPA's most important substantive and procedural provisions, which had the effect of lessening the role of science in the process and eliminating certain environmental impacts from consideration. In effect, these exemptions laid the groundwork for HLW decisions to be based on less informed reasoning than they would have been under the full provisions of NEPA. The use of science as a basis for decision making in the site selection process was reduced further by NWPAA, which identified Yucca Mountain as the only site to be characterized as a potential HLW repository. The NWPAA represented a congressional determination that Yucca Mountain likely was suitable as a HLW site, as opposed to a determination being made with the use of more sound science. Such a determination combined with the fact that NWPAA violated the original NWPA provisions calling for participatory decision making in selecting potential repository sites have generated endless controversy.

A performance assessment of a HLW repository site has never before been conducted. Performance assessment methodologies, models, and procedures for their verification and validation are only beginning to be developed, and it is not known whether some geohydrologic and tectonic models are capable of true validation. The early developmental stage of performance assessment makes it difficult to resolve the pervasive scientific uncertainties regarding geohydrologic, tectonic, volcanic, and other conditions that might affect the integrity of the Yucca Mountain site for 10,000 years or beyond. In addition, a large part of the information used in performance assessment contains numerous value-laden assumptions, judgments, inferences, and evaluations that exacerbate problems of scientific uncertainty. All of the uncertainties make it questionable whether performance assessment at Yucca Mountain will yield reasonably certain scientific information to satisfy the EPA standards and NRC regulations for radionuclide exposure levels. If reasonably certain information is not available, then it is likely that repository licensing will be delayed or denied by litigation under existing standards. Alternatively, EPA's quantitative standards can be made more qualitative so that they would more appropriately reflect the more limited capabilities of performance assessment to yield reasonably certain scientific information about future repository performance at Yucca Mountain.

The U.S. approach to HLW disposal differs from every other country grappling with HLW problems. Other nations have neither a detailed plan nor rapid schedule for developing a HLW repository, nor requirements for quantitative probabilistic predictions of future repository behavior. In addition, the U.S. criteria for formal protection from HLW differ from those promulgated by the IAEA, ICRP, and many other countries. In contrast to the U.S. criteria, the international criteria and those of other nations provide greater protection to future generations, protect individuals and populations outside of repository boundaries, and restrict radionuclide exposure within repository boundaries. In part, the less stringent criteria of the U.S. presumably are necessary because of its rapid repository development schedule which, given the pervasive scientific uncertainties about repository performance, precludes reasonably certain predictions about future repository performance. In contrast, the approaches of other countries reflect an acknowledgment of the pervasive scientific uncertainty surrounding performance assessment of a potential HLW site and they allow opportunities for future decisions to be based on more reasonably certain scientific information as it becomes available.

The U.S. approach to repository development corresponds to a positivistic role for science. In other words, it assumes that scientific procedures exist and will be used that will allow the government to determine the environmental and human health impacts of actions in a reasonably certain and objective manner. However, because the methodologies and tools of performance assessment include value-laden assumptions, judgments, inferences, and evaluations, it is not possible to guarantee a neutral or objective description of the facts about the adequacy of the Yucca Mountain site. This creates the potential for the values positions held by fact finders and decision makers in governmental agencies to become part of the analysis of the scientific derivation of the facts. While most of the value-laden aspects of performance assessment cannot be avoided, it is important for public-policy and ethical reasons that the value-laden assumptions, judgments, inferences, and evaluations be disclosed so that decision makers and the public are able to differentiate better between facts and values. Further, distributional problems of risks are posed by a HLW repository. Accordingly, public policy should be based on criteria of fair decision making and distribution of burdens. Ideally, science has a role to play in distributional problems because it can provide factual information about risks to individuals and populations in space or time for use in public-policy discussions regarding whether and to what extent risks should be placed on individuals or populations, and what levels of compensation, if any, should be provided for those subjected to risks. Public-policy decisions about risk distributions are difficult enough under conditions of scientific certainty, but they are compounded under conditions of uncertainty.

Given the likelihood that the scientific uncertainties surrounding a potential HLW repository at Yucca Mountain will remain for the foreseeable future, one question which should be discussed is whether the U.S. HLW repository program should continue to be guided by present plans and policies or whether fundamental changes should be made. Any proposal to change HLW plans and policies may be infeasible at this historical juncture because the NIMBY syndrome would emerge again, and because the litigation and political controversy would be endless. However, the environmental goals enunciated in NEPA may be sufficiently important to justify a different course of action, which would be to leave the HLW in adequate above-ground holding facilities until the scientific uncertainties are better resolved or until better technologies are developed. This approach is a feasible long-term option only if the nation's nuclear weapons and nuclear energy programs are scaled back or eliminated. Shrader-Frechette has presented an extensive analysis supporting the establishment of temporary negotiated, monitored, retrievable storage (NMRS) facilities to handle HLW until the scientific uncertainties and inequities associated with the potential repository at Yucca Mountain are capable of being resolved more fully (21). Consistent with more stringent international criteria, such an approach would allow promulgation of radionuclide standards that would better protect future individuals and populations. In this sense, it would be ethically more justified.

References

1. Emel I, Cook B, Kasperson R, Renn O, Thompson G. Nuclear waste management: a comparative analysis of six countries. NWPO-SE-034-90. Carson City: State of Nevada, Agency for Nuclear Projects, 1990.
2. International Commission on Radiological Protection. Radiation protection principles for the disposal of solid radioactive waste, ICRP Publication 46. New York: International Commission on Radiological Protection, 1985.
3. Carter LJ. Nuclear waste imperatives and public trust: dealing with radioactive waste. Washington, DC: Resources for the Future, 1987.
4. Lemons J, Malone C, Piasecki B. America's high-level nuclear waste repository: a case study of environmental science and public policy. Int J Environ Stud 1989;34:25–42.
5. Brown DA, Lemons J. Scientific certainty and the laws that govern the location of a high-level nuclear waste repository. Environ Manage 1991;15:311–319.
6. Nuclear Waste Technical Review Board. Sixth report to the U.S. Congress and the U.S. Secretary of Energy. Washington, DC: US Government Printing Office, 1992.
7. Lemons J, Brown DA, Varner G. Congress, consistency, and environmental law: nuclear waste at Yucca Mountain, Nevada. Environ Ethics 1990;12:311–327.
8. Murchinson K. Does NEPA matter? An analysis of the historical development and contemporary significance of the National Environmental Policy Act. Univ Richmond Law Rev 1988;18:557–614.
9. Stryker's Bay Neighborhood v. Karlin. 1978. 44 US 223, at 227.
10. Vermont Yankee Nuclear Corporation v. Natural Resources Defense Council. 1978;435 US 519, 558.
11. Administrative Procedure Act 5 USC 702.
12. State of Nevada. A role in environmental compliance for the state of Nevada during site characterization of the proposed high-level nuclear waste repository site at Yucca Mountain, Nevada. NWPO-TR-008-88. Carson City: State of Nevada, Agency of Nuclear Projects, 1988.
13. Burton ES. Statutory environmental assessments under the Nuclear Waste Policy Act of 1982: process, content, and status. In: Proceedings of the 1983 civilian radioactive waste management information meeting, CONF-83127. Washington, DC: US Department of Energy, 1984:245–247.
14. Department of Energy. 1984. General guidelines for the recommendation of sites for the nuclear waste repositories. Federal Register 1984;49:47714–47770 (19 CFR 960).
15. Montagne CH. The initial environmental assessments for the nuclear waste repository under section 112 of the Nuclear Waste Policy Act. J Environ Law 1985;4:187–229.

16. Sierra Club v. Morton, 510 F2d 813, 825 (5th Cir. 1975).

17. Sierra Club v. Feaster. 1969. 410 F2d 1354, 1364 (5th Cir. 1969), cert. denied, 396 US 962 (1969).

18. Bernero RM. Performance assessment: the bridge between regulatory decision-making and scientific uncertainties. In: High level radioactive waste management, Proceedings of the Second Annual International Conference. Las Vegas, NV, 1991:869–873.

19. Campbell JE, Cranwell RM. Performance assessment of radioactive waste repositories. Science 1988;239:1389–1392.

20. McCombie C, Papp T, Coplan S. The development and status of performance assessment in radioactive waste disposal. Paris: International Symposium on the Safety Assessment of Radioactive Waste Repositories, 9–13 October, 1989.

21. Shrader-Frechette KS. Burying uncertainty: risk and the case against geological disposal of nuclear waste. Berkeley: University of California Press, 1993.

22. Winorad IJ. Radioactive waste disposal in thick unsaturated zones. Science 1981;212:1457–1464.

23. National Research Council. A study of the isolation system for geologic disposal of radioactive wastes. Washington, DC: National Academy Press, 1983.

24. Ross B. A conceptual model of deep unsaturated zones with negligible recharge. Water Resour Res 1984;20:1627–1629.

25. Department of Energy. Environmental assessment: Yucca Mountain site, Nevada research and development area, Nevada. DOE/RW-0073. Washington, DC, 1986.

26. Department of Energy. Site characterization plan, Yucca Mountain site, Nevada. DOE/RW-0199. Washington, DC, 1988.

27. Gertz CP. Yucca Mountain, Nevada: is it a safe place for isolation of high-level radioactive waste? J Inst Nucl Mater Manag 1989;17:10–12.

28. Malone CR. Geologic and hydrologic issues related to siting a repository for high-level nuclear waste at Yucca Mountain, Nevada, U.S.A. J Environ Manag 1990;30:914–929.

29. Malone CR. Environmental performance assessment: a case study of an emerging methodology. J Environ Syst 1990;19:171–184.

30. Hunter RL, Barr GE, Bingham FW. Preliminary scenarios for consequence assessments of radioactive-waste repositories at the Nevada Test Site, SANDS82-0426. Albuquerque, NM: Sandia National Laboratories, 1982.

31. Nuclear Waste Technical Review Board. Tenth report to the U.S. Congress and the U.S. Secretary of Energy. Washington, DC: US Government Printing Office, 1993.

32. Rogers AM, Harmsen SC, Carr WJ, Spence W. Southern Great Basin seismological data report for 1981 and preliminary data analysis, open-file report 83-669. Washington, DC: US Geological Survey, 1983.

33. United States Geological Survey. A summary of geologic studies through January 1, 1983, of a potential high-level radioactive waste repository site at Yucca Mountain, Southern Nye County, Nevada, open-file report 84-792. Washington, DC: US Geological Survey, 1984.

34. National Research Council. Probabilistic seismic hazard analysis. Washington, DC: National Academy Press, 1988.

35. Scott RB, Bank J. Preliminary geologic map of Yucca Mountain, Nye County, Nevada. U.S. open-file report 84-494. Washington, DC: US Geological Survey, 1984.

36. Pearthree PA, Wallace TC. Evidence of temporal clustering of large earthquakes in central Nevada. Seismological Research Letters 1988;59:17.

37. Crowe BM, Johnson E, Beckman RJ. Calculation of the probability of volcanic disruption of a high-level radioactive waste repository within southern Nevada, USA. Radioact Waste Manage Nucl Fuel Cycle 1982;3:167–190.

38. Crowe BM, Vaniman DT, Curtis D, Bower N. Volcanism: research and development related to the Nevada nuclear waste storage investigation, LA-10006-PR. Las Alamos, NM: Los Alamos National Laboratory, 1984.

39. Crowe BM, Self S, Vaniman DT, Amos R, Perry F. Aspects of potential magmatic disruption of a high-level radioactive waste repository in southern Nevada. J Geol 1983;91:259–276.

40. Knight JH, Philips JR, Waechter RT. The seepage exclusion problem for spherical cavities. Water Resour Res 1989;25:29–37.

41. Philip JR, Knight J, Waechter RT. Unsaturated seepage and subterranean holes: conspectus, an exclusion problem for circular cylindrical cavities. Water Resour Res 1989;25:16–28.

42. Buscheck TA, Nitao JJ. Repository-heat-driven hydrothermal flow at Yucca Mountain. I: Modeling and analysis. Nucl Technol 1993;104:418–448.

43. Beatley JC. Effects of rainfall and temperature on the distribution and behavior of *Larrea tridentata* (creosote bush) in the Mojave Desert of Nevada. Ecology 1974;55:245–261.

44. Quade J, Tingley JV. A mineral inventory of the Nevada Test Site, and portions of Nellis Bombing and Gunnery Range, Southern Nye County, Nevada, open-file report 84-2. Carson City, NV: Nevada Bureau of Mines and Geology, 1984.

45. Department of Energy. General guidelines for the recommendation of sites for nuclear waste repositories. Code of Federal Regulations 10, Part 960. Washington, DC, 1988.

46. Jones GM, Blackey ME, Rice JE, et al. Survey of geophysical techniques for site characterization in basalt, salt and tuff, NUREG/CR-4957. Washington, DC: US Nuclear Regulatory Commission, 1987.

47. Buxton BE. Geostatistical, sensitivity, and uncertainty methods for ground-water flow and radionuclide transport modeling. Columbus, OH: Battelle Press, 1989.

48. National Academy of Sciences. 1989. Ground-water models: scientific and regulatory applications. Washington, DC: National Academy Press.

49. McCarthy JF, Zachara JM. Subsurface transport of contaminants. Environ Sci Technol 1989;23:496–502.

50. Hunter RL, Mann CJ. Techniques for determining probabilities of events and processes affecting the performance of geologic repositories: literature review, NUREG/CR-3964. Washington, DC: US Nuclear Regulatory Commission, 1989.

51. Gillis D. Preventing human intrusion into high-level nuclear waste repositories. Underground Space 1985;9:35–43.

52. Knepp AJ, Dahlem DH. Role of uncertainty in the basalt waste isolation project. In: Buxton BE, ed. Proceedings of the conference on geostatistical, sensitivity, and uncertainty methods for ground-water flow and radionuclide transport modeling. Columbus, OH: Battelle Press, 1987:123–131.

53. Younker JL. Report of early site suitability evaluation of the potential repository site at Yucca Mountain, Nevada. (SAIC-91/8100). Las Vegas, NV: US Department of Energy, 1992.

54. Crowe BM. Volcanic hazard assessment for disposal of high-level radioactive waste. In: Studies in geophysics: active tectonics. Washington, DC: National Academy Press, 1986: 247–260.

55. Malone CR. Environmental performance assessment: a case study of an emerging methodology. J Environ Syst 1989;19:171–184.

56. Lemons J, Malone CR. High-level nuclear waste disposal and long-term ecological studies at Yucca Mountain, Nevada. BioScience 1991;41:713–718.

57. Roxburg IS. Geology of high-level nuclear waste disposal. London: Chapman and Hall, 1987.

58. Kirchner G. A new hazard index for the determination of risk potentials of disposed radioactive wastes. J Environ Radio 1990;11:71–95.

59. Environmental Protection Agency. Environmental standards for the management and disposal of spent nuclear fuel, high-level and transuranic radioactive waste. 40 CFR 191, Federal Register 1985;50:38066–38089.

60. Lemons J, Brown DA. The role of science in the decision to site a high-level nuclear waste repository at Yucca Mountain, Nevada, USA. Environmentalist 1990;10:3–24.

61. Pitman S. Corrosion and slow-strain-rate testing of type 304I stainless in tuff groundwater environments. In: Corrosion '87. San Francisco: Pacific Northwest Lab, 1986 (Item 172 in US DOE, DE88004834).

62. Broxton D. Clinoptilolite compositions in diagenetically altered tuffs at a potential nuclear waste repository, Yucca Mountain, Nevada. In: Hoffman, P, ed. Interface science and engi-

neering. Columbus, OH: Battelle Memorial Institute, 1987 (Item 90 in US DOE, DE90006793).

63. Environmental Protection Agency. Draft environmental impact statement, 40 CFR, Part 191, environmental standards for management and disposal of spent nuclear fuel, high level, and transuranic radioactive wastes. Washington, DC, 1982.

64. Department of Energy. Performance assessment strategy plan for the geologic repository program. Washington, DC, 1990.

65. Sinnock S. Preliminary estimates of groundwater travel time and radionuclide transport at the Yucca Mountain repository site. (SAND85-2701). Albuquerque, NM: Sandia National Laboratory, 1986.

66. Oreskes N, Shrader-Frechette KS, Belitz K. Verification, validation, and confirmation of numerical models in the earth sciences. Science 1994;264:641–646.

67. General Accounting Office. Nuclear waste. Quarterly report as of December 31, 1989. GAO/RCED-90-130. Washington, DC, 1989.

68. National Research Council. Rethinking high-level radioactive waste disposal. A position statement of the board of radioactive waste management. Washington, DC: Commission on Geosciences, Environment, and Resources, National Academy Press, 1990.

69. Camp v. Pitts. 1973. 411 US 138, 36 L Ed 2d 106, 93 S Ct 1241.

70. Natural Resources Defense Council v. Securities and Exchange Commission. 1977. (DC district court) 432 F. Supp. 1190, revd on other grounds 196 App DC 124, 606 F2d 1031.

71. Plaza Bank of West Port v. Board of Governors of Federal Reserve System. 1978. (CA 8) 575 F2d 1248.

72. Friends of Endangered Species, Inc. v. Jantzen. 1984. (ND Cal) 589 F Supp 113, affd, (CA 9 Cal) 760 F2d 976.

73. Davis JG. Letter to the Environmental Protection Agency transmitting Nuclear Regulatory Commission staff comments on the proposed high-level waste standard. In: Rethinking high-level radioactive disposal, a position statement of the Board of Radioactive Management. Washington, DC: National Academy Press, Washington, 1990:13.

74. Natural Resources Defense Council v. Environmental Protection Agency. 1987. 26 EPA 1233.

75. Alaska v. Andrus. 1978. 580 F.2d 465 (D.C. Cir. 1978), vacated in part as moot, 439 U.S. 922.

76. Yost N. Administrative implementation and judicial review under the National Environmental Policy Act. In: Novick S, ed. The law of environmental protection. New York: Clark Boardman, 1987:1–48.

77. Pollack S. Reimagining NEPA: choices for environmentalists. Harv Environ Law Rev 1987; 9:359–393.

78. Partridge E. Responsibilities to future generations. Buffalo, NY: Prometheus Books, 1981.

79. Clary B. Beyond Yucca Mountain and environmental gridlock: an alternative future for nuclear waste policy. In: Herzik EB, Mushkatel AH, eds. Problems and prospects for nuclear waste policy. Westport, CT: Greenwood Press, 1993:15–30.

80. Bryan RH. The politics and promises of nuclear waste disposal: the view from Nevada. Environment 1987;29:8–13.

Scientific Uncertainty as a Constraint to Environmental Problem Solving: Large-Scale Ecosystems

Carl F. Jordan and Christopher Miller

Background

Environmental problems have existed throughout the history of humankind. The ancient city-states of Sumaria between the Tigris and Euphrates Rivers provide an example. In the early dynastic period, approximately 3000 B.C., the city-states had a food surplus that enabled them to build their bureaucracies and armies and to extend their influence. They maintained this surplus, despite the region's hot, dry climate because of water storage and irrigation projects. Records of the declining amount of wheat cultivation and its replacement by the more salt-tolerant barley indicate that, over the centuries, irrigation resulted in salinization of the region's soil.

As the land in Subir was abandoned, populations immigrated southward at the same time that southern irrigation agriculture was suffering from reduced flow from the Euphrates. The growing population of Sumer necessitated cultivation of new areas. But the amount of new land that could be cultivated was limited, even with the more extensive and complex irrigation works that were becoming common at the time. Consequently, the size of the bureaucracy and the army that could be fed and maintained fell rapidly, making the state vulnerable to external conquest. The decline and fall of Sumer closely followed the decline of its agricultural base (1).

Other environmental catastrophes, as well as plagues and epidemics were common throughout the ages, and still frequently occur today. Humans often have dealt with such problems by moving away or by carrying out rituals which sometimes worked and sometimes did not.

Hypothesis Testing and Scientific Advancement

Modern reductionist science changed the approach to the problems of mankind. It replaced fatalism and superstition with analysis that enabled humans to understand and sometimes control the mechanisms which govern nature. The idea that all phenomena may be explained by decomposing a system and then reassembling it began in the sixteenth century with Copernicus and, later, Kepler, Galileo, Bacon, and culminated in the works of Descartes and Newton. Reductionist science has been the dominant approach in physics, chemistry, and biology and continues to be the primary means of explaining phenomena in the natural world.

Reductionist science made possible the Industrial Revolution. Newtonian physics encompasses a body of theory that enabled engineers to build machines, factories, and other infrastructures of the Industrial Revolution. Reductionist science, particularly the theory that bacteria and viral infection caused many human diseases, also revolutionized medical practice. Once scientists and physicians had an understanding of the mechanisms by which a disease affected the body, they were much more effective in controlling the disease.

Because of the success of science in engineering and medicine, in the late nineteenth century science also began to be applied to agriculture. Agricultural colleges were established whose mission was to increase understanding of crop production.

However, agricultural science, and associated applied sciences such as forestry and fisheries, differed in a fundamental way from basic sciences such as physics. In applied sciences such as agriculture, there were few central, unifying theories on which scientific predictions could be made. While ''laws'' such as Liebig's Law of the Minimum helped direct researchers toward problems of soil fertility, advances usually were made on a trial-and-error basis. Techniques empirically found to be effective were described by agricultural researchers, and recommendations were carried to farmers by extension agents. It was what has sometimes been called the ''let's throw another ton of fertilizer on the cornfield and see what happens'' approach to scientific advancement.

Selective breeding has been an important tool of farmers, horsemen, and cattlemen, and the outcome of selective breeding depends on genetics, which in the last century has become based on evolutionary theory. However, until very recently, plant and animal breeding, like agricultural crop production, was almost entirely empirical.

Once major advances had been made in increasing crop production, some agricultural scientists became interested in the side effects of farming, particularly erosion. This led to important recommendations, such as planting rows of trees as windbreaks and using contour plowing to help prevent soil from washing into rivers. Advances in erosion control also came about as a result of empirical testing rather than as a result of predictions based on theory.

Aside from such activities as erosion control, in the first half of the twentieth century there was little of what today we call environmental science. Ecology was mainly natural history, a descriptive activity. While there were environmental problems, there was little recognition that they were amenable to scientific solutions based on a basic understanding of the mechanisms of nature.

All this changed since mid-twentieth century. After World War II, with the rapid swelling of world populations and increases in standards of living, there was pressure to boost resource production. To do this, more research was needed, especially in relation to increasing agricultural yield and timber production, heightening the efficiency of pest control, and other practical problems related to production of food and fiber.

Because the hit-or-miss nature of the empirical approach was proving increasingly unsatisfactory for answering resource questions, researchers began testing hypotheses about the mechanisms that controlled ecosystem dynamics. There was a major increase in emphasis on studying the basic mechanisms hypothesized to control resource production. Instead of throwing another ton of fertilizer on the cornfield, scientists began a reductionist analysis of the corn and the soil in which it grew. Physiologists analyzed the rate of nitrogen uptake by corn. Soil scientists studied the ability of the soil to retain the nutrients from fertilizer, and calculated the proportion available for uptake and the proportion that would be leached away. Agricultural ecologists studied the competition between corn and weeds for nutrients in the soil. Mathematicians modeled the interactions between all these components to facilitate predictions. As a result of these studies, we gained a fair ability to predict the consequences of adding another ton of fertilizer to the cornfield, without actually having to add the fertilizer.

Predictions were never 100% perfect. Climate, pests, and other stochastic variables made completely accurate predictions impossible; nevertheless, this reductionist approach to science was clearly superior to the empirical approach for agricultural problems. In the 1950s, the approach began to be applied to environmental problems resulting from man's increased influence on the environment.

Modern Environmental Problem Solving

The first "modern" environmental problem treated in a systematic manner was radioactive fallout from atmospheric nuclear testing. Because of the possible danger from this fallout, the U.S. Atomic Energy Commission began an extensive and intensive series of environmental studies of radioactivity in the environment. Studies were made at the whole system level, at the process level, and at the genetic level. At the system level, whole ecosystems were irradiated, and effects on ecosystem process were studied at an oak-pine forest on Long Island, a tropical forest in Puerto Rico, a southern hardwood forest in Georgia, a northern hardwood forest in Wisconsin, and several other ecosystems on a smaller scale.

An important result was that populations of organisms with large physical structure and long life spans such as trees were more affected by radiation than those with small structures and short life spans such as insects. Within the plant kingdom, species such as pine that have a small photosynthetic area relative to total size were more affected than those such as grasses where the ratio was smaller. At the genetic level, many scientists concluded that sensitivity of species to radiation depended on volume of chromosomes, and that species with large volumes would be more susceptible to damage from radiation. At the process level, important gains were made at measuring isotope movement through the ecosystem, which enabled a prediction of the environmental half-time of the isotope (the length of time that it takes for half of the radioactivity to disappear from the ecosystem in question; it combines radioactive decay, and loss processes such as leaching into groundwater).

From the 1960s onward, the reductionistic approach to environmental problems spread to areas outside of radioactive fallout. The number of studies on effects of management or pollution on natural or man-managed ecosystems increased dramatically. Examples of some of the studies are given in Orians et al (2).

While reductionism has dramatically increased our understanding of nature, our ability to increase the accuracy of predictions concerning the consequences of environmental insults has tapered off in recent years. Scientists and the general public had expected that as the new, reductionistic ecology began to mature, its capability for prediction would rival that of the scientists who could predict the position of Mars with enough accuracy that a spaceship could be guided to land on that planet.

However, the hopes for an improved ability to forecast the effects of oil spills, wetlands drainage, pesticide leakage, and so forth in the environment have generally been unfulfilled. Despite a voluminous increase in the literature of resource management, our uncertainty of the effects of pollution in the environment has not been much reduced in recent decades. For example, if an oil spill results from a tanker breaking apart along the coast, how many seabirds will be killed? How will the pollution affect the local fishery this year? next year? Ten years from now? What will be the cumulative effect of a dozen oil spills along a coastline? Ecologists usually cannot predict within an order of magnitude, a margin of error so large that many scientists and citizens alike would hesitate to label the predictions "scientific."

What is the problem here? Why can ecologists and other environmental scientists not do a better job of advising government and business about what the effects of some new development project will be or whether projected increases in amounts of pollutants will harm human health or global climate?

Some say that the problem is that we simply need a greater effort: more scientists, more funding, more studies, and when these new studies are completed, our ability to forecast the effects of man's activities will be increased. Reductionist science holds that a central, absolute "truth" exists in science and that with the collection of enough data we can construct a model that perfectly reflects reality. Descartes's vision is still held by many scientists and most edu-cated citizens that "All science is certain evident knowledge. We reject all knowledge which is merely probable and judge that only those things should be believed which are perfectly known and about which there can be no doubts" (3, p. 57).

This chapter develops the idea that for many ecologic situations, science already has reached its predictive capacity and can scarcely further improve its capability to predict the effects of environmental insults. We do not reject ecol-ogy as a science, nor do we reject the idea that science is useful in solving many environmental and conservation problems. Rather, we seek to explore why sci-ence is inherently incapable of more precise environmental predictions on a scale outside the laboratory.

Facts and Probabilities

Most people believe that a scientific "fact" is a hypothesis that has been proven to be true. The educated public generally understands that an experimental test is usually necessary to evaluate a scientific hypothesis. What is less recognized is that no test is capable of proving a scientific hypothesis to be true (4). All that a test can do is show that a scientific theory is not false. The more tests that are run and the more times a theory is shown to be not false, the more confidence scientists have in that theory; however, alternative hypotheses can never be entirely ruled out.

The educated public generally believes that what scientists do is carry out experiments to "discover facts." The perception is that if scientists are unsure about the effects of some sort of environmental insult, they should go out and do more studies and collect more data that will enable them to better predict how a stress will affect an ecosystem. Many scientists, in an effort to obtain more research funds, foster this idea in their statements to the general public, and sometimes even in grant proposals.

Facts

If, indeed, modern reductionist science—sometimes called the hypothetico-deductive method (5) or the method of inductive inference (6)—can never prove a hypothesis to be true, then what is a scientific fact? To explain the difficulty that ecologists have in producing "hard facts," it is important to inquire into the nature of a "fact."

Philosophers of science have recognized two types of facts: institutional facts and natural facts (7). Institutional facts are statements about the relationship between things in space and time and depend for their existence on human institutions and conventions. *I had coffee with Mike yesterday* is an institutional fact verifiable by asking Mike or by asking the waiter in the restaurant where we purportedly had coffee. Institutional facts, however, are useless for predictions. The fact that I had coffee with Mike yesterday does not help predict with whom I will have coffee tomorrow.

Facts in a scientific sense are used more restrictively. A scientific fact is a statement about some fundamental quality or quantity of the universe that is true regardless of when or where that observation is made. The first law of thermodynamics—that energy cannot be created or destroyed—is a scientific fact.

Facts and Data

A datum is not the same as a fact. A scientist may be studying the population of squirrels in Georgia and determine that, on a particular day at a particular place, there were 17 observable squirrels per acre. This datum is valid only for the time, place, and conditions under which the observation was made. It cannot be assumed that at another place or at another time there will be 17 observable squirrels per acre. Many such data collected at different times and places could be used to predict the squirrel population of Georgia, but the prediction obviously would not be a "fact," merely an "estimate."

Facts in Ecology

It is difficult to find examples of facts in ecology. Some of what the public might think of as facts are actually just definitions, labels, or tautologies. An example of a definition or tautology is "birds fly south in the winter because of instinct." What is instinct? Instinct is that which causes birds to fly south in the winter. Instinct is a definition of all the physiologic and neurologic processes that go on that cause the birds to fly south. We do not really know how all these processes work, but we give them the label "instinct."

Peters (8) has criticized ecologic "laws" because variables are regressed or plotted against some function of themselves. For example, the −3/2 thinning "law" states that as the density of a population of plants increases, intraspecific competition will begin to affect the weight of the individuals in the populations. As density increases, plant weight decreases and the two variables can be plotted on a log-log scale along a line with a slope of −3/2.

The famous logistic equation $dN/dt = rN(K - N)/K$ has been used in ecology since the nineteenth century, when it was first formulated. However, many ecologists have described populations with this simple equation and its variants such as the Lotka-Volterra competition equations, even though the populations they are studying may not be growing exponentially but fluctuating around a long-term mean. The data either fit the logistic or they do not; when the data do fit, the equation is appropriate, but when they do not the equation is not appropriate. This tautology, claims Peters, has existed for over 150 years.

The final, and perhaps most damning, tautology identified by Peters (8) includes the central organizing idea for much of biology, the theory of evolution by natural selection. There are three conditions necessary for evolution by natural selection.

1. *Variation* in organisms: But the absence of variation does not disprove natural selection.
2. *Heredity:* Differences in traits are passed from one generation to the next. However, some traits are *not* heritable (phenotypic plasticity).
3. *Selection:* Differential survivorship and reproduction is necessary for evolution, but similar survivorship may also occur.

Each premise to the three conditions entails its converse, and therein is the tautology. The converse of each condition can be met (and often is met), but evolution occurs only if the converse conditions are *not* met.

Peters also criticizes ecology for its dependence on an esoteric jargon that includes terms so obscure they may not exist in the real world. Nonoperationalized terms such as "niche," "guild," "competition," "community," and "carrying capacity" have so many definitions in the literature that authors often are discussing ideas with many, and sometimes divergent, characteristics.

Ecology is criticized because its theories cannot make accurate predictions. May (9) plotted net reproductive rate (births minus deaths) against type of competition in a population growth model and found no apparent repetition of a pattern. Even in populations with a moderately high reproductive rate and density-dependent competition, the population under study exhibited chaotic behavior. May concluded that this is an entirely intrinsic response of the system, and no external control of the population is necessary to generate this behavior. "Even if the natural world was 100% predictable," he said, "the dynamics of populations with 'density-dependent' regulation could nonetheless in some circumstances be indistinguishable from chaos."

Facts in Taxonomy and Systematics

Taxonomy is concerned with the process of naming organisms. Often the names of plants and animals are considered by the public to be "scientific facts." However, taxonomy is basically a labeling exercise or a tautology. Giraffes are mammals with very long necks. If we encounter a mammal with a very long neck, it must, ipso facto, be a giraffe.

Taxonomy is based on systematics—the study of evolutionary relationships between organisms. Much indirect evidence seems to illustrate the process of speciation. However, evolutionary theory has never been proven because the length of time for a new species to evolve is too long to be amenable to a scientific experiment.

Facts in Ecosystem Science

Ecologic succession is one of a few central ideas of ecology. It is an orderly process of community development that is reasonably directional and, therefore, predictable (10). While there have been many theories to explain ecologic succession (11, 12), none have been accepted as fact by all ecologists, due to the inability of the theories to accurately predict successional trends under *all* situations.

Odum (10) hypothesized a series of trends to be expected in the development of ecosystems through the process of succession, such as an increase in biomass, an increase in species diversity, and an increase in the efficiency of nutrient cycling. The predictions have been criticized by Drury and Nisbet (13), Vitousek and Reiners (14), and others on the basis that other trends, or the reverse of the trends, sometimes are encountered. While the criticisms are valid, they do not address the fundamental problems. One is that Odum labeled the changes "trends," not "hypotheses," thereby acknowledging that the changes might not always occur. A second problem is that even if the trends *were* encountered under all situations, the "trends" are merely tautologies. Thus, an expected trend in succession is an increase in biomass; if one goes out and samples an old field over a period of years and determines that biomass per unit area is in fact increasing, then one can conclude that succession is occurring. If biomass does not increase, this does not refute the "hypothesis" that succession is an accumulation of biomass. It merely means that in this particular instance, succession is not occurring.

If succession is a hypothesis, it must be amenable to testing and rejection. Because the preceding definition of succession is a tautology, it cannot be falsified and therefore is not truly a theory. The ability for falsification is what distinguishes science from pseudoscience (4).

Facts in Ecophysiology

Some statements at the level of ecophysiology appear to be facts. For example, plants capture solar energy and reduce carbon dioxide to carbohydrates, thereby storing energy in a form usable to the plants and herbivores. This is a "fact" in ecophysiology. But if we examine this at lower hierarchical levels, then the "fact" becomes a probability. The problem is that we do not know, exactly, by what mechanism the chlorophyll captures solar energy, and therefore we cannot predict, except in a very general way, the effect of a given level of sunlight on the photosynthetic ability of plants.

Even more difficult is predicting how a given level of pollution will effect photosynthesis because that depends not only on sunlight but on all the other environmental variables at the time of exposure. The best that we can do is make a probability statement that a given increase in pollution, under given soil, water, and climatic conditions, will decrease photosynthesis a certain percentage plus or minus a rather large standard error. That which is a "fact" at one hierarchical level—ecology (pollution affects photosynthesis)—is a probability at a lower level—physiology (plants suffer a 25% reduction in photosynthesis at the 95% level of confidence).

This problem is not trivial but basic to all branches and levels of reductionistic science—that is, of "true" science. Observations at one hierarchical level are explained by mechanisms at a lower hierarchical level. As a result, that which is a "fact" at one level is a probability statement at the next lower level.

Facts in Physics

Are there any examples at all of scientific facts? We have not been able to find one in ecology. However, there is one physical law that always has been valid, when observations are restricted to time spans greater than a millionth of a second and to locations within our solar system. Under these conditions there have never been any known observations of an exception to the first law of thermodynamics, which states that energy can be neither created nor destroyed. The first law is the closest thing we have to a scientific fact, a fact being defined as something that is "always true."

Are there any other facts?

No!

Even the second "law" of thermodynamics is a statistical statement. Boltzmann showed that because certain processes do not seem to occur—for example, the spontaneous conversion of heat energy into mechanical energy—does not mean that they are impossible but merely that they are extremely unlikely. In microscopic systems, consisting of only a few molecules, the second law is violated regularly, but in macroscopic systems which consist of vast numbers of molecules the probability that the total entropy will increase becomes a virtual certainty (3, p. 73).

Much of what the general public considers to be physical and chemical "facts" are merely conventions and definitions. For example, "The sky is blue" is not a scientific fact but a definition, since children are taught that the sensation that they experience when looking at the sky is called "blue" (in Spain, the sensation is called "azul"). Physicists say that the sensation of blueness from the sky results from the differential refraction of wavelengths within a certain bandwidth. While this is not a tautology, neither is it a universal truth, in that sometimes the sky is red or gray and some people cannot see the color blue.

Probabilities

One way to view facts is to consider them as hypotheses with a high probability of occurring. Where can we find examples of high-probability hypotheses? A good place to begin looking might be in the "hard" sciences.

Hard Science and Soft Science

Within the fields of science, there is a series of hierarchical levels from "hard" to "soft." Hard science is often considered to be more rigorous; that is, it has more predictive power than soft science. Although individuals might make minor changes in classification, most would agree that the series from hard to soft would be approximately as follows:

- Physics
- Chemistry
- Biology
- Environmental sciences (including ecology and earth sciences)
- Economics
- Social sciences (political science, sociology, anthropology)

Classical Newtonian physics has been considered the "hardest" science, in that theories and hypothesis have been amenable to the most rigorous testing with results that have a high level of confidence. This is where most people gain their impression of the "rigor" of science.

Although physics is harder than ecology, and ecology is harder than anthropology, there is nothing innate about physics that makes it better than ecology, or ecology better than anthropology. It is the practical aspects of testing theories that distinguish between hard science and soft science. It is statistically and logistically easier to carry out experiments in the hard sciences than in the soft sciences.

Statistics and Logistics

Newtonian physics is the most rigorous of the sciences simply because it has the good fortune to have a very good statistical base. Each physical object used in an experiment consists of billions and billions of experimental objects (atoms), all of which are virtually identical. For example, let us test Newton's hypothesis that every body continues in its state of rest unless it is compelled to change that state by forces impressed upon it. We conduct the test by laying a pencil on the desk, watching it for a while, and observing that it does not move. However, we have not proven the hypothesis. We have merely failed to falsify the hypothesis.

Is this a silly experiment? In a practical sense it is, but in a theoretical sense it is not, because it illustrates how easy classical physics is in terms of logistics and statistics. The theoretical aspect of the pencil experiment is well illustrated by Modis (15, p. 189), who as a physics student had to calculate how long one has to wait in order to see a pencil fly. He describes his calculations as follows:

> Since atoms in solids are vibrating in random directions, it is conceivable that there may be a time when all the atoms of one pencil direct themselves upward, in which case the pencil will lift off on its own. My calculation showed that had someone been watching a pencil without interruption from the day the universe was created, he most certainly would not have

witnessed such a levitation. In fact, it is much worse; in order to have a fair chance for such a sighting, one would have to wait another period of that length, but this time with every second itself stretched to have the age of the universe! Yet the possibility is there, and it is only a question of time.

Many pencil experiments lasting 1 minute each could be conducted by a class during a laboratory period, with a result that has a very high degree of statistical confidence. The experiment has easy logistics and good statistics.

A physics experiment with more difficult logistics was carried out by Billas et al (16). They asked how many metal atoms are needed in a cluster for it to start exhibiting the properties of the bulk metal. They measured the magnetic moment of different sizes of clusters of iron, cobalt, and nickel. Ferromagnetism was observed for even the smallest clusters, but for 30 atoms or fewer the cluster's magnetic moment is atomlike. Bulk magnetism becomes the norm for clusters of several hundreds of atoms. Predictability, at least in this branch of science, appears to begin when sample size is several hundred.

At the level of the individual physical particle, predictability becomes impossible. Heisenberg expressed the limitations of classical concepts with his discussion of the impossibility of simultaneously describing atomic phenomena as either wave-like or particle-like. The more we emphasize one aspect in describing the behavior of an object, the more the other becomes uncertain (3, p. 79). The more precisely one can determine the momentum of a particle, the less the observer will know about its position.

In the field of chemistry, the logistics remain relatively easy. Experiments can be carried out in a laboratory, regardless of where in the world that laboratory might be located. The statistics, while not as good as in Newtonian physics, still are good enough that many hypotheses are considered to be facts. Take, for example, the periodic table of elements, considered by most scientists to be "confirmed." In reality, the periodic table is a probability table. In an ordinary mixture of hydrogen gas, 0.0156% of the atoms are deuterium and 10^{-15}% are tritium. As a result, the atomic weight of hydrogen is not really 1.000000, but 1.00797. Deuterium and tritium have different properties than hydrogen, but since there are so few in relation to stable hydrogen, their effect is so small that it is not detectable without the use of highly sensitive analytical instruments.

When we move into the realm of biology, things become more difficult, but they are still manageable. For example, let us hypothesize that a certain drug will have a certain effect on mammals, and we wish to test that hypothesis. We decide to administer the drug at four levels to four groups of genetically identical mice, 100 in each group, and keep 100 mice as a control. To eliminate crowding as an uncontrolled variable, the mice must be kept in individual cages, and the room containing the cages must be climatically controlled and the air well circulated. Still there can be some effects due to uncontrolled variables, such as some cages being closer to the walls than others. This is where statistics comes to the rescue. With an adequate statistical design (cages placed randomly in the room, dosages assigned randomly to cages), the effects of uncontrolled variables can be pretty much eliminated.

The logistics of this experiment are messy, in that a room that allows for feeding and cleaning of the mice is difficult to keep at exactly uniform conditions. The statistics (only 100 experimental subjects for each dosage level) are not nearly as good as in chemistry. Nevertheless, it still is possible to get results that are repeatable 95% of the time, which statisticians consider to be good enough.

When we move into ecology, things become very difficult. Let us say that we want to test the hypothesis that conversion of a watershed from trees to grass will result in an increased water yield. To test this, we must find watersheds that are "identical." Since no two watersheds are identical, how similar can they be? In the mouse experiment, any mouse more than a few percent above or below the average weight would probably be discarded. However, it is impossible to find two watersheds that are very similar to each other in terms of water yield, length of streams, or biomass of vegetation. What can we reasonably expect in terms of similarity? If one watershed had a stream 1 km long and another 1.2 km long, this is probably as close as we can find. Ecosystems are so complex and are subject to so many variables, it is impossible to find two that resemble each other within 5% in any quantifiable way.

Logistics are also very difficult at the ecosystem level. Given the cost of constructing the weirs for the watersheds, measuring the runoff over the decade necessary to do the experiment, and the associated management costs, we can scarcely expect to have more than a few experimental watersheds and one control. This number is unacceptable to statisticians. Watershed scientists attempt to avoid criticism by comparing an experimental and a control watershed before and after treatment.

Other ecologic questions, while theoretically amenable to experimental testing, in fact are not directly testable due to the length of time that a test would take. Hypotheses regarding the minimum critical area necessary to prevent extinction of species is a case in point. While experimental tests of biogeography theory (17, 18) readily demonstrate that reserves of a certain size are too small, an experiment must be run over a period of several hundred years at least to prove what size reserve is large enough to prevent extinctions. The effect of inbreeding upon rates of extinction cannot be assessed in a matter of decades. Very rough estimates can be made based on an understanding of the mechanisms underlying the process of extinction, but these are really "order of magnitude" estimates.

When we move up the hierarchy into the social sciences, experimental tests are generally not possible. Tests of sociologic theory, such as the effects of homelessness, while theoretically possible, are not ethically possible. It is not legally possible to take a homogeneous group of people and randomly select half to become experimentally homeless. Similarly, in anthropology, experiments on cultural groups are theoretically but not practically possible.

The statistical nature of the differences between "hard" and "soft" science has resulted in a series of other characteristics associated with the extremes (Table 3.1). These differences are not innate to each science, but are merely a reflection of the statistical base on which that science rests.

Influence of Observer on Observed

Regardless of the hierarchical level of investigation, the observer in any science will affect the observed. At higher hierarchical levels, such as anthropology, the effect will be greater than at lower levels, such as Newtonian physics. Regardless of the level, however, the false notion that there is a segregated and distinct observer who independently notes the world is a carryover from the old man-nature dichotomy propagated by Cartesian-Newtonian science. Heisenberg stated very nicely, "What we observe is not nature itself, but nature exposed to our method of questioning" (19, p. 114).

Predictions in Ecology: How Good Can They Be?

Given the variability of subjects studied in ecology, compared to those studied in biology, chemistry, and physics, and given the logistical difficulty of outdoor ecologic experiments in comparison with controlled conditions in the laboratory, what levels of prediction can we expect from ecologic experiments? There are two types of ecologic field experiments.

Table 3.1 Characteristics of "Hard" and "Soft" Sciences

HARD SCIENCE	INTERMEDIATE	SOFT SCIENCE
Newtonian Physics, Chemistry, Biology, Ecology, Economics, Anthropology, Sociology		
Theory reflects "real world"	Theory reflects limited vision of world	Theory is site and time dependent
Deterministic	Partially deterministic	Nondeterministic, empirically based
Universally accepted	Not always accepted	Many competing theories
Mathematically expressed models	Not always expressed mathematically	Most theories qualitative
Deviation from model is experimental error	Deviation may be due to changes in environment	Much covariance of variables, not easily separable

Designed Experiments

In designed experiments the ecosystem to be studied is prepared or selected to be as homogeneous as possible so that plots can be set up to statistically determine the effect of a treatment on ecosystem properties. When proposals to carry out such experiments are submitted, it is expected that a suitable statistical design such as analysis of variance (ANOVA) will be incorporated to determine if the effect of the natural or human-caused perturbation is really significant.

However, it is impossible to find a natural ecosystem that is homogeneous enough to permit a statistically significant number of "identical" plots within that ecosystem. Scores of graduate students have complained about the impossibility of finding a homogeneous plot. Regardless of the size or location of the "ecosystem," the students all have the same complaint: "My ecosystem is too heterogeneous to implement a good statistical design."

Sometimes ecologic field sites are prepared beforehand in an effort to make them homogeneous. One graduate student at the Institute of Ecology, University of Georgia, wanted to compare the effect of "alley cropping" (a type of agroforestry) with conventional green manuring on the productivity of sorghum. We hunted long and hard to find an old field that was big enough to accommodate a statistically meaningful number of plots and that was also homogeneous throughout. The best that we could find was a field about 50 × 100 meters that had a slope in one direction that resulted in about 2 meters difference in elevation over the width of the field. We plowed and disked the field several times to homogenize it as much as possible. Still, there was little we could do about the slope. In our initial survey, we detected a difference in soil chemistry between the highest and lowest elevations. As a consequence, we settled on a stratified design, with five plots along a topographic level at each of four levels down the slope.

At the end of 3 years of field research, the student could show no statistical difference between the plots at even the 90% level of confidence. While there were differences in soil fertility and sorghum production, variability between plots within the same treatment was so great that any differences between treatments was masked. We speculated that to show differences at a 95% level of confidence, the experiment might have to be run again with a greater number of plots for at least 5 years.

In order to achieve the type of homogeneity required for standard statistical tests, and within a time frame that is achievable, experiments within a controlled environment such as a greenhouse are much more feasible. While such tests have scientific validity within the bounds of the experimental design, they usually have limited value for predictions under uncontrolled conditions. When results of such experiments are introduced into legal procedures, they are easily criticized as not being relevant, since they cannot make accurate predictions outside of laboratory conditions.

Natural Experiments

In the other type of ecologic field study, the objective is to determine or to predict the effect of some type of perturbation, such as an oil spill, on some type of ecosystem, such as a lake. In this type of "natural" experiment, the ecologist does not have the opportunity to set up any kind of statistical design. He or she must work with whatever ecosystem is in question, usually with the number of experimental treatments being 1. Further, in contrast to watershed experiments, there usually is no opportunity to compare the experimental with a control site before and after treatment. What kind of predictions can be made in such a case? How good will the predictions be?

Studies supported by the U.S. Atomic Energy Commission in the 1960s provide an illustration. One study was to predict the environmental half-time of radioactive strontium in a tropical rain forest in Puerto Rico. At the time such studies began, it was not known if the half-time would be on the order of minutes, days, years, decades, centuries, or millennia.

The first step was to measure the stocks and fluxes of stable strontium in the forest and use the data to construct a model (Fig. 3.1; 20). Measurements of stable strontium concentrations in the ecosystem are straightforward, and concentrations of strontium multiplied times biomass of ecosystem compartments, or times volume of flux between compartments resulted in the average stocks and fluxes in Fig. 3.1.

Next, the transfer coefficients between each of the compartments were expressed mathematically. For each compartment, there was a differential equation that predicted the amount of strontium in that compartment as a function of time, based on flows into and out of that compartment. The equations for each compartment were solved simultaneously for a series of time intervals. The environmental half-time of strontium in the forest was predicted to be about 22 years.

To validate the model, actual strontium-90 fallout data was used (Fig. 3.2) as input for the model. The curves in Fig. 3.3 show the predictions of strontium-90 in each compartment, and the circles show the observed concentrations in the litter from the forest floor.

To what extent did inaccuracies in measurements of the stocks and fluxes affect the predictions of the model? In subsequent years a detailed study of the throughfall component of our model was carried out. It showed that throughfall data was off by about 10%. A sensitivity analysis indicated that a 10% error in throughfall estimates could affect the prediction by about 2 years.

Given that when the experiment began it was not known whether the half-time would be 1 year or 1000 years, an error of 2 years does not seem very significant. The error is not great enough to influence one way or the other any policy decisions, whereas an error of 100 years or more could have important implications.

Given the limited time and budget for the original study, it seemed more

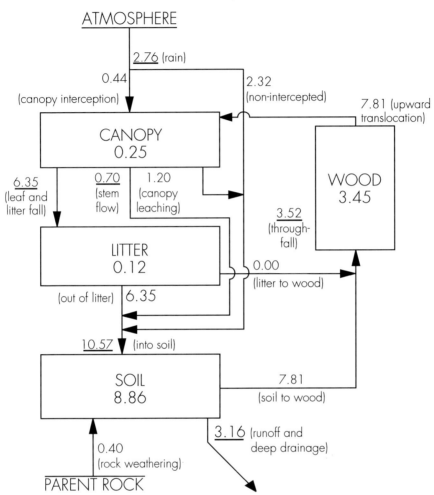

Figure 3.1 Stocks and fluxes of stable strontium in Luquillo forest, Puerto Rico (20).

reasonable to do a study that gave an approximate but useful result rather than an intensive study that yielded an accurate result for part of the system but which did not answer the initial question about response of the entire eco-system.

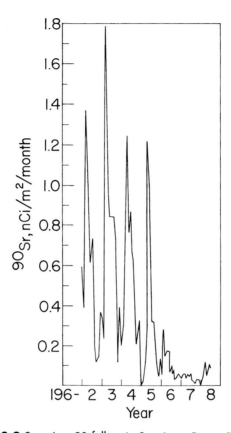

Figure 3.2 Strontium-90 fallout in San Juan, Puerto Rico (20).

What Should We Expect?

Pomeroy et al (21) suggest that ecologists should not expect to establish universal laws but should attempt to state what normally or usually happens in a given system. "Normic statements" are not probabilistic, but they "attempt to state what will happen in a particular case, given a series of conditional statements around circumstances." Normic statements are not as predictive as physical laws, but they provide explanations for phenomena under a given set of conditions.

The problem here is that we again are faced with a tautology. If the normic statement fails to be accurate, the explanation must be that the given set of conditions was not met. How feasible is it to specify initial conditions? In May 1994, at a workshop dealing with chaos theory and the limits to scientific knowledge, the subject was whether the world was too complex for many important problems to be solved through science. Crutchfield estimated that the

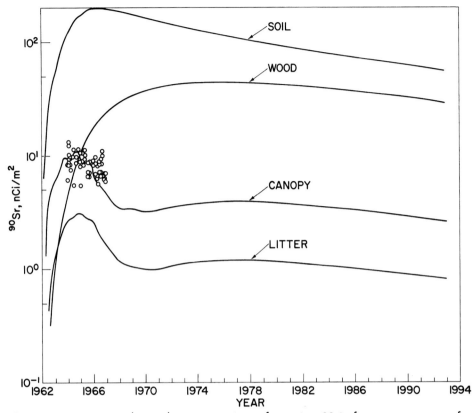

Figure 3.3 Curves predicting the concentrations of strontium-90 in four compartments of a Puerto Rican rain forest (curves) resulting from the input shown in Figure 3.2, and validation data (circles) showing observed concentrations in the litter layer (20).

gravitational pull of an electron, randomly shifting position at the edge of the Milky Way, can change the outcome of a billiard game on Earth (22).

When Can Predictions Be Better?

We have emphasized that the natural variability in ecosystems is so great that it is not possible to predict the results of some environmental insult within the limits of commonly accepted statistical confidence. To criticize the ecologist for not being able to predict with a high level of confidence is like criticizing a scientist for not being able to predict the toss of a coin with any greater than 50% accuracy. One cannot predict the result of the coin toss because of the random nature with which it lands. Because of the random nature of factors that influence ecosystems, a high degree of confidence is not possible in ecology either.

Predictions in Ecology: How Good Need They Be?

How accurate do predictions have to be to be useful? In the strontium study, it would seem that only if the error were greater than an order of magnitude would the implications be serious. For example, if in reality the half-time were 1 year or 100 years instead of 22 years, the difference could be meaningful. However, for practical purposes, the difference between 20 and 22 years is meaningless.

Let us consider the implications of an oil spill in a bay. If an ecologist were to predict that the spill might kill one or two ducks, most people might not consider this serious. In contrast, if the prediction were that the spill might kill thousands of ducks, this would be a serious situation. However, to discredit an ecologist because he or she could not predict within a 95% level of confidence how many ducks would be killed reflects a lack of understanding of the probabilistic nature of ecology, of the practical implications of what an ecologist can do, and of the meaning of the results. What would be the purpose, for example, of trying to do a prediction with 95% predictive accuracy? What difference would it make if 940 rather than 960 ducks would be killed by a potential oil spill?

Let us consider the effects of a situation such as the Yellowstone fires in the summer of 1988. The best an ecologist could do would be to say that due to the accumulation of fuel resulting from fire suppression, there would be a high probability that some people would be killed or injured by an accidental fire and that surrounding businesses would suffer some losses. Because an ecologist could not predict the exact number of people that would be killed or the exact value of the economic damage, does this lessen the importance of the warning that a dangerous fuel accumulation had occurred?

However, we do not advocate abandoning all further efforts to forecast environmental effects. We agree with Ravetz (23), who suggests that in cases where systems uncertainties are high and where the decision stakes also are very high, further efforts at improving forecasts are warranted. The most notable example is the case of global warming due to increases in carbon dioxide in the atmosphere. In this case, the contents of the compartments (sensu Fig. 3.1) are so poorly quantified that further study could significantly improve the predictive capacity of the global models.

Scientific Uncertainty as an Excuse, When the Real Problems Are Economic, Social, and Political

Calls for more scientific studies and more ''facts'' by politicians and bureaucrats are usually a stalling technique, a way to postpone or avoid politically unpopular action.

The call for further studies, a common conclusion in scientific papers, is mostly a self-serving tactic to justify further grants. In most cases, more studies are not going to increase the accuracy of predictions, any more than further study will increase the accuracy of predicting a coin toss.

Scientists should be careful about feeding the illusion that they can increase their predictive capability. Rather they should emphasize that, except at the global biogeochemical and climatologic level, our predictions are about as good as possible. Scientists should emphasize the fact that they will never be able to produce perfect knowledge. By so admitting, they can shift the responsibility for action to the economic and social sphere. Ecologists should no longer allow politicians to use "lack of certainty" as an excuse for not taking action.

Case Studies

In this section we give examples of applied situations where further study and data collection in theory should help improve ecological predictions, but where in practice, improvement in predictions is unlikely.

Florida Everglades

Since the early 1900s, the exploding urban population and farming in South Florida have resulted in diversion of more and more water that originally fed the Everglades ecosystem. Over 1400 miles of canals, dikes, and levees have been built, and the resulting drainage has led to loss of more than half of the Everglades. By the 1980s, the amount of water flowing through the Everglades was just one-fifth of the amount that used to reach the ecosystem at the turn of the century. Only 5% of the wading birds that used to nest in the wetlands were still doing so (24).

Because of the increasing awareness of the importance of wetlands and the species that they harbor, efforts at restoration of the Everglades began in the 1980s. Several approaches are being tried. One has been to reroute and redistribute some of the water from the canals into the ecosystems. Another has been to improve water quality by filtering the agricultural nutrients through marshlands. A third is to eliminate exotics such as melaleuca, Australian pine, and Brazilian pepper which outcompete the native species.

One major problem in deciding what actions to take is predicting how successful each of the various actions might be. Is removing nutrients, a costly procedure, cost-effective in promoting reestablishment of desirable species? How much water must be returned to the Everglades in order to rehabilitate biodiversity and ecosystem function to desirable levels?

It is impossible for scientists to answer these questions with any degree of accuracy. In fact, trying to answer the question—that is, in the sense, if we double the amount of water, we will increase populations by X%—is meaningless. In natural ecosystems, populations fluctuate wildly, and we can say with certainty that the number and distribution of species in the Everglades before human interference fluctuated greatly from year to year. Natural fires also greatly influenced populations. Given the natural variability of the ecosystem, predictions of the effect of restoration activities can have meaning only with a range of several orders of magnitude.

Given this inability to make accurate predictions, some scientists believe that "rehabilitation" is a better term for such activities than "restoration." Usually there is a good sense of what the original environment looked like. To begin to work toward that without precisely specified goals, such as "15 species of wading birds per hectare," is more reasonable.

"Adaptive management" is the term frequently used these days for a strategy undertaken to determine the outcome of a management technique when basic science is incapable of the task (25). In the case of the Everglades, modelers at the South Florida Water Management District are writing computer programs to reconstruct how water traveled before the region was crosshatched by canals and levees. Once finished, this "natural system model" will be superimposed on an existing model of how water currently flows. The results will be used as a guide to reconstruct the water flow to restore or rehabilitate the Everglades (26). This is probably about the best that we can do.

Hydrologists, ecologists, and other scientists working in the Everglades recognize that numerous factors may complicate prediction, but that delaying action in restoring the ecosystem will result in greater loss of function and structure, and rehabilitation will become increasingly difficult. We cannot wait until all the data have been collected, for instead of gaining information needed to heal the Everglades, we will have documented its death (27). Scientists' knowledge of the system is incomplete, but action must be immediate (28).

Once a plan is tried, ecologists evaluate the results and proceed with recommendations as to whether more water is needed, and if so, how it should be distributed. After a second round of trials and results, a third approximation can be attempted. To many scientists, this "adaptive" approach is the most reasonable strategy. Actually, "adaptive management" is the old "let's throw another ton of fertilizer on the cornfield and see what happens" approach, updated for the 1990s.

Oil Spills

Wrecked tankers and broken pipelines have resulted in oil spills in aquatic and terrestrial ecosystems. Freedman (29) has described the results of some of the more spectacular accidents. For example, the wreck of the Torrey Canyon on the Cornish coastline resulted in the death of at least 30,000 seabirds. The damage resulting from such spills is resulting in new regulations regarding safety precautions in ship's hulls and pipeline maintenance. Requirements for double-hulled ships and for better pipeline inspections may reduce the frequency of environmentally damaging oil spills. Nevertheless, we can certainly expect that in the future there will be at least a few more oil spills.

Let us hypothesize that a developing country, Vietnam for example, is considering building a new port for handling crude petroleum pumped from recently discovered oil fields. To obtain a loan from an international lending agency, the country might have to prepare an environmental impact statement, predicting what could happen if a tanker were damaged in the proposed harbor and oil seeped across the surrounding ecosystems.

The problem faced by ecologists in making such predictions is illustrated by Wheelwright (30) in his assessment of the results of the *Exxon Valdez* oil spill in Prince William Sound, Alaska. The short-term effects of the spill, such as blackened carcasses of seabirds and otters were extremely obvious. However, any long-term effects were not discernible, he said, because they were drowned out by background "noise." In the years after the spill, the salmon and herring returned strongly. Pink salmon catches broke records in 1990 and 1991, and herring catches broke records in 1991 and 1992. However, the last two seasons were very poor, and both populations fell sharply. The problem is that fishery science and marine ecology cannot account for a delayed response from an oil spill. All that can be said is that the population fluctuations must have resulted from other causes such as shifts in the weather, ocean temperature, food supply, predation and disease, and inconsistent spawning. Against such a stochastic, chaotic background, how is it possible to predict with any certainty at all what the results of an environmental insult might be?

Management of National Parks

Despite extensive research in Yellowstone National Park, the park and surrounding ecosystems have suffered greatly due to logging, tourism, oil and gas exploration and exploitation, and private land-use practices. Lack of agency coordination, intense public pressure, and an unfavorable political climate in recent years has led to general ecosystem decline and to problems with the reintroduction of wolves and attempts to boost the number of buffalo in the park.

The failure to properly manage the ecosystem is not due to lack of research. Instead, the inability to coordinate activities between federal agencies such as the National Park Service and the Forest Service has created problems in defining management goals. The Park Service is oriented toward encouraging public use without compromising ecosystem integrity. In contrast, the Forest Service has attempted to maximize sustained yield of natural resources, particularly timber, and juggle numerous private and public interests. Furthermore, current federal laws further deepen the differences between the agencies. For example, the National Forest Management Act protects sensitive areas from logging. The

Forest Service, however, often acknowledges potential destruction due to timber practices and then proceeds to log an area anyway (31).

The Yellowstone Park fire of 1988, perhaps the largest and most intensive to have ever occurred in the region, has brought urgency to the question of how much more research is needed, as opposed to putting into practice what is already known. Elfring (32) has pointed out that it was a lack of understanding by the *public* and by *politicians* of the natural disturbance regime in the Yellowstone ecosystem, and not ignorance on the part of scientists, that resulted in criticism of management policy. Ecologists had been aware of the importance of fire in forested ecosystems since the 1950s. Recognition of this fact ended the fire suppression policy (for remote natural fires) that Yellowstone maintained from 1872 to 1972 (33).

Even within the ranks of scientists, it is philosophical differences that are the problem rather than lack of data. One school argues that Yellowstone has been subjected to so many adverse human impacts that allowing natural disturbances to occur without proactive management can lead only to further declines in park resources (34, 35, 36). In general, advocates of proactive management want plans with specific goals and a "scientifically sound program for assuring that progress is being achieved" (37, p. 98).

In contrast, scientists advocating a passive or semipassive approach argue that precise goals, even if they could be achieved, preempt the opportunity to learn about ecologic phenomena that have occurred for millennia. Not only is it extremely difficult to predict what the results of proactive management strategies, such as prescribed burning might be, they say it has been impossible to thoroughly evaluate *after the fact* what the effects of the 1988 fires have been. Knight (37) summarizes his review of the Yellowstone fire controversy by saying (p. 99):

> Because definitive statements about adverse ecological effects are not yet possible, significant changes in fire management policies at this time seem unnecessary. (Policies at this time are suppression of human-caused fires only.) Moreover, adoption of a policy that attempts to eliminate uncertainties, even if that were possible, creates an inherently unnatural situation. If a change in the policy does occur, the reason will probably be adverse public opinion rather than adverse ecological effects.

Biological Reserves

Island biogeographic theory predicts that the probability of a species becoming extinct in a reserve is inversely proportional to the size of the reserve. The larger the reserve, the less likely a species will disappear from that reserve. Although scientists working in the field may not agree exactly as to the minimum area required to prevent a species from disappearing, virtually all agree that a single large reserve is more effective than a single small reserve in preventing species extinctions. Some of the disagreements between scientists stem from uncertainties about how quickly the genetic inbreeding within small populations in small reserves will affect the ability of that species to survive. For a predator, estimates of the minimum area required could vary between 5000 and 50,000 hectares, depending on assumptions in the genetic model used.

To actually test the theory, it would be necessary to measure the length of time required for a species to become extinct on reserves of a certain size. Several experiments have been carried out that show what size reserve is too small. For example, in the Minimum Critical Ecosystem Size experiment near Manaus, Brazil, scientists found that as a forest is felled around a reserve, many birds find sanctuary in the remnant patch of forest. The elevated numbers, which are inversely correlated with fragment size, persist for approximately 200 days, at which point the number of birds in the reserves falls sharply to levels below preisolation values. The rate and extent of this collapse are greater in smaller reserves than in larger ones (17).

What the experiment was unable to show was how large a reserve was necessary to prevent a species from becoming extinct. The reason is that in order to state with assurance that the population of a species and the area that it occupies are sufficiently large enough to prevent extinction, the experiment might have to be run for several hundred years.

Because testing of the theory requires such a long time, it has not been tested. Because it has not been tested, in legal litigation between the Sierra Club and the U.S. Forest Service, a former chief of the Service argued that: "the island biogeography theory is not new, but until there is conclusive empirical evidence that the conclusions, hypotheses, or predictive capabilities for terrestrial ecosystems are valid, it is proper to acknowledge it as untested theory." The district court voiced a similar concern: "whatever their theoretical validity, considerable uncertainty seems to surround the question of how exactly these conservation biology principles should be applied," and the court cited "scientific uncertainty" as a reason to uphold the chief's decision (38).

Implications

The response of the (then) chief of the Forest Service to the Sierra Club suit is a common one for bureaucrats and politicians who wish to avoid action that might be unpopular among their constituencies. It reflects a lack of comprehension as to the level of scientific certainty that can be expected. There can be no "empirical test" of island biogeography theory without running an experiment for several hundred years over an area of tens to hundreds of thousands of hectares, clearly an impossibility. The best that can be reasonably expected is the opinion of scientists who have observed rates of species extinctions in small areas over short periods of time.

Given that "proving" such ecologic hypotheses is a practical impossibility, decision makers must resign themselves to making decisions based on less than perfect data. They must accept the burden of decision and not try to weasel out by telling the (often willing) scientists to go out and collect more data to "prove" their hypothesis.

Recommendations

Ravetz (23, 39) has discussed the problem of ignorance in science and its implications for the role of scientists in environmental problem solving. He says that scientists have to admit that science alone cannot solve the problem. Science can help, providing nuggets here and there that can illuminate some aspects of the issues, but overall it is an iterative process, with experience and practical outcomes influencing policy until one side becomes so discredited that it either admits defeat or becomes irrelevant. The nuclear industry and the tobacco industry provide two contrasting examples. The nuclear industry has generally admitted its problems and is therefore maintaining some of its credibility. The tobacco industry, on the other hand, is clinging desperately to a position that, as more and more circumstantial evidence becomes available, appears to be ultimately wrong and thereby is losing all credibility with the public.

To maintain credibility, scientists must fit into a new paradigm of environmental problem solving. They must admit that science cannot solve it all. The role of scientists is to issue warnings, concerning chlorofluorocarbons for example, where our predictive ability is low, but where risk, should the worst scenario be true, is unacceptable. At the local and regional levels, it is up to the scientists to make the public aware that they cannot do much better than they already have done, and it is at the political and economic levels that further actions must be taken.

Scientists, by continuing a charade that if they were only given more money and more facilities they can solve the environmental crises, will only contribute to a greater discreditation of science. The longer they hold on to this myth, the greater will be the crash when the public finally realizes the limitations of science. It would be better for scientists to publicize the limitations of science. While such candor might result in decreased funding, in the long term it will be beneficial because it will enable scientists to retain some measure of credibility and not be entirely destroyed when the truth is finally understood by all.

In the words of Ludwig et al (40), we must

> Confront uncertainty. Once we free ourselves from the illusion that science or technology (if lavishly funded) can provide a solution to resource or conservation problems, appropriate action becomes possible. Effective pol-

icies are possible under conditions of uncertainty, but they must take uncertainty into account.

References

1. Ponting C. Historical perspectives on sustainable development. Environment 1990;32(9): 4–33.
2. Orians G, Buckley J, Clark W, et al. Ecological knowledge and environmental problem solving. Washington, DC: National Academy Press, 1986.
3. Capra F. The turning point. New York: Simon and Schuster, 1982.
4. Popper K. Science, pseudo-science and falsifiability. In: Tweney RD, Doherty ME, and Mynatt CR, eds. On scientific thinking. New York: Columbia University Press, 1981:92–99.
5. Hanson NR. The origin of hypothetico-deductive explanations. In: Tweney RD, Doherty ME, Mynatt CR, eds. On scientific thinking. New York: Columbia University Press, 1981: 305–312.
6. Platt JR. Strong inference. Science 1964;146:347–353.
7. Doeser MC. Can the dichotomy of fact and value be maintained? In: Doeser MC, Kraay JN, eds. Facts and values: philosophical reflections from Western and non-Western perspectives. Dordrecht: Martinus Nijhoff, 1986:1–19.
8. Peters RH. A critique for ecology. Cambridge: Cambridge University Press, 1991.
9. May R. Biological populations obeying different equations: stable points, stable cycles and chaos. Theoret Biol 1975;49:511–524.
10. Odum EP. The strategy of ecosystem development. Science 1969;164:262–270.
11. Golley FB (ed.). Ecological succession. Stroudsburg, PA: Dowden, Hutchinson and Ross, 1977.
12. McIntosh RP. The relationship between succession and the recovery process in ecosystems. In: Cairns J, ed. The recovery process in damaged ecosystems. Ann Arbor, MI: Ann Arbor Science, 1980:11–62.
13. Drury WH, Nisbet ICT. Succession. J Arnold Arboretum 1973;54:331–368.
14. Vitousek PM, Reiners WA. Ecosystem succession and nutrient retention: a hypothesis. BioScience 1975;25:376–381.
15. Modis T. Predictions: society's telltale signature reveals the past and forecasts the future. New York: Simon and Schuster, 1992.
16. Billas IML, Châtelain A, de Heer WA. Magnetism from the atom to the bulk in iron, cobalt, and nickel clusters. Science 1994;265:1682–1684.
17. Bierregaard RO, Lovejoy TE, Kapos V, dos Santos AA, Hutchings RW. The biological dynamics of tropical rainforest fragments. BioScience 1992;42:859–866.
18. Wilson EO. The diversity of life. New York: Norton, 1992.
19. Zukav G. The dancing Wu masters; an overview of the new physics. New York: Bantam Books, 1979.
20. Jordan CF, Kline JR, Sasscer DS. A simple model of strontium and manganese dynamics in a tropical rain forest. Health Phys 1973;24:477–489.
21. Pomeroy LR, Hargrove EC, Alberts JJ. The ecosystem perspective. In: Pomeroy LR, Alberts JJ, eds. Concepts of ecosystems ecology: a comparative view. New York: Springer-Verlag, 1988:1–17.
22. Johnson G. Cosmic noise: scaling lofty towers of belief, science checks its foundations. *The New York Times* 1994; Section 4 (The Week in Review:1).
23. Ravetz JR. Usable knowledge, usable ignorance: incomplete science with policy implications. In: Clark WC, Munn RE, eds. Sustainable development of the biosphere. Cambridge, U.K.: Cambridge University Press, 1986:415–434.
24. Holloway M. Nurturing nature. Sci Am 1994;270(4):98–108.
25. Walters CJ, Holling CS. Large-scale management experiments and learning by doing. Ecology 1990;71:2060–2068.

26. Fennema RJ, Neidrauer CJ, Johnson RA, MacVicar TK, Perkins WA. A computer model to simulate natural Everglades hydrology. In: Davis S, Ogden J, eds. Everglades: the ecosystem and its restoration. Delray Beach, FL: St. Lucie Press, 1994:249–289.

27. Mazzotti FJ, Brandt LA. Ecology of the American alligator in a seasonally fluctuating environment. In: Davis S, Ogden J, eds. Everglades: the ecosystem and its restoration. Delray Beach, FL: St. Lucie Press, 1994:485–505.

28. Davis S, Ogden JC. Toward ecosystem restoration in Everglades: the ecosystem and its restoration. In: Davis S, Ogden J, eds. Everglades: the ecosystem and its restoration. Delray Beach, FL: St. Lucie Press, 1994:769–796.

29. Freedman B. Environmental ecology: the impacts of pollution and other stresses on ecosystem structure and function. San Diego: Academic Press, 1989.

30. Wheelwright J. Exxon is right, alas. *The New York Times,* July 31, 1994, Section 4:15.

31. Goldstein B. The struggle over ecosystem management at Yellowstone. BioScience 1992; 42:183–187.

32. Elfring C. Yellowstone: Fire storm over management. BioScience 1989;39:667–672.

33. Romme WH, Despain DG. Historical perspective on the Yellowstone fires of 1988. BioScience 1989;39:695–699.

34. Bonnicksen T. Fire gods and federal policy. Am Forests 1989;95:14–16,66–88.

35. Chase A. Playing God in Yellowstone: the destruction of America's first national park. New York: Atlantic Monthly Press, 1986.

36. Bonnicksen T, Stone EC. Managing vegetation within U.S. national parks: a policy analysis. Environ Management 1982;6:101–102,109–122.

37. Knight DH. The Yellowstone Fire controversy. In: Keiter RB, Boyce MS, eds. The greater Yellowstone ecosystem: redefining America's wilderness heritage. New Haven: Yale Univ. Press, 1991:87–103.

38. Lawrence NSW. Brief of AMICI CURIAE the Society for Conservation Biology and the American Institute of Biological Sciences, Case No. 94-1736 in the United States Court of Appeals for the Seventh Circuit, Sierra Club et al Plaintiffs v. Floyd J. Marita et al Defendants, 1994.

39. Ravetz JR. The merger of knowledge with power. London: Mansel, 1990.

40. Ludwig D, Hilborn R, Walters C. Uncertainty, resource exploitation, and conservation: lessons from history. Science 1993;260:17,36.

Uncertainty in Managing Ecosystems Sustainably

4

Richard A. Carpenter

Humans are in charge of the biosphere. This incontrovertible fact precludes any discussion here as to whether that is good or bad news. Being in charge does impose, in my opinion, a requirement to act rationally with the most timely and complete information obtainable, although at the same time accepting that subjective values affecting actions are important and also inescapable.

Solving the problems of using renewable natural resources to provide a sustainable flow of goods and services is one of the major tasks before environmental scientists today. Our measurements, in terms of biology, geology, physics, and chemistry, are *fundamental* to the analyses of colleagues in other disciplines and to policies and decisions made by laypersons. Scientific knowledge of cause-effect relationships and of quantification of status and trends guides the preparation of strategic and programmatic alternatives, undergirds value judgments, and informs the political choices that combine social, cultural, and ethical factors. To dramatize the point, environmental justice, for instance, may well be an important issue in land-use planning, but first we have to understand the hydrology and chemistry of the local drinking water quality.

The sustainability that is now embraced in "sustainable development" (SD) means intensive use of, and high yields from, natural systems without damage to their continuing productivity. In fact, however, around the world there are few examples of sustainable practices. Most exploitation is guided by short-term economics, and most managers do not know if they are ruining the underlying resource. Nor do they know whether an alternative practice or harvest objective would ultimately be more or less sustainable.

A pre–Earth Summit report on South America and the Caribbean, for instance, notes (1):

> The link between economic agents and natural resources does not pose a problem in the region, since it is generally recognized. The problems arise when it is a question of understanding how natural ecosystems function

and the nature of their relationship with economic systems. Unfortunately, the amount of information we have about these relationships and the extent to which they are monitored are negligible, and as a result certain circumstances go unrecognized when the time comes to take decisions.

The well-publicized closing of the Atlantic cod fishery by Canada in 1993, and the U.S. shutdown of Georges Bank more recently, illustrates the surprises that can occur even in resources that have been intensively studied for many years. The population of commercial ground fishes has rapidly dropped to near extinction levels. The causes may include overfishing, but the catch was already strictly regulated, supposedly to be sustainable, according to scientific understanding and frequent trawler monitoring data. The cost of this management error is billions of dollars per year in compensation and lost harvest, great social disruption, and perhaps, permanent reduced potential from the fishery.

Exploitation of the "natural endowment" is strongly motivated by direct political, social, and economic benefits. Table 4.1 is a selected list of typical projects in developing countries where decisions about how to manage ecosystems, including the achievement of sustainability, depend on statistically reliable biophysical information. Even monoculture retains, and depends on, a considerable biodiversity in the soil and in border areas.

A host of scientific uncertainties about the behavior of ecosystems under

Table 4.1 Typical Managed Ecosystem Projects in Developing Countries

1. Irrigation systems for intensified agriculture; e.g., a third rice crop per year.
2. Implementing soil conservation on sloping lands; choosing among methods such as mechanical terracing, vegetative strips, intercropping, agroforestry.
3. Establishing plantation forests; e.g., raw material for pulp and paper mill, tree crops such as oil palm.
4. Selective logging and enrichment planting in natural forests.
5. Introduction of exotic species in croplands, rangelands, and forestry.
6. Cultivation of marginal lands, stabilization of shifting agriculture.
7. Biological pest control programs.
8. Aquaculture and mariculture.
9. Establishing fishing fleets and gear; amount and kind; e.g., fish attraction buoys.
10. Setting allowable fish catch, spatial and temporal limits.
11. Choosing coastal preserve sites as compensation for coastal development sites.
12. Setting grazing lands stocking patterns and paddock rotation.

anthropogenic and natural perturbations combine to frustrate statistically reliable biophysical measurements and ecologic understanding. As a result, the prudent strategy for the future is one of exploitation-as-experiment, with ongoing research and monitoring sufficient for expected mid-course corrections to adapt to inevitable surprises. Explicating this complex problem to all parties involved (i.e., to everyone) is an urgent task for environmental professionals.

Many of the insights in this chapter were obtained when I was a member of the steering committee for the June 1992 United Nations University International Conference on the Definition and Measurement of Sustainability: The Biophysical Foundations, held in Washington at the World Bank (2). I have compiled over 200 recent literature references to the inadequacies of biophysical measurements and ecologic applications. I also draw on personal experiences in the course of various projects and field work in Southeast Asia and the Pacific islands from 1977 to 1993 under the auspices of the East-West Center (3, 4).

Categories of Uncertainty

This chapter first examines the context for ecosystem management as the crucial strategy in sustainable development. An overview of uncertainties is organized into three groups on the basis of opportunities for them to be reduced, a slight adaptation from Hilborn (5). The first category is uncertainties that cannot be eliminated or reduced but whose magnitude and relative importance can be estimated. These include the so-called unknowable responses, or true surprises, that arise from the self-organizing, ever-changing character of ecosystems and their response to perturbations that are unprecedented (at least to the current ecosystems). For example, surprising consequences in ecosystem behavior are likely as a result of rapid climate change, carbon dioxide fertilization, and enhanced ultraviolet (UV) radiation. Ocean warming may bring larger and more frequent typhoons, previously unexperienced by tropical coastal zones. Another source of surprises are the rare stochastic events such as large meteor impacts, major earthquakes, and large volcanic eruptions. Multiple causes and nonlinear response are also sources of unknowable outcomes. The concepts of complexity theory and chaos may have manifestations in ecosystem behavior that allow explanation of deterministic relationships but not prediction.

The second category is uncertainties arising from lack of ecologic understanding and principles on which dependable predictive models can be constructed. Reduction of these uncertainties is occurring and more progress is possible, but ecologic research is inherently difficult and long term. Control and replication may be practically impossible, and nonlinear temporal and spatial scaling makes transferring results difficult. Even the formulation of appropriate testable hypotheses is a barrier to ecologic research, especially when ethical arguments (about the subtle shifts of the burden of proof between exploiter and conservator) enter the experiment design.

The category of uncertainty that can most readily be reduced has to do with data quality. Determining which parameters are most relevant to decisions affecting sustainability should focus on field investigations. Better monitoring program design can make data collection more efficient and cost-effective. More use can be made of prior similar studies and existing information. Decision makers should be involved in data quality management in order to express their acceptable decision error rate—in other words, the probability of making an incorrect decision based on data that inaccurately estimate the true state of nature (6). This chapter presents examples of the practical problems in biophysical measurements in the field.

The demand for single indicators of sustainability should be resisted, but there are possibilities for selecting "minimum data sets." These would capture directly related measures, such as harvest yield, and the underlying condition of the ecosystem that reflects its continued *potential* for producing desired goods, services, and environmental amenities. Minimum data sets, if constructed and implemented through international cooperation, could provide continual monitoring guidance. All levels and stages of management would be aided in adapting practices when the inevitable surprising responses occur in the intensive exploitation of ecosystems.

What Is Sustainable Development?

It is clear what is *not* sustainable: the currently continuing increases in population growth, natural resource use, and environmental degradation. The most profound aspect of SD is the rejection of continued growth of throughput of materials and energy; that is, Malthus was right. Development, however, defined loosely as change for the better, may go on indefinitely within the constraints of thermodynamics and insolation.

Economists are now anxious to expand their analyses and ledger sheets beyond economic sustainability to include environmental "externalities" such as pollution damage costs and the depreciation of natural capital. Nobelist Robert Solow asserts, ". . . talk without measurement is cheap. If we—the country, the government, the research community—are serious about doing the right thing for the resource endowment and the environment, then the proper measurement of stocks and flows ought to be high on the list of steps toward intelligent and foresighted decisions" (7). For example, soil erosion not only decreases crop yields but the sediment is delivered to clog stream channels, which then must be cleared at some cost. The relative effectiveness of different soil conservation techniques must be measured quantitatively in biophysical terms. Then the costs of their implementation can be compared to the damage costs of lost crop production and dredging to provide a basis for management action. The economic valuation or monetization techniques cannot proceed except on the basis of biophysical quantifications (8).

Sustainability Defined

Sustainable development has multiple meanings with diverse roots in ecology (both "deep" and conventional), resources, carrying capacity, antitechnology, and ecodevelopment (9). Operational definitions, and indicators of implementation achievement, are required if SD is to be anything more than an attractive, but empty, phrase. There are many substantial and varied ongoing efforts around the world to supply policy and decision makers with quantitative measures related to SD. The World Bank, the United Nations University, the United Nations Environment Programme, the Organization for Economic Cooperation and Development, the World Resources Institute, the U.S. Environmental Protection Agency's Environmental Monitoring and Assessment Program, and the governments of Canada and the Netherlands are among the leaders in this work. Indicators of sustainability being suggested are numerous and varied.

The International Institute for Environment and Development offers a long list covering energy use, biologic wealth, policy, economics, institutions, society, and culture. But only one criterion of sustainability seems to deal with the root definition: "Renewable resources are increasingly used and harvested at rates within their capacity for renewal" (10). Other attributes of sustainability are important but they must be based on an accurate picture of the biophysical production potential. Most natural scientists who are managing ecosystems, such as agriculture, are skeptical about their capability to measure sustainability (11, 12, 13).

My working definition: sustainability is when the productive *potential* of a managed ecosystem *site* will *continue* for a long time under a particular management *practice*.

A managed ecosystem is a fairly homogeneous region demarcated at its boundaries by abrupt changes in some biophysical characteristics, such as an upland agricultural area, production forest, lake, river basin, coastal zone, or island. A long time means several decades or a generation, not indefinitely because of the likelihood of new technologies and changed values.

Nature is constantly varying and changing as to its output of harvestable surplus for consumption. Human societies are far beyond the hunter-gatherer stage and seek to impose a constancy of production in order to have some security of food and other materials. Management practice is the intensity and type of technical and social input activities, such as energy, nutrients, genetic variety, harvesting procedures, and their planned variations over time.

The productive potential relates to the quantity and mix of goods and services from the environment. For example, ". . . forest lands—including soils, plants, animals, minerals, climate, water, topography, and all the ecological processes that link them together—[are viewed] as living systems that have *importance beyond traditional commodity and amenity uses*" (14) (emphasis in the original). This set of goods and services is, ideally, participatively chosen by the society that owns, or should control, the natural system at the site under consideration.

Sustainable development depends largely on renewable natural resources of air, water, soil, sunlight, and communities of plants and animals. These must provide the replacements for consumed nonrenewable resources when the latter cannot be recycled. The utility, capacity, or potential of these natural systems for producing goods and services is what is to be continued, and even enhanced.

Sustainability of High Yields is the Objective

Donella Meadows and her colleagues recently revisited the "limits to growth" models and confirmed their earlier warnings, now widely apparent in reality, that the world has too many people too aggressively exploiting its resources (15). They recommend choosing to reduce ". . . the burden of human activities upon the earth . . . [to] aim for an average industrial output per capita of $350 per person per year . . ." by 2050 with the population leveling off at just under 8 billion through ". . . perfect birth control effectiveness . . ." starting in 1995. This universal equitable level of affluence would be about one-fifth of the present world average and about $1/25$ of the present U.S. average.

Significantly (for the thesis of this chapter) in achieving this remarkable conversion of civilization to their version of a sustainable future, the authors call for *extensive and reliable natural science information* about local and planetary resources and sinks of pollutants, prompt awareness of changes in the environment, forecasts of the consequences of stress on ecosystems, harvests only at regeneration rates, and use of all resources with maximum efficiency.

An interesting similar calculation is the "ecologic footprint," or that aggregate area of ecologically productive land required to provide the entire world's population with the present North American lifestyle. It turns out that two additional Earths would be necessary (16).

Robert Goodland of the World Bank exemplifies the well-meaning, simplistic conclusions of many megaplanners that the industrialized nations must unilaterally reduce their "overconsumption" (17). He says, "The North is responsible for the overwhelming share of global environmental damage today, and it is unlikely that poor countries will want to move toward sustainability if the North doesn't do so first."

To judge the credibility of these proposals remember that the famous 1987 Brundtland Commission report opened with the phrase "The Earth is one but the world is not." Lynton Kieth Caldwell (18) reminds us that,

> Relatively few people yet see the earth as an object of concern, respect and responsibility. . . . Many contemporary values, attitude, and institutions militate against international altruism. As widely interpreted today, human

rights, economic interests, and national sovereignty would be factors in opposition. The cooperative task would require behavior that humans find most difficult: collective self discipline in a common effort.

Operationalizing Precaution

Another proposed solution to the dilemma of getting the most out of nature without ruining the underlying resource base is the "precautionary principle." Certainly, when there are high risks to human health (exposure to ionizing radiation) or of irreversible environmental change (e.g., species extinction), great precautions and safety factors are warranted. When the consequences are less profound, although adverse, the degree of precaution is not so obvious. We have the technological means to exhaust most resources. Economically, even modest discount (interest) rates encourage overexploitation in ecosystems, such as forests and top predator fisheries, where natural growth of the harvestable product is slow (19):

> Even within a solely economic context, we can see the importance of uncertainty. Fluctuations in economic factors such as the interest rate will cause perspectives on conservation to change: About a dozen years ago, when U.S. interest rates were ~14%, a sole owner of a resource growing at 8% would have great incentive to eliminate it and reinvest the capital. The situation is very different today.

The precautionary principle now in vogue, or its 40-year-old predecessor, Ciriacy-Wantrup's "safe minimum standard" of conservation (SMS), asks that whenever there is *uncertainty* about thresholds of degradation, the burden of proof should be shifted and exploitation be voluntarily reduced, "to the extent that it is cost-effective or economically feasible" (20). SMS, in practice then, tries to distinguish somehow between conservation costs that are intolerable and those that are merely substantial, and it implies a confident prediction of worst-case consequences which has seldom been documented (21). Minimax and maximin are similar pseudorigorous exercises in logic and ethics.

These formulations do not produce objective operating guidelines and simply require another value judgment, by present generation technocrats, of the proper balance of risk between taking immediate gains and losing long-term potential. Regardless, if applied in the real world, they depend on some measure of the proximity to thresholds because the cost usually varies strongly with the degree to which exploitation is foregone. ". . . The [precautionary] principle offers no guidance as to what precautionary measures should be taken. It . . . does not tell us how much resources or which adverse outcomes are most important" (22).

Uncertainty can cut two ways and the real costs due to inadequate biophysical measurement of sustainable practices are in two forms of management error:

a. Urgent demand may increase risk taking and drive overexploitation that subsequently results in irreversible degradation.
b. Conservative precautions may turn out to needlessly lower the intensity of inputs so that harvests are less than they could be, and some people are denied a portion of basic human needs. Precaution may mean unnecessary hunger or prolonged poverty. And so again, *the need for reliable biophysical data is central to decision making.*

The expectations of the public and layperson policy and decision makers are that environmental sciences can provide straightforward advice. They mistake myths popularized by pseudoecologists in the 1970s as established theories; that is, everything is connected, you cannot change just one thing, nothing goes away, and there is no free lunch (23). There is, unfortunately, ample evidence that the accuracy and statistical reliability of measurements of managed ecosystems under field conditions are inadequate for establishing the sustainability of ongoing management practices or predicting the sustainability of proposed alternative practices. It is necessary to sort these uncertainties out, communicate them forthrightly to consumers of biophysical data, and to prioritize our efforts in reducing them to get the most cost-effective improvement in the overall management information package.

Holling characterizes five belief systems that are creating future scenarios and promoting actions to deal with each: (1) the cornucopian exponential growth with technology; (2) a minimal demand on nature by humankind (deep ecology); (3) a clever but turbulent transition to a sustainable plateau; (4) discontinuous growth and collapse; and (5) radically new opportunities through integrating disciplines by theories of complexity and nonlinear dynamics (24).

My own belief is that the absolute amount of production from ecosystems will be increased in order to (a) maintain per capita flows, since the human population is going to grow, and (b) improve the welfare of poor people, since they will demand it. The alternative, that the affluent will drastically, and voluntarily, reduce their consumption, is not credible. At least several decades of increasing pressures and stresses on presently exploited natural systems, intensively managed for a stable maximum annual harvest, is the most plausible future and these are the conditions with which we must try and cope. This fine tuning of sustainability at high yields is where the concept will be tested and where uncertainty manifests itself in frustrating the development of alternatives and the choice of best practices. These are the implications of sustainable development for the environmental science community, regardless of how arguments are settled about equities between or within generations, conversions among financial and natural capital, value judgments as to which goods and services are desired, or risk aversion as a function of economic status.

The Nature of Uncertainty

Science is the activity of understanding the regularities of the universe and revealing the simple laws that produce them. Prediction to guide human actions is also a primary goal of science and uncertainties interfere causing what actually occurs to differ from what was expected. Scientific truth is always somewhat uncertain.

The scientific method constructs hypotheses that are then rejected (falsified) or conditionally accepted on the basis of existing evidence from experiments or empirical observations. A hypothesis may always be rejected in the future if additional information is seen to falsify it. Within this truth-seeking methodology, information is characterized by kinds and degrees of doubt, change, and availability.

Ecologic uncertainty is of two basic kinds: what we do not know at all, and errors in what we do know. The latter is quantitative departure from the truth, and can sometimes be expressed statistically as a distribution, around a mean value, of a number of repeated measurements. Other forms of quantitative uncertainty are incomplete data, "anecdotal" data which are not gathered with a statistical design, inappropriate extrapolation, and temporal and spatial variability of the measured parameter (25).

The U.S. National Research Council (26) separates uncertainty as measurement problems of insufficient observations and natural variability; problems of extrapolating from the scale in time and space of observation to the scale of management; and inadequacy of models and fundamental knowledge of underlying mechanisms. More appropriate and better measurements can sometimes be obtained, and extrapolation errors can at least be quantified. Uncertainties due to general scientific ignorance are hard to even quantify, much less reduce. Most important environmental problems, however, suffer from *true* uncertainty (i.e., indeterminacy) or events with an unknown probability, according to Costanza (22).

Surprise is a manifestation of uncertainty that, by definition, cannot be predicted. Totally unanticipated rapid and adverse changes in ecosystems may arise from apparently unrelated policy or social events. Very-long-term effects sometimes become evident only long after the original cause, and explanation may be confounded (27).

One example of surprise is the potential for what are called "chemical time bombs." Long lasting chemicals such as heavy metals and refractory organic compounds may not be toxic or bioavailable in the chemical combination or form in which they first are introduced into the biosphere. They are ignored in conventional ecotoxicology. But, accumulating in soils, sediments and groundwater, they can undergo transformations due to changes in land use, climate, acidification, redox potential, and cation-exchange capacity. The responses of biota to these pollutants is time delayed and nonlinear. The June 1992 meeting of the Society for Environmental Toxicology and Chemistry in Potsdam was devoted to chemical time bombs. One conclusion was ". . . that it will never be possible to match the data and the results from chemical risk assessment and ecosystem responses in a truly scientific manner" (28).

Uncertainties have relative importance depending on their size. If stochasticity is large, then less error in the knowledge base cannot help much and predictive power is limited. Stochasticity is the variation in response of an ecosystem due to random, uncontrolled factors such as weather, and not due to the stressor being studied.

Bias is a source of uncertainty due to measurement and sampling errors of the parameters in a model. These may be reduced at some cost. But field operations, especially in developing countries, make improvements difficult (see below).

Natural Variation

A more familiar source of uncertainty is natural variation. Any "signal" that a change in condition is due to human action is often hidden in the "noise" of natural changes in the value measured. For example, no statistically significant change in the volume of water discharged by the Amazon river, or the amount of sediment delivered from the deforested Rondonia region, has yet been detected. A signal that deforestation has altered the hydrologic cycle or soil erosion rate in that basin is obscured by the high natural variability of rainfall. The El Nino events explain most of the occasional trends in the "noise" that are evident in the hundreds of river gauging stations (29).

> Our ability to distinguish the signal from the noise depends directly on our understanding of the system under question. As an example, we recall that a fraction of the physical and biological variability in the tropical Pacific and adjacent regions that was previously assigned to noise is now interpreted as an El Nino signal. Scientific research cannot decrease natural variability, but can certainly improve our ability to detect the over exploitation signal.

Early detection of management failures, however, remains difficult due to natural variability.

In my own work associated with colleagues in Thailand, a well-designed and well-conducted experiment was performed to test the hypothesis that replacing present monoculture of casava with agroforestry would create sustainable land management systems and provide farmers with more income. After 5 years, no significant difference in soil erosion could be detected between any of the different tree-crop combinations, casava, or volunteer weed cover. The only statistically significance was between any and all of the vegetation plots and the bare plots (3).

Agricultural production strategies may try to incorporate, rather than overcome, natural variation of biomass productivity and crop yields. A satisfactory overall average harvest is the goal. But such strategies must be at large scales of time and space to allow the fluctuations to take place. Thus, they are more costly, politically more complex, and the reduction in uncertainty that they might provide does not become available.

Statistical Analysis of Ecological Uncertainty

Shrader-Frechette and McCoy argue (somewhat philosophically, in my opinion) that ecology does not and will not yield general and deterministic laws useful for environmental management. They recommend, therefore, an emphasis on natural history and accurate description of particular situations as case studies (31).

Further, they argue that ecologists should pay more attention to researching null hypotheses in such a way that the risk of type II error is preferentially avoided (over type I). Thus the burden of proof is shifted to the exploiter. The null hypothesis is, usually, that human intervention causes no effect on the ecosystem. A type I error is to reject a true null hypothesis. A type II error is to accept a false null hypothesis. When uncertainty prevents *both* types of error from being avoided, science traditionally prefers to minimize the risk of making a type I error. For example, if a regulatory agency makes a type I error (that adverse change has occurred when actually none has), unnecessary protective or corrective expenses may be incurred (32).

Shrader-Frechette and McCoy, however, say that, ethically (31)

> There are a number of reasons for holding that assessors' and ecologists' prima-facie duty is to minimize type II errors, viz., to minimize the risk of not rejecting a dangerous development/environmental action, or to minimize public risk. . . . the philosophy of scientific method underlying the application of ecology to environmental problems, a philosophy tied to both scientific and ethical rationality, provides a basis for using ecology in many practical situations because it is a conservative rationality. It is conservative in that it puts the burden of proof on the ecologist who wishes to posit no detrimental effects from human manipulation of the environment. It puts the burden of proof on the ecologist who does not want to reject the null hypothesis.

In Chapter 1 Shrader-Frechette calls for use of a "three-valued epistemic logic" involving a third hypothesis (in addition to the usual null and alternative); that is, that there are not adequate data to either reject or not reject the null hypothesis. This may be the most honest finding but there is a danger that decision makers will be turned off and no longer seek the assistance of ecologic science. More dangerous, however, is the usual total disregard of variance and the presentation of a single (expected, or mean) value as a go–no go advisory from the impact assessor. This is misleading at best, and quite often will be disputed by subsequent events, again earning the distrust of the decision maker. Uncertainty must be embraced and preserved by the analyst and communicated clearly to the manager (33).

And, the shift of burden of proof is not without its costs (see below). Statistical "power" of a hypothesis test is the probability of avoiding the type II error—that is, accepting the null-no effect when change has actually occurred. In contrast with the more familiar "confidence," or the probability of avoiding the type I error (rejecting the null when in fact there has been no change), power is seldom considered in designing monitoring programs. "If the size of an impact is small relative to natural variability, it will be difficult to detect with any degree of confidence. Therefore, it is critical to consider statistical power in planning and interpreting environmental impact assessment studies" (34). So, the probability of correctly concluding that change has occurred is equally as important as knowing the confidence level, and may be preferred from an ethical standpoint.

Incidentally, the so-called 95% confidence limit (that often seems to be the minimum statistical significance required for scientific data to be taken seriously) needs to be understood by scientists and their clients for what it is: an artifact of arithmetic in normal distributions of sets of repeated measurements. The probability is 0.95 that the next measured value will lie in an interval on the distribution between 1.96 standard deviations on either side of the mean (25). Then 95% of samples taken in the same way will yield confidence intervals that contain the true mean. The coincidence of this interval being about $\pm$ 2 standard deviations and the "comfort" that decision makers seem to feel in being right 19 times out of 20 has given a false authority to this arbitrary formulation that is now widely accepted as a standard of certainty. Some benchmark is required for expediency, however, and this one is useful.

Hypothesis Testing of Sustainability

Let us try to apply these arguments to sustainable management of production ecosystems. The following is adapted from a recent paper by Smith and Shugart (35). Say that a parameter of degradation, or deviation from, sustainability (D) exists such that a sustainable practice produces $D \le 0$. Then the null hypothesis is $D = 0$. A truth table may be constructed as shown in Table 4.2. Whether a management practice at a site is sustainable or not is akin to the question of safety, and safety (a negative concept—the absence of harm) is not provable in a statistical study. "In such studies, it is desirable that the null hypothesis not be rejected [if it is true]. . . . The difficulty arises because the rejection of a hypothesis is a strong statement, while evidence that favors the null hypothesis is regarded as confirmatory evidence and not proof.'' Thus the decision that the practice is sustainable really means that there is no evidence to indicate otherwise, and the decision should be viewed with some suspicion. It is much more difficult to calculate the power, or probability of a type II error, than the confidence, or probability of a type I error.

An actual example of the confusion of error types occurred in 1983. Despite substantial, but inconclusive, evidence of stock declines, the allowable whale harvest was set by the International Whaling Commission at a high level because of their reluctance to make a type I error (claiming an effect when none had occurred) (36). ''It is important, therefore, that gross uncertainties about hypothesis tests be controlled by proper statistical control of studies. Small sample sizes, poorly designed experiments, ignorance of the proper variables to measure or proper times to sample may lead to acceptance of a no-effect assessment when in fact there is an effect and high uncertainty'' (35).

If one were to follow the above suggested ethical guidance, it would be necessary to calculate the power of the monitoring program designed for the practice and site in question. In contrast to confidence, power *does* depend on the magnitude of the hypothesized change to be detected. The larger the change, the larger is the power. Of course, increasing sampling stations and occasions can also increase power but at some cost in time and effort. Analysis of variance in typical environmental problems shows that the number of samples required to give a power of 0.95 increases rapidly if changes smaller than 50% of the standard deviation are to be detected. Thus, the desired emphasis on avoiding type II error must be balanced against other opportunities to use limited scientific resources to reduce uncertainty in achieving sustainability.

There are fundamental objections by some ecologists to using the hypothetical deductive approach and these standard frequentist statistical methods for hypothesis testing in ecology. There is an assumption that data are sampled

Table 4.2

	STATE OF NATURE, TRUTH	
SCIENTIST'S DECISION	NULL IS TRUE—NO CHANGE (PRACTICE IS SUSTAINABLE)	NULL IS FALSE—CHANGE (PRACTICE IS UNSUSTAINABLE)
Do not reject null (practice is sustainable)	No error	Type II error
Reject null (practice is unsustainable)	Type I error	No error

repeatedly to give means and standard errors, and that the distribution is normal. The conclusions depend on data that could exist in theory but usually do not in practice. Ecologic data may be unique, not replicable. There are often low probability–high consequence events that skew the distribution and should not be ignored. The use of Bayes' theorem, a method of statistical analysis that includes prior information as well as the current data set being analyzed, is an alternative (37). Bayesian probability is a "belief" of what is true, based on combining prior information with the observations or measurements of the current study. Power, the probability of making a type II error, is not necessary or relevant because there is no accept-reject framework. The distribution of *D* (in the above example) is what is important. The sensitivity of the resulting distribution to prior information may be tested by selecting exploitive, neutral, or conservative values for the calculation. The decision maker still must state what probability of unsustainable *D* is acceptable. Bayesian analysis is growing in popularity.

Limits to Reducing Ecologic Uncertainties Through Ecological Research

Ecology is obviously the science needed to undergird management of ecosystems. There is considerable controversy within the profession of ecology about the capabilities of this discipline to be of practical assistance. The self-examination is commendable, but some criticisms seem snide and unconstructive. For example, Peters attempts to prove that ecology is a weak science by blaming it for the extent and continuation of environmental degradation. "It is that the problems that ecology should solve are not being solved. They are worsening, growing more imminent, more monstrous" (23). He seems to confuse science with religion, as if any scientific discipline would be able to radically change selfish, shortsighted human nature.

The Ecological Applications Forum

A 45-page forum in the November–December 1993 issue of *Ecological Applications* debates whether basic ecologic research, with its long-term experiments, difficulties of replication, control, and randomization, and evolutionary change in the systems under study, can test hypotheses and produce useful results quickly enough to affect the course of human perturbations of the biosphere (38).

The discussion was initiated by an article in *Science* by Ludwig, Hilborn, and Walters (39). These authors, experienced in fisheries management, conclude that "... assigning causes to past events is problematical, future events cannot be predicted, and even well-meaning attempts to exploit responsibly may lead to disastrous consequences." They advise, "Distrust claims of sustainability. Because past resource exploitation has seldom been sustainable, any new plan that involves claims of sustainability should be suspect." In a mild rebuttal, Rosenberg et al (40) note progress in the use of probabilistic advice to management through risk assessment in stock evaluations and alternative management actions. Fisheries may be uniquely vulnerable ecosystems because of the tragedy-of-the-commons syndrome but a lack of scientific consensus as to their condition and trends is an added barrier to sustainable management.

Commenting on the research plan in the Sustainable Biosphere Initiative of the Ecological Society of America (41), Ludwig et al (39) caution:

> Such a claim that basic research will (in an unspecified way) lead to sustainable use of resources in the face of a growing human population may lead to a false complacency: instead of addressing the problems of population growth and excessive use of resources, we may avoid such difficult issues by spending money on basic ecological research.

This line of thought appears to be similar to the religious-like "deep ecology" where humankind is only another species. A related objection to ecology-based management is "environmental therapeutic nihilism" (42) with its goal of completely avoiding human perturbations on the grounds that we have no reliable basis for action and nature should be allowed to proceed alone. For biologic conservation, this is the extreme "hands-off" policy.

But, there is no way to learn how an ecosystem will respond to stress except to stress it. Properties of an ecosystem cannot be predicted from its components. It may not be possible to achieve the 95% confidence level that is generally regarded as essential for science. The usual means of dealing with uncertainty (through decision theory and models) may not work in such situations. Most thoughtful ecologists accept an anthropocentric view and believe that they had best get on with management, exhibiting due humility and purpose.

The more optimistic of the ESA forum discussants argue that ecology can (only, or at least?) guide management in conducting economic development as a well-planned, but risky, *experiment.* Holling effectively restates his adaptive environmental management message of the 1970s that ". . . there is an inherent unknowability, as well as unpredictability, concerning these evolving managed ecosystems and the societies with which they are linked . . . [so that] . . . uncertainty and surprises become an integral part of an anticipated set of adaptive responses" (43).

Responses in nature to perturbations involve lags, thresholds, and rapid transformations from one state to another. For example, despite considerable knowledge of atmospheric chemistry and years of monitoring the gradual buildup of chlorofluorocarbons (CFCs), the appearance of the Antarctic ozone hole was a surprise. Mixtures of positive and negative feedbacks, each poorly understood, or perhaps not even known, make predictions difficult.

John Cairns emphasizes (45)

The capacity of ecosystems to adapt to anthropogenic stress is presently poorly understood. Unfortunately, there are few places in the world where human activities do not have a major influence on natural systems. The capacity of natural systems to generalize an adaptation from one stress to new stresses is virtually unknown.

Mechanisms of adaptation at the community level and the influence of the type of stress, either common in evolutionary history, such as organic enrichment, or unprecedented, such as synthetic pesticides, are only partially understood, and their reversibility is not assured once the stress is removed. We should improve our ability to predict the effects of anthropogenic stress on natural systems.

Some environmental data are both uncertain (difficult to measure precisely) and variable (changeable). For instance, emissions vary temporally. Transport of pollutants varies spatially with wind speed and direction. Human exposures vary with the time spent in a particular place. Some effects occur long after the cause. For example, only recently has forest management recognized the danger of building up accumulations of fuel on the ground through suppression of fires at the local level. These variabilities should be noted and explained. Variability does not always lead to the familiar symmetrical bell-shaped normal distribution, and so the mean value may not be a good communicator of the true situation. There may be infrequent but catastrophic events that should be considered now. The parameter of interest may actually be a random variable, so that only a distribution, and not a point value, is pertinent.

These fundamental uncertainties are the root of the difficulties of monitoring and predicting sustainability but there are other, perhaps more tractable, problems as well. Ecologic research has not emphasized large area investigations, and that scale (regions such as the American midwest agricultural heartland, large marine ecosystems, or the boreal forests) is most important in monitoring and predicting sustainability of management practices in economic development. What few studies that have been made of ecoregions or biomes reveal some of the sources of natural variation.

Concurrently, new research shows that combining large-scale experiments with long-term and comparative studies may obviate the necessity to tackle each ecosystem stress problem as if it were unique. "The challenge is not to eliminate uncertainty, for that is impossible, but to assign priorities to the tractable uncertainties and reduce them as rapidly as possible through targeted research" (45). Another view (46) is that risk expressions are useful; for example, specifying a 95% probability of persistence of a development strategy is a reasonable criterion for sustainability. Then one can identify and evaluate the comparative risk of various threats (biotic and abiotic, stochastic and deterministic) to the system. In some applications to population biology, such an exercise reveals the relative risk of different management strategies. Improved understanding may be on the way, but ecologists are cautioning against raising false expectations.

Chaos in Ecosystems?

The headline in a 1989 news article in *Science* read "Ecologists Flirt with Chaos" (47). The dalliance continues and it may, in part, be due to a facetious thought that *with* chaos there would be more hope for understanding ecosystems than there is today. At least there would be some "attractors" along the way instead of unremitting randomness. It is beyond my competence to summarize the state of application of chaos theory to biology, but a few observations are pertinent to this chapter's purpose.

Self-organizing systems develop in a nondeterministic manner that changes their nature, not merely their behavior. Life is quite apparently a self-organizing complex system, far from thermodynamic equilibrium, rich in negentropy at the expense of more disorder in its surroundings. The inference is that ecosystems, involving feedbacks, and with their future states highly dependent on random perturbations of initial conditions, are chaotic (48). Self-organized criticality is the term for a state in which a minor event can lead to a major, possibly catastrophic consequence. For example, ". . . natural systems may evolve toward increasing adaptability to environmental change and that adaptability is maximized when the system has evolved to the verge of chaos. This condition is analogous to that of a mature prairie or forest which is experiencing fires. The conditions for the support of occasional fire become optimal as the system reaches maturity" (49). Compare this idea with the creative destruction proposed by Holling in adaptive environmental management (below).

Chaos is jargon for one type of behavior of the dynamic behavior of nonlinear systems (50). This scientific meaning is often confused with the everyday use of the same word to connote pure probabilistic randomness. Chaos is ultimately mathematics, and the real physical systems that have been studied are very simple, observable with negligible errors, and possess only a few degrees of freedom (e.g., a pendulum, an electrical circuit, fluid flow, chemical reactions). The *in*determinism of nature at the quantum level is not relevant to these macrosystems, and they can be deterministic yet unpredictable. Almost all real systems, and certainly all ecosystems are nonlinear (a small change in a parameter can lead to a sudden large change in behavior), but not all nonlinear systems are chaotic. Chaos is an alternative explanation for noise or random-like *behavior* that might be caused by uncontrolled outside influences.

There is no doubt that it can be actually observed in simple physical-chemical systems but its main introduction into biology has been in *calculations* of simple population dynamics models containing a feedback term and, therefore, nonlinearity (e.g., the logistic equation) (51). Due solely to a peculiarity of arithmetic, the trajectory of sequential calculations (where each solution is the basis for the next) changes unpredictably with slight variation of the initial state, and the solution of the equation cannot be calculated when the feedback exceeds certain numerical values. "This is the signature of chaos. . . . there are systems, even within mathematics, that are both deterministic and unpredictable. We cannot blame this failure on the presence of unknown factors because there are none. It is rather the result of our own terminal inability to measure or represent the present with infinite precision (52). The implication for ecology is that the outcome of a stress-response experiment can never be predicted; the only way to learn what the consequence is is to let the experiment run. "Thus, even God [Laplace's calculating Intelligence] must allow these chaotic systems to evolve to see what will happen in the future. There is no short cut to prediction for chaotic systems" (50). In a sense, nature is its own simplest and fastest simulation model.

Furthermore, studies to distinguish between chaos and random noise require very large (long term) data series and absence of multiple causes, unknown degrees of freedom, and true noise. Ecosystems do not usually yield data of the sort required, and the inexactness of ecological measurements, the focus of this chapter, is an additional barrier to establishing chaos. For example, there are no single-species ecosystems. It may not be possible to transfer the emerging principles of chaos theory to aid understanding of ecosystem behavior. And, if ecosystems are chaotic, they may also be so noisy (true randomness) that for all practical purposes it does not make any difference.

The Value of Additional Certainty

The value of additional information must be weighed against the cost of obtaining it because all management enterprises are limited in resources. For example, a resource may be harvested cautiously because of uncertainty and the desire not to overexploit and ruin the resource base. The question is whether, with perfect knowledge so that the harvest could be exactly at the maximum sustainable yield, would the additional economic return offset the cost of the research to remove uncertainty. How much more resource would be harvested? In some investigations, the answer has been "not much more" (53).

It is my opinion that in most ecosystem developments we are a long way from the point where the cost of the next increment of knowledge exceeds its worth. Holling and Walters (24, 54) explain:

Two kinds of science influence renewable resource policy and management. One is a science of parts, e.g., analysis of specific biophysical processes that affect survival, growth, and dispersal of target variables. It emerges from traditions of experimental science where a narrow enough focus is chosen in order to develop data and critical tests that will reject invalid hypotheses. The goal is to narrow uncertainty to the point where acceptance of an argument among scientific peers is essentially unanimous. It is appropriately conservative, unambiguous, and incomplete. The other is a science of the integration of parts. It uses the results of the first, but identifies gaps, invents alternatives, and evaluates the integrated consequence against planned and unplanned interventions in the whole system that occurs in nature.

Typically, alternative hypotheses are developed concerning the integrated properties of the whole to reveal the simple causation that often underlies the time and space dynamics of complex systems. . . . Often there is more concern that a useful hypothesis will be rejected than a false one accepted: 'don't throw out the baby with the bath water.' . . . Since uncertainty is high, the analysis of uncertainty becomes a topic in itself.

Ecologic Risk Assessment

Risk assessment is a way to explicitly deal with uncertainty (55). The U.S. National Research Council, in its 1993 report, notes (26)

> Ecological risk assessments have no equivalent of the life time cancer-risk estimate used in health risk assessment. The ecological risks of interest differ qualitatively between different stresses, ecosystem types, and locations. The value of avoiding these risks is not nearly so obvious to the general public as is the value of avoiding exposure to carcinogens . . . the function of risk assessment is to link science to decision-making, and that basic function is essentially the same whether risks to humans or risks to the environment are being considered.

Comparative environmental risk assessment is more doable and more useful than estimation of risk in an absolute probabilistic sense. As applied to ecosystems, the idea is (a) to rank (relatively, into broad groupings of risk) a comparable set of threatened ecosystem sites, and (b) to target response actions toward those ecosystem sites that are at greatest risk. The Risk Forum of the U.S. Environmental Protection Agency (EPA) (35) and the Oak Ridge National Laboratory (ORNL) (56), among others, have sponsored conferences, reports, and research to foster a consistent approach for the developing field of ecologic risk assessment (ERA). There is as yet no widely applicable, established procedure. Several approaches have been used.

Reductionist ecologic methods attempt to compartmentalize ecologic processes and effects into a myriad of understandable units and linkages. Generally these lack, and do not allow, evaluation of synergism. An example is tracking the flow of nutrients from their sources through the ecosystem. Quantification of the flow is possible, but drawing implications about effects at the regional level is complex. Reductionist approaches are useful in defining how individual stressors affect individual species and (sometimes) biologic communities, and how they can be detected and monitored. With two or more stressors operating on the same system, the analysis and interpretation become increasingly more difficult.

Bottom-up methods rely on the use of models and laboratory data to quantify biologic and ecologic processes and impacts, primarily at the species and community levels. This can be useful at site specific locations, but extrapolating the results to ecosystem and regional levels is more difficult, especially if two or more ecosystems and stressors are involved. A standard water column model comprising many biogeophysical parameters is used at ORNL "... to extrapolate the results of laboratory toxicity data into meaningful predictions of ecological effects in natural aquatic ecosystems" (56).

Top-down methods evaluate structural and functional changes at the ecosystem and regional levels and are most easily applied where there is large-scale homogeneity in both the ecosystem and the stressor that affects it. Conversely, these methods break down when a region is a mosaic of many stressors and ecosystems. Normally there is a lack of sufficient data from a broad region to allow quantification.

"Practical" methods recognize the need to design and accomplish practical comparative ecologic risk assessments useful to decision makers, politicians, and nonscientists. To meet this goal, comparative ERA need not be quantitative, and it may actually be preferable to keep it qualitative. A combination of the best judgment of ecologists and professional land/water managers with on-site experience and the systematic evaluation of risks from available information is pursued. Effective communication to decision makers is accomplished through use of maps, simplified scoring systems, clearly defined evaluative criteria, and a manageable set of ecologic stressors. Defining the specific problem areas and classifying the ecosystems of the study region are important early steps in this approach to comparative ecologic ERA.

Health risk assessments (with heavy emphasis on public health) differ from ecologic risk assessments in several significant ways. For ecosystems, the ERA must consider effects beyond just individual organisms or a single species. No single set of ecologic values and tolerances applies to all of the various types of ecosystems. Stressors are not only chemicals or hazardous substances. They also include physical changes and biologic perturbations. For public health purposes all humans are treated equally, but with ecosystems some sites and types are more valuable and vulnerable than others.

Risks to ecosystems are based on the values (intrinsic and anthropocentric) of actual individual sites and the probability that stressors from human activities will significantly degrade these values in the near future. Just as the individual human being is the focus of health risk assessment, the individual ecosystem site is evaluated in ecologic risk assessment. Uncertainties about value, frequency of adverse impacts, and severity of response to stress are identified and evaluated as a part of the ERA. The ability of the ecosystem occurrence (site) to recover is also considered. Accommodating these factors complicates comparative ecologic risk assessments and renders them more subjective. Nevertheless, a number of state and local governments are using this technique as a basis for allocating funding in conservation and environmental protection (55).

Biophysical Measurements in the Field

Meanwhile out in the forest, on the farm, at the fishery, decisions affecting the sustainability of renewable resources are being made daily on a short-term, localized basis with little regard for collecting any measurements at all beyond elementary economic calculations of input costs and rough estimates of harvest. The contrast of this reality with the sophisticated statistical analysis and ethics discussed above could hardly be more stark. But, it is here that improved biophysical measurements could make a difference in the future condition of production ecosystems.

Even where there is a willingness to take precautions, data inadequacies are frustrating. A seven-country study on SD for the Asian Development Bank (57) concludes:

> In the preparation of this report no task proved more daunting than the assembly of reliable statistical indicators of recent trends and the current state of the environment. Aside from the most glaring cases where officials and researchers are aware of what is happening and can describe conditions in general terms most eloquently, the lack of quantitative environmental information comparable with the statistics available regarding economic parameters is a major obstacle to integrated economic-cum-environmental planning. Statistical information on the environment is scarce, often inaccurate, seldom comparable from country to country, and rarely available in a time series covering a sufficient number of years to indicate trends in a reliable way. Thus, descriptions remain anecdotal and lack the hard edge of quantification which is necessary for analysis and policy formulation.

Nevertheless, whatever the management practices, they are subject to influence by governments and development assistance agencies through incentives and demonstrations of improved benefits to practitioners. Alternative practices are sought in applied ecologic research that goes on all over the world in field stations and other somewhat controlled experimentation. The Consultative Group on International Agricultural Research (CGIAR) Centers, for example, has added sustainability to its productivity goal (58). The empirical observations and experimental results in these developing country studies are the basis for planning large-scale, high-budget projects (recall Table 4.1) involving the welfare of millions of people and the viability of huge natural systems. The quality of much of these data is inadequate (59).

The watershed is a preferred unit for study. That water runs down hill is one of the principles of ecology (the other one is that animals move around— ''biota disperse''). Knowledge of the origin, transport, storage, and delivery of nutrients, suspended solids, and pollutants (hydrology and water chemistry) is fundamental to monitoring and predicting sustainability in the watershed ecosystems. Paired watershed studies such as the classic Hubbard Brook site in New Hampshire have never been accomplished in tropical developing countries. A setting devoid of residents (and thus free from sociocultural and economic factors) or that can be protected from unwanted human perturbations is almost impossible to find.

Collecting statistically reliable data in the field in developing countries over a long period of time can be problematic (4). Access to remote sites may be disrupted during rainy seasons. Political unrest, corrupt officials, and guerilla activities intimidate researchers. Vandalism and theft deplete project equipment and supplies. Heat and humidity, unreliable electric power, poor or nonexistent repair and maintenance facilities, and untrained manpower reduce the effectiveness of sophisticated instruments. A typical account of field research problems concerns a tsetse fly control trial in Ethiopia. Cloths soaked with pesticide and odor-baited with small containers of cow urine and acetone are hung on wood frames. The experiment failed because most of these targets were stolen. ''People used the cloths for bedding and to make clothes and bags. They used the urine and acetone containers as household containers and even fashioned them into coffee cups'' (60).

Natural variation of rainfall, soil characteristics, crop yields, biomass productivity, and pest population dynamics are a reality in managing ecosystems. The range of natural variability is usually from 100 to 1000%. Spatial variation can be reduced by sampling a larger area, using larger representative plots, more sample points, or more transects. Temporal variation can be reduced by more frequent sampling or sampling over longer time periods. Both types of variation can be reduced by using composite sample or by concentrating sampling in the area or period of greatest variability. The detection of a change in one of these parameters due to some management practice or action depends on the magnitude of the change compared with the magnitude of the natural variation; the signal-to-noise ratio. If this ratio is not greater than 1 the large number of replicate samples required to give a 95% confidence level is often too expensive or time consuming to be obtained.

Sophisticated measurements or indicators are not needed to find that a management practice is *un*sustainable when gross and obvious damage occurs to the environment, such as gullies deep enough to hide the overstocked cattle in Lesotho fields, fertilizer runoff eutrophicating reservoirs in Southeast Asia, or salt appearing on soil surfaces in irrigated northern Thailand. In contrast, it is easy to agree that, when an ecosystem is robust (e.g., with thick soils), thinly populated by humans, relatively inaccessible, shielded from external effects, and lightly harvested, then it is probably in a sustainable condition. Sustainable agriculture, by any reasonable definition, is exemplified by centuries-old sites such as Asian terraced rice systems, and sheep grazing near Stonehenge. Sanborn Field at the University of Missouri at Columbia has grown grain crops for well over 100 years, although obviously with external inputs that eventually may not be sustainable (61).

Amidst the uncertainties of global and international environmental problems there are some components that are well documented, understood, and which we are confident about: for example, the increase in atmospheric carbon dioxide, alterations in the biogeochemical cycle of nitrogen, and gross changes in land use and vegetative cover (62). Others are the loss of topsoil, and contamination of groundwater. Also, much is known with adequate certainty to proceed (e.g., conditions for green revolution agriculture). The next stage of the cause-effect sequence from these changes into impacts on human health and welfare are less certain. Unsustainable practices occur for reasons other than inadequate biophysical measurements, such as the lack of any alternative available to poor farmers and fishermen. But there is continuing documentation of failure and surprise in exploitive projects that were expected to be sustainable but were not. The following sampling is organized by major types of managed ecosystems.

Agroecosystems

Managers are not waiting for new measures of sustainability; they need constant guidance. Most often, they use the economic returns from annual harvest yields, with suitable regressions to account for variation in weather, pests, inputs of fertilizer, equipment, irrigation, market demand, and other short-term factors. But the economic calculus will not usually give a correct signal that the practice they are following is unsustainable. There are too many subtle, non-monetizable influences on any natural system. "Historical references to degradation of agricultural lands and forests can be found in the writing of Plato and even earlier. Yet many centuries later, we are far from consensus as to the identity of a minimal but sufficient set of indicators by which to measure changes in the state of nature" (63).

The kinds of decisions about agroecosystems that are made every year, with whatever information is available, include choice of soil conservation techniques, pest control, amounts of fertilizer and irrigation, and cropping patterns. It is difficult to establish whether the choices are cost-effective, much less whether they result in a sustainable system. Crosson notes concerning U.S. agricultural practices, "Sustainable agriculture could be defined in many ways, including the ability to meet indefinitely the demand for agricultural output at socially acceptable economic and environmental costs. . . . But quantitative measures of environmental costs needed to complete this assessment are lacking" (64). Much economically oriented information is available about agroecosystems, but less is known about noncrop components such as soil organisms that ultimately affect sustainability. Faeth et al note that "in the field, erosion-induced productivity changes are almost impossible to isolate and measure accurately" (65).

In developing countries additional uncertainties arise because of the difficulties of field research. "In the area of understanding and evaluating environmental degradation in Africa, the following causes of uncertainty emerge. First, there is the problem of data—its scantiness, unreliability, irrelevance and ambiguity. Statistics are seldom in the right form, are hard to come by, and even harder to believe let alone interpret" (66).

Henderson (67) writes about conditions in Rajasthan:

> Environmental variability is a major problem for agricultural populations. In arid regions where droughts are frequent, annual variability may be so great that it is difficult to speak of a "normal" year's production. . . . This situation reflects the fact that research is often carried out within a single year's time. . . . Quite simply, there is no such thing as a cyclical pattern that can be identified by examining differences in annual rainfall totals. . . . The problem with using a mean value is that it tends to mask the nature and extent of variation and raises the expectation that behavior will be geared to this central tendency rather than to other factors. . . . In this regard, an agricultural expert who was visiting India during my stay there told me it was impossible to grow sorghum in Renupur, because the average rainfall was below the water limits for this crop. At this time the residents of Renupur were busily sowing sorghum, though there was no irrigation for the fields where the crop was sown and the year was one of drought. . . . The difficulty for the investigator is that, in highly variable environments, "normal" conditions are unlikely to occur very often. In general, the probabilistic aspect of hazards is misunderstood.

Some new agronomic techniques pose unique questions about their long-term effects. For example, plastic film technology (PFT) is widespread even in developing countries (68).

> During recent years the Earth's surface has been covered increasingly with plastic mulch (usually polythene film) for plant cultivation. . . . [It] can

help to increase temperature, retain water, promote the germination and emergence of seeds, accelerate the growth and development of roots (and even whole plants) and, finally, can encourage higher yields and better quality products. . . . PFT developed concomitantly with the petrochemical and high polymer chemical industries.

It is interesting to speculate what will happen to the soil system after many years of application of this essentially nonbiodegradable film, and what predictive measurements could be made now.

High harvest yields can mask loss of soil organic matter and nutrients, impending salinization and water logging from irrigation, pest resistance to chemicals, and strained social institutions, for example, through inequitable labor rates. Further, stable, high harvests are where ecologic knowledge seems to warn of inherent problems. Holling notes that "[stability]. . . . emphasizes the equilibrium, the maintenance of a predictable world, and the harvest of nature's excess production with as little fluctuation as possible" (69). But nature is cyclic at best and usually is changing unpredictably but fundamentally, so that considerable inputs are needed to assure sufficient harvests, and these inputs themselves may not be sustainable.

For example, in Vietnam ". . . rice farming is already very intensive and fully modernized. Farmers exclusively plant high yielding rice varieties (HYVs) and make heavy use of chemical fertilizers and pesticides. . . . the best farmers in the north are already getting yields representing 80% of the genetic potential of the available HYVs. . . . There is no yield gap to exploit in the Red River Delta" (70).

Harvests vary considerably and the reasons are not always obvious. Figure 4.1 is based on data from experimental plots (high input, three crops/year) in Thailand, and actual farms in Vietnam. The intensively managed Chiang Mai University experiment station yield dropped steadily with no obvious explanation until an increasing boron deficiency in the soil was finally detected. The Duy Tien District yields varied widely, and the reasons could be reliably ascribed to unforeseen floods and insect outbreaks. In developing countries rice harvests are often never quantified by measuring weight or volume but merely estimated and negotiated between farmers and buyers. False, low yields are reported when taxes or government shares are related to harvests.

Forestry

The time between harvests in silviculture is usually many decades so there are few places where sustainability, even just of saw logs, has been demonstrated. Duncan Poore, in a worldwide survey for the International Union for Conservation of Nature and Natural Resources (IUCN), concluded: "It is not yet possible to demonstrate conclusively that any natural tropical forest anywhere has

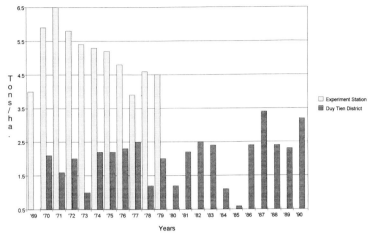

Figure 4.1 Variation in rice yield.

been successfully managed for the sustained production of timber" (71). In contrast with temperate forests, almost all of the nutrients in a tropical forest are in the trees, either natural or plantation stands. When cut, the nutrients are hauled away in the timber and must be replaced or the site is degraded. Botkin and Talbot could not find documentation of sustainability in any original forest under commercial harvest (72). "Clearly, defining sustainability in terms of the continuous yield of timber alone is a trap . . . timber yield is not a primary indicator of a forest ecosystem's health" (73).

For example, the block clear-cutting practice in the U.S. Pacific Northwest is now seen, after 40 years, as unsustainable. For the future, American society is now also demanding spotted owls and water for salmon, in addition to timber, from these forest lands. A long rotation alternative management approach would reduce impacts on soil and water quality and maintain greater biodiversity. Franklin (74) suggests that a reasonable assurance of sustainability of such a mix of forest products and services might require reducing the frequency of final harvest cut to 160 to 200 years from the current 80 to 100 years. This is indicative of (1) the substantial sacrifices of economic production in order to achieve sustainability and (2) the considerable uncertainty involved. While the lower return on investment would only be acceptable on publicly owned forest lands, even there the sustainability outcome would not be known for perhaps 200 years.

Concerning a quite different but important forest product, a Forest Service ecologist trying to regulate recreational uses says, "We don't even know how abundant the resource is; are we picking 95% of the mushrooms? Are we picking 5%?" (75).

No statistically significant change in the volume of water discharged by the

Amazon river, or the amount of sediment delivered from the deforested Rondonia region, has yet been detected. A signal that deforestation has altered the hydrologic cycle or soil erosion in that basin is obscured by the high natural variability of rainfall, and the El Nino events explain most of the occasional trends in the noise that are evident in the hundreds of gauging stations (76).

Grazing Lands

Grazing lands are the ecosystems where the tragedy of the commons is most obvious, but even when the land and the ungulates have the same owner, it is difficult to set the right utilization rate. Pickup and Morton (77) describe a method using remote sensing in arid lands to separate grazing-induced changes in forage biomass from rainfall-driven variations. "Grazing gradients" are patterns developed because sheep and cattle graze out only so far from water sources, and the areas where the vegetation is consumed can be detected. If the grazing gradient does not disappear (with the recovery of vegetation) after large, infrequent rainfalls, the land has been more or less permanently damaged. If recovery does occur once, it does not mean that practice at that site is permanently sustainable. Large, unpredictable fluctuations in rainfall, and explosions of weeds and mammalian pests may yet ruin the pasture. They conclude that for Australia, "meshing of wealth-generating uses of land with ecological sustainability remains an unachieved goal."

Fish habitat is affected adversely by sediment delivered from erosion induced by riparian zone livestock grazing (78):

> The Forest Service has made land management decisions without data quantifying relations between land use and aquatic habitat quality applicable to the water sheds in question. Given the scientific uncertainty, these land use decisions will inevitably reflect values and management bias. The present policies are biased towards grazing. . . . Grazing has been so widespread in western North America that it is difficult to locate ungrazed sites to serve as controls in studies of grazing impacts.

Lusigi (79) points out that African tropical range lands are not actually a managed production system but are responding to population pressures and political disruptions by shifting from a nomadic (plausibly sustainable) mode to a sedentary situation which is obviously rapidly degrading entire regions.

The U.S. National Research Council advised the government, in 1994, that the ecologic condition of western range lands is so poorly understood, due to inconsistent and fragmented data, that a determination of whether they were degraded, and how they should be managed cannot be made. Despite reports of widespread range degradation, the utility of current methods and data for classification and inventory was deemed unsatisfactory (80).

Fisheries

Even catch data are inadequate according to the International Center for Living Aquatic Resources Management in the Philippines. Underestimation occurs because as much as twice the marketable catch may be consumed or processed locally without being measured. In developing countries, data on the species composition of the catch are rarely available. "Measurements relating stock size and composition to environmental stresses such as pollution, or to loss of spawning habitat, are not well developed" (81).

Sherman calls for standardized sampling by trawlers of the large marine ecosystems around the world (82). He notes that a consistent data base over many decades is necessary to judge whether overfishing is the cause of changes in species composition. Few areas have such records as yet, the North Sea and Georges Bank being notable exceptions.

The Kemp's Ridley sea turtle on the Gulf Coast of Mexico was being extirpated by shrimp fishing nets. A massive effort involving regulation of fishermen was instigated, but without scientific controls. "As a result, fourteen years and $4 million later, no one can tell if the program has worked" (36).

The difficulties in gaining international cooperation toward sustainable fisheries include a measurement problem. Regier reports that, historically, "Networks of scientific researchers routinely shared information, sometimes even when such sharing was discouraged by a scientist's country. On occasion researchers informally provided relatively accurate data to their foreign peers while their countries were officially declaring inaccurate data for political purposes" (83).

The number of whales of various species is going up, according to recent assessments and the census data may lead to renewed fishing, a controversial topic internationally. But, "Researchers also warn that today's higher estimates could partly reflect better methods of locating the animals, and that human activities and natural events such as El Nino may be altering the geographic range of some whales, bringing them within reach of the scientists. What's more, incomplete and inaccurate historical records make it nearly impossible to establish a baseline from which to assess the relative health of current populations" (84). For example, in the Hawaiian Islands Humpback Whale National Marine Sanctuary, "Independent studies suggest that while the population is slowly recovering, the magnitude of recovery is unknown because of a number of factors, including both the lack of a uniform methodology for assessing humpback whale populations and the absence of an adequate compilation of existing data" (85).

A study to advise the management of the Great Barrier Reef finds: "The sobering reality is that there appears to be no large-scale example of a Marine and Estuarine Protected Area wherein management has demonstrably or predictably led to resource sustainability over a reasonably long time frame" (86).

GESAMP has reviewed the adequacy of chemical and biological data for the oceans and concludes: "If environmental assessments are to be valid, the chemical measurements on which they are based must be reliable and adequate for their intended use. . . . It has recently become clear that many chemical measurements in the sea made more than 10 years ago are dubious, making it difficult to establish time trends . . ." (87).

Measurements for Adaptive Ecosystem Management

This, then, is the state of the science (biophysical measurements and ecologic understanding) as summarized by Walters and Holling (54):

> . . . in no place can we claim to predict with certainty either the ecological effects of the activities, or the efficacy of most measures aimed at regulating or enhancing them. Every major change in harvesting rates and management practices is in fact a perturbation experiment with highly uncertain outcome, no matter how skillful the management agency is in marshaling evidence and arguments in support of the change.

Adaptive Environmental Management

It is remarkable that the way to deal with uncertainty in managing ecosystems was figured out in the mid-1970s by the ecologists working with C. S. Holling at the University of British Columbia (and later at the International Institute for Applied Systems Analysis near Vienna). That their method, Adaptive Environmental Management (AEM), has not been widely implemented is more of a comment on institutional failures and the stubbornness of conventional economics than the merit of the idea; AEM is not so much a confrontation, or embrace, of uncertainty as it is a wise capitulation to the inevitable. We cannot know precisely what we are doing so surprise should be expected (the semantics of expecting the unexpected seems diverting, to me). Ecosystems are not static, but continually change, and this is the source of their resilience. Avoid irreversibilities. Design projects with built in monitoring and remedial mechanisms.

Kai Lee, in the recent book *Compass and Gyroscope,* observes that, "The adaptive approach is needed if scientific uncertainty is not to thwart socially timely action. Sustainability turns on the ability to manage large ecosystems, and thus requires understanding their behavior in practically useful ways—that is, while recovery from overexploitation is still within practical reach" (36). But he goes on to say that data are sparse and inexact, theory is limited, and surprise is unexceptional. Furthermore, AEM must cross jurisdictional and political boundaries, and cover time scales longer than terms of elective office or budget and business cycles.

Walters and Holling explain (54):

Almost by definition, the impacts will be the consequence of disturbances that are unlike any the natural system has yet experienced. . . . the post-project system is a new system, and its nature cannot be deduced simply by looking at the original one. If the project planning and development sequence fundamentally incorporates adaptive assessment throughout all of its stages, then the ecological response of both the old and the new systems will be studied . . . assessment merges into environmental management. . . . Environmental assessment should be an *ongoing investigation into*, not a *one-time prediction of*, impacts [emphasis in the original].

Managers usually arrange for measurements that relate directly to their primary goal. Holling (24) has studied a variety of managed ecosystems that failed or were unsustainable. He reports a sequence of events common to each of the systems studied. The management goal was always to reduce variability (for economic advantage) of some target parameter such as cattle grazing stocking density, fish population, or occurrence of fire or insects in forestry. In each case control was achieved as measured by the target variable. Natural capital increased and was stored. But slowly the total system changed to become more homogeneous, brittle, less resilient, an accident ready to happen. At the same time, management, having succeeded in its primary purpose, switched its concentration to improved efficiency in shifting cattle among paddocks, hatching fish, fighting fires, and spraying insects. Measures of the status and trends of conditions in the whole ecosystem were too difficult or too expensive. The stabilized managed systems were attractive to investors and were further developed. These economic/social/ecologic systems became increasingly vulnerable; to drought in pastures, inbreeding and disease in fish hatcheries, contiguous forests that spread insect outbreaks and fires. Also, the other goods and services from the now simplified ecosystems declined. Ultimately, each system could not sustain production of the target commodity, and the rigidified management and investment made reorganization and remediation more difficult. Measurements that might have guided adaptive management were not available.

In these examples, the old hypothesis of a stable nature fails to match reality. Holling proposes a discontinuous change hypothesis instead. Four ecosystem functions interact at different rates and times. Exploitation dominates at first and slowly yields to conservation at the climax stage. Then creative destruction, such as fire or pest outbreak, moves the system rapidly through a renewal stage using the readily available nutrients and then back to exploitation (88). Knowing where the ecosystem is in this cycle allows adaptive management.

AEM is active *learning*, learning by doing. While there are nontechnical impediments to learning, such as loss of institutional memory, and inherent ecological problems of replication and control, learning is the most rational procedural response to uncertainty (91). The rate of learning is reflected in the reduction of the standard deviation of the prediction error. This has been demonstrated in the manipulation of whole lake ecosystems. "As experimentation creates greater variance in the forcing variable, results are increasingly informative about responses of the total biomass. Manipulation is a very effective way of improving the capability to predict ecosystem change in experimental systems" (90). For learning to occur, well-designed data collection systems must be in place as the development begins, and the data must be relevant and statistically reliable.

The Project Bias

AEM runs counter to the "project" nature of development. Intensified exploitation of renewable natural resources in developing countries is a major sector (along with building infrastructure and industrialization) in the "sustainable development" to improve human health and welfare. The money for development comes from in-country revenues, foreign private investment capital, and the international development assistance agencies such as the multilateral banks (e.g., the World Bank, Asian Development Bank) and bilateral aid programs (e.g., U.S. aid). The money comes largely in project-sized amounts.

It is obvious that sustainable management of production ecosystems must be planned and implemented on a biome or regional scale, far larger than most projects. Although environmental impact assessment (EIA) is being successfully moved to higher levels of the decision hierarchy (program, sectoral, and regional master planning), it is difficult to overcome the bias (and tyranny) of the project. Decisions as to which projects are funded depend heavily on individual calculations of benefit-cost ratio, and internal rate of return on investment for each candidate. Even when environmental damages are quantified and monetized for internalization into the economic analysis, it is on a project basis.

In fact, an even worse distortion of rational decision making often occurs. Banks have a pressure to lend money even when environmental warnings appear, if the calculated return on investment meets some arbitrary, internally established threshold. Some environmentally dubious, or even obviously disastrous, landscape interventions are set in motion by this politically dominated process of moving money to developing countries. And there are "soft loan" windows where lower interest rates, no-interest grace periods, and long bank times make a mockery of any calculations. This process goes on even as more conventional loans create a debt burden that may eventually have to be forgiven.

A further barrier to holistic adaptive management is the decoupling of the payback of development loans from the success of the project. Most development assistance loans are guaranteed by the central bank of the recipient country. If the project fails (perhaps or in part due to adverse consequences to the environment), the lender is repaid anyway. So-called structural adjustment loans are blocks of money transferred to central governments for generalized development strategies with no specification of what projects are going to be undertaken. These decouplings seriously detract from the detailed scrutiny of environmental effects that might otherwise be undertaken by the lender for purely self-serving financial reasons. But the Third World asks: why should foreign financing dictate, to a sovereign nation, environmental considerations and constraints on exploiting a natural endowment? The reply about a mutually shared biosphere is easily rejected, and this sensitivity is another barrier to implementing AEM.

The "project" is also a cause of resistance to AEM, or candidly conducting development as experiment. Economic and financial analyses assume that projects will be carried forward as planned, even though contingencies for conventional uncertainties (e.g., labor troubles, weather patterns, and market fluctuations) are recognized. Large investments have their own irreversible character. The midcourse corrections envisioned by AEM may well require radical departures from original project objectives. Retrospective evaluation of most multilateral bank projects shows that they failed to meet original objectives and time tables for production and return on investment, and had to be substantially altered as they proceeded. But even this admitted experience does not allow for AEM-type (anticipated) changes to be formally considered in project appraisal and allocation of funds among competing proposals. Financiers are just unable to compute an investment that is forthrightly labeled an experiment.

AEM will have to be *sold* to financial decision makers; it is not reasonable that they will easily accept what appears to be a blind-faith underwriting of ecologic research. My approach is to work with their own retrospective evaluations to show that the continuing lack of scientific knowledge and statistically reliable monitoring data is a hazard to even the environmentally benign projects. There are always times in the operational course of a project when changes can be made relatively easily, and these opportunities could well lead to increased profit, longer life, and enhanced reputation of all concerned. Responsiveness to more accurate and timely biophysical information is also needed to judge the cost-effectiveness of mitigation measures mandated by environmental regulations and to check predictive models of the project. The benefit-cost ratio of improvement in ecologic understanding and monitoring should be substantially greater than one.

The second selling point is to gain collaborative financial support from the investors for regional environmental master planning, particularly in major watersheds, estuaries and coastal areas. Contributions to study these natural systems from all development interests could be organized by the recipient country. Ecologically sensitive areas could be identified, research begun on the major ecosystems, cumulative effects of past, current and future development analyzed, and rough carrying capacity established. The studies are more difficult because they must be done in the midst of a crowded changing landscape that is already substantially simplified and disturbed. Long-term ecologic research sites are not analogous, although they too must be established worldwide for their own worthwhile purposes. There are already a few examples of environmental master plans: The Asian Development Bank funded a study of the Klang Valley downstream from Kuala Lumpur, Malaysia, and the entire island of Palawan in the Philippines (57). The research planned by the World Bank to support the recovery of Lake Victoria may be another. But these studies must be long-term commitments and not the on-off type as with EIAs.

Data Quality

As illustrated, the sources of uncertainty and error, variability and bias, include field instruments, laboratory analyses, sample collection, and population variability. Acceptable data quality depends on the decision for which it is to be used and the comfort level demanded by the decision maker. Accuracy is a combination of precision (standard deviation) and systematic errors (bias in measurements). Other elements of quality are mean square error, representativeness, comparability, and completeness (91).

The practical management decisions in keeping production ecosystems sustainable dictate the required certainty. For example, if soil-building processes at a site suggest a tolerable soil loss rate of 7 ± 3 tons/ha^{-1}/year^{-1} and soil erosion measurements under the existing agricultural practice are 20 ± 7 tons/ha^{-1}/year^{-1}, then a clear signal of unsustainability is given even with the uncertainty.

There are also institutional aspects to data quality such as budget disruptions and policy changes that prevent time series from accumulating. But all of these quality aspects are amenable to standardization and control. More important is what to measure.

No Simple Indicator

The "economics envy" of some natural scientists is seen in the urge to concoct some ecologic equivalent of gross national product (GNP). Proposals for a GNP-like single indicator of the sustainability status of a natural system include net primary productivity (NPP) as measured by remote sensing, and various biodiversity indices (BI). Such integrating concepts could hide important qualitative differences among ecosystems that might show the same NPP or BI (and in fact, the GNP also is defective in this manner). For example, in forests there are "immense technical difficulties associated with measuring below-ground productivity" (74). And yet, the biomass of roots may have a high turnover rate. Aggregation of biophysical measurements at the country level, as characterized by the reports of the United Nations and other international agencies, may be quite misleading because political boundaries seldom coincide with natural demarcations of the landscape.

Hannon and his colleagues suggest a strictly physical measurement of sustainability (49). Life is, in a sense, an alternative to radiating away the free energy from insolation. The dissipation in forming highly complex biologic structures eventually uses all this energy. They argue that evolution has produced the climax ecosystem as the *sustainable* upper limit on the negentropy that can be produced therein. With this as a reference point, the total respiratory heat generated by a managed ecosystem (including that from fossil fuels used) can be compared with that generated by a natural climax system at the same site. If the managed ecosystem heat (e.g., an agricultural crop) is less then it is sustainable. Their calculations depend on many assumptions and extrapolations that introduce substantial uncertainties into this thermodynamics based indicator of sustainability. A purely physical approach is to compare the total respiratory heat generated (as an expression of entropy) by an agricultural system with that of a natural climax vegetation.

"Health" is another proposed analogy: that is, a medical doctor checks a few "vital signs" and judges the health of an individual; so some similar data should succinctly describe the health of an ecosystem. But the human body is much more tightly integrated than any ecosystem. The use of the term "ecosystem health" by Costanza (91), Schaeffer (92), and others is more qualitative and vague, and perhaps more realistic, than the temperature–blood pressure concept.

The Minimum Data Set

Ecology is unlikely to develop any simplified single indicator of sustainability. However, a *minimum data set* for each type of managed ecosystem may be possible. The most important scale for managing (and therefore for measuring) sustainability is at the landscape or regional level of perhaps thousands to millions of hectares. Local harvest-related data cannot be aggregated to give the needed information. Land-use changes appear to be a fundamental cause of reduced ecosystem function and reduced sustainability of production of the goods and services desired. But ecosystems are naturally changing and adapting, so measures must discriminate effects of human activity against this background of change.

Over half the Earth's land surface has been transformed from its natural state by agriculture, forestry, urbanization, desertification, and other interventions. The altered landscape affects adjacent aquatic systems. One third of the loss of species is ascribed to land-use change. For conservation, this implies that stringent restrictions on land use are necessary.

One new measurement approach is to combine growing capabilities in remote sensing, geographic information systems, and landscape ecology. The product is a set of data depicting land-use change that is practical, sensitive over time, and interpretable in terms of sustainability (93, 94). Maplike outputs readily communicate to decision makers about habitat coverage, ecotones, patch configuration, economic activity, water quality, and vegetation.

Supporting these measurements of landscape composition and pattern, a selected biophysical data set could be established for the particular type of managed ecosystem under consideration. Site-specific measurements (although individually unreliable for determining regional sustainability) can be gathered into status and trend reports. Integrating organisms and keystone species can be monitored (e.g., lake trout in Lake Superior). Biologic indicator organisms, such as salt-sensitive plants, will be important. The research and monitoring agendas of management agencies and scientific organizations are beginning to focus on the broader conditions of ecosystems that relate to their sustainable utility.

The CGIAR has recognized that measurements are not available at the regional level to evaluate the extent to which agricultural practices degrade, maintain, or enhance the total ecosystem. Major research areas are the reversibility of degradation, thresholds of decline, and the biodiversity necessary for the future genetic base of agriculture. "Several quantifiable indicators taken over time can provide data along crucial dimensions that help to indicate the sustainability of most agricultural production systems. These include, especially, soil organic matter, soil acidity, crop yields or biomass yields per hectare, and net value added to production" (58).

A site with high "biologic integrity" is, supposedly, able to withstand natural or human disturbances (95). The components of an index of biologic integrity (IBI) are species-abundance counts and ratios, water quality, habitat structure, flow regime, energy source, and biotic interactions. This is essentially a resilience measurement and, although valuable in protection and restoration management, does not relate directly to productivity or sustainability.

The U.S. Tennessee Valley Authority (TVA) uses five indicators of the "health" of a lake or impoundment: algae, dissolved oxygen, fish species diversity, benthic diversity, and sediment.

Indicators for sustainability in agroecosystems of sub-Saharan Africa at the cropping system scale are suggested by Izac and Swift (96) as follows:

1. Ratio of annual yield for all products to potential and/or farmer's target yield
2. Soil pH, acidity and exchangeable aluminum content
3. Soil loss and compaction
4. Ratio of soil microbial biomass to total soil organic matter
5. Abundance of key pest and weed species

Specific criteria for assessment of soil quality are listed by Cole (97) as follows:

1. Biodiversity: soil organisms and plants
2. Physical: annual erosion rate, topsoil depth, drainage rate, depth to water table, aggregate structure
3. Chemical: pH, nutrients, soluble salts, toxic chemicals, organic matter
4. Biologic: biomass production, extent of vegetative cover, condition of carbon and nitrogen cycles

A promising example of a minimum data set is suggested by Risser (98) for grasslands. He proposes that "selected indicators should be small in number and relatively easy to measure, should reflect useful characteristics of the ecosystem as perceived by human values, and should be quantitatively conservative so as to accommodate changes in driving variables such as changing climate." The following values would indicate a sustainable situation: a high percentage (70–80%) of herbaceous cover species palatable to livestock; seasonal peak standing crop of more than 300 gm/m^2; plant species diversity exp. (H^1) not under 5.0; soil organic carbon in the top 20 cm will be 3.0 kg/m^2 for sandy soils and 5.0 in silt-loam soils; and the nitrogen content of above-ground herbage is at least 0.6 gm/100 gm dry biomass. "In other words, the proposal is that if these five conditions are met, the essential grassland properties should persist indefinitely regardless of the use made of the grassland." The actual values of the parameters would vary for grassland ecosystems throughout the world.

Table 4.3 is an attempt to select a minimum set of data about environmental conditions in each type of managed ecosystem that would, taken together, inform management as to whether an ongoing practice was sustainable. This draft "minimum data set" derives from the United Nations University International Conference (2).

Further development by panels of experts and managers should lead to a consensus on measures and their critical values that could be standardized for collection around the world. Trial application in the field for several years could bring refinements and establish statistical reliability. Eventually, a practical, and at least partial, approach to monitoring and predicting sustainability in these intensively managed ecosystems may result. Timely and relevant biophysical measures can support adaptive management with mid-course corrections, treating sustainable development as an experiment, which it most certainly is. Then the continuing international discussions and negotiations on all of the other aspects of achieving SD will be on a firmer and more rational basis.

Conclusion

I have shown that *even* sustainable development means the production ecosystems of agriculture, forestry, grasslands, and fisheries will be severely stressed

Table 4.3 Biophysical Measurement Categories Relevant to Sustainable Managed Ecosystems

CATEGORY	AGRICULTURE AND FORESTRY	FISHERIES
Land-use patterns	Rate of change into and out of present use Patchiness, connectivity, size of parcels	Spawning/nursery habitats changes
Production harvest	Total biomass, usable harvest—all products Extent of pest damage and trends	Catch size and composition, per unit effort
Biological diversity	Species abundance—crop, pests, predators, soil organisms	Top predator, keystone species
Water quality	Pollutants, sedimentation, nutrients	Dissolved oxygen, toxics
Soil properties	Erosion rate, organic matter, nutrients	Sedimentation rate
Atmospheric composition	Acidic precipitation, carbon dioxide concentration	Toxic deposition, UV radiation
Climate	Precipitation and temperature—mean and seasonal variation	

in the decades to come. This is due to the demand for more goods and services by an increasing population, and by the poor. High yields of food, fiber, and secondary products, as well as continued provision of environmental amenities are expected. Unsustainable practices will not only degrade nature but will increase human suffering and social disruption. Fortunately, there is a growing political will to sustain the underlying productive potential of these natural systems, including their biodiversity. Supporting the management of ecosystems for *sustainable high yields* is the most important task for ecology and related environmental sciences.

The state of these sciences is characterized by uncertainties, some inherent and intractable, and others reducible at a cost. The sum of these uncertainties at the present time is frustrating the achievement of sustainability in most managed ecosystems. Information to guide policy and decision making is inadequate because of inexact and statistically unreliable measurements of biophysical parameters, and lack of fundamental understanding of the behavior of ecosystems under stress. It is difficult to determine scientifically whether existing management practices at a given site are sustainable and whether a different alternative management practice would be more or less sustainable.

Limits to replication and control of experiments are barriers to the reduction of these uncertainties through ecologic research, but improvements are on the way (e.g., findings from whole ecosystem manipulation). The measurement problem is worst under field conditions, especially in developing countries where intensified exploitation (short of causing irreversible degradation) is proceeding rapidly, and is most needed.

Dealing with uncertainty is best addressed by adaptive ecosystem management, a learning-while-doing approach in which projects are accepted and treated (by governments and investors) as being experiments. A merging and concurrence of applied ecologic research with management is what is being called for. Natural variation cannot be reduced but can be more accurately represented and interpreted. Actual measures of the productive potential, the condition that is to be sustained, can be found. Sampling and analytical errors can be reduced through better training and supervision of data quality. The monitoring of essential relevant biophysical parameters can be designed into projects, and larger-scale environmental master planning, at their outset. International collaborative efforts are necessary to devise, test, and perfect minimum data sets for each major production ecosystem in each climatic zone.

The expectations of other actors in the sustainable development for certainty and predictive capability from environmental science should be tempered and made realistic. The great stakes in the achievement of sustainability are strong motives for financial and institutional support to improve the means of dealing with continuing uncertainties.

References

1. Economic Commission for Latin America and the Caribbean. Sustainable development: Changing production patterns, social equity and the environment. Santiago, Chile: United Nations, 1991.
2. Munasinghe M, Shearer W, eds. The definition and measurement of sustainability: the biophysical foundations. Washington, DC: Joint publication of the United Nations University and The World Bank, 1995.
3. Rerkasem K, Rambo A. Agroecosystem research for rural development. Chiang Mai, Thailand: Multiple Cropping Centre, Faculty of Agriculture, Chiang Mai University, and Southeast Asia University Agroecosystem Network, 1988.
4. Carpenter RA, Harper D. Towards a science of sustainable upland management in developing countries. Environ Manage 1989;13:43–54.
5. Hilborn R. Living with uncertainty in resource management. N Am J Fish Manage 1987; 7:1–5.
6. Quality Assurance Management Staff. Guidance for planning for data collection in support of environmental decision making using the data quality objectives process. Washington, DC: EPA/G-4. U.S. EPA, 1994.
7. Solow R. An almost practical step toward sustainability. Washington, DC: Resources for the Future, 1992.
8. Dixon J, Carpenter R, Scura L, Sherman P. Economic analysis of environmental impacts. 2nd ed. London: Earthscan, 1994.

9. Kidd C. The evolution of sustainability. J Agric Environ Ethics 1992;8.

10. Dalal-Clayton B. Modified EIA and indicators of sustainability: first steps towards sustainability analysis. London: International Institute for Environment and Development, 1992.

11. Carpenter RA. Monitoring and predicting sustainability. In: IUCN symposium proceedings, Buenos Aires, Argentina, January, 1994. London: Earthscan, forthcoming.

12. Carpenter RA. Can we measure sustainability? Ecol Int Bull 1994;21:27–36.

13. Carpenter RA. Biophysical measurement of sustainable development. Environ Prof 1990; 12:356–9.

14. Kessler WB, Salwasser H, Cartwright CW Jr, Caplan JA. New perspectives for sustainable natural resources management. Ecol Appli 1992;2:221–225.

15. Meadows DH, Randers J. Beyond the limits: confronting global collapse, envisioning a sustainable future. Post Mills, VT: Chelsea Green, 1992.

16. Rees W, Wackernagel M. Ecological footprints and appropriated carrying capacity: Measuring the natural capital requirements of the human economy. In: A-M Jansson et al, eds. Investing in natural capital, Washington, DC: Island Press, 1994.

17. Goodland R. Environmental sustainability and the power sector. Impact Assess 1994;12: 275–304.

18. Caldwell LK. Between two worlds. Cambridge: Cambridge University Press, 1992.

19. Mangel M, et al. Ecol Appl 1993;3(4):573–575.

20. Ciriacy-Wantrup SV. Resource conservation. Berkeley: University of California Press, 1952.

21. Bishop R. Endangered species and uncertainty: the economics of a safe minimum standard. Am J Agric Econ 1978;60:10–18.

22. Costanza R. Developing ecological research that is relevant for acheiving sustainability. Ecol Appl 1993;3(4):579–81.

23. Peters R. A critique for ecology. Cambridge: Cambridge University Press, 1991.

24. Holling CS. New science and new investments for a sustainable biosphere. In Ref. 2.

25. Cothern C, Ross P. Environmental statistics, assessment, and forecasting. Boca Raton: Lewis, 1993.

26. U.S. National Research Council. Issues in risk assessment. Washington, DC: National Academy Press, 1993.

27. Magnuson J. Long-term ecological research and the invisible present. BioScience 1990;40: 495–501.

28. Hekstra G, Stigliani, W. Report of the closing session at the SETAC Conference, Potsdam, Germany, June 24, 1992: Chemical time bombs. Land Degrad Rehab 1993;4(4):199–206.

29. Richey J. Tropical water resources management. In Ref. 2.

30. Mooney H, Sala O. Science and sustainable use. Ecol Appl 1993;3(4):564–66.

31. Shrader-Frechette K, McCoy E. Method in ecology: strategies for conservation. Cambridge: Cambridge University Press, 1993.

32. Mar B, et al. Cost-effective data acquisition. Honolulu: East-West Center, 1987.

33. Carpenter R. Communicating environmental science uncertainties; what we do know; don't know; could know; should know. Environ Prof (in press).

34. Osenberg C, et al. Detection of environmental impacts: natural variability, effect size, and power analysis. Ecol Appl 1994;4(1):16–30.

35. Smith E, Shugart H. Uncertainty in ecological risk assessment. In: Ecological risk assessment issue papers. Washington, DC: The Risk Forum, U.S. EPA, 1994.

36. Lee KN. Compass and gyroscope. Washington, DC: Island Press, 1993.

37. Hilborn R, Ludwig D. The limits of applied ecological research. Ecol Appl 1993;3(4):550–552.

38. Levin S. Forum: Science and sustainability. Ecol Appl 1993;3(4):544–589.

39. Ludwig D, Hilborn R, Walters C. Uncertainty, resource exploitation, and conservation: Lessons from history. Science 1993;260(17):36–39.

40. Rosenberg A, Fogarty M, et al. Achieving sustainable use of renewable resources, Science 1993;262:828–9.

41. Lubchenco J. The sustainable biosphere initiative: an ecological research agenda. Ecology 1991;72:371–412.

42. Hargrove E. Environmental therapeutic nihilism. In: Costanza R, Norton B, Haskell B, eds. Ecosystem health. Washington, DC: Island Press, 1992.
43. Holling CS. Investing in research for sustainability. Ecol Appl 1993;3(4):552–55.
44. Cairns J, Niederlehner BR. Adaptation and resistance of ecosystems to stress: a major knowledge gap in understanding anthropogenic perturbations. Specul Sci Technol 1989;2: 23–30.
45. Carpenter SR, et al. Complexity, cascades, and compensation in ecosystems. In: Yasuno M, Watanabe MM, eds. Biodiversity: its complexity and role. Tokyo: Global Environmental Forum, 1994.
46. Wilcox BA. 1992. Institute for sustainable development, personal communication.
47. Pool R. Ecologists flirt with chaos. Science 1989;243:310–13.
48. Hollick M. Self organizing systems and environmental management. Environ Manage 1993;17(50:621–628.
49. Hannon B, Ruth M, Delucia E. A physical view of sustainability. Ecol Econ 1993;8(3):253–268.
50. Hilborn RC. Chaos and nonlinear dynamics: an introduction for scientists and engineers. New York: Oxford University Press, 1994.
51. May R. The chaotic rhythms of life. In: Hall N, ed. The new scientist guide to chaos. London: Penguin Books, 1992.
52. Vivaldi F. An experiment with mathematics. In: Hall N, ed. The new scientist guide to chaos. London: Penguin Books, 1992.
53. Walters C. Adaptive management of renewable resources. New York: MacMillan, 1986.
54. Walters C, Holling CS. Large-scale management experiments and learning by doing. Ecology 1990;71:2060–2068.
55. Carpenter RA. Environmental risk assessment. In: Vanclay F, Bronstein D, eds. Environmental and social impact assessment. Chichester: Wiley, 1995.
56. Bartell SM, Gardner RH, O'Neill RV. Ecological risk estimation. Boca Raton: Lewis, 1992.
57. Asian Development Bank. Economic policies for sustainable development. Manila: Asian Development Bank, 1990.
58. Consultative Group on International Agricultural Research (CGIAR). Report of the Committee on Sustainable Agriculture. Consultative Group Meeting May 21–25, 1990. The Hague: CGIAR, 1991.
59. Carpenter R. Do we know what we are talking about? Land Degrad Rehab 1989; 1(1):1–3.
60. Swallow B, Woudyalew M. Evaluating willingness to contribute to a local public good: application of contingency valuation to tsetse control in Ethiopia. Ecol Econ 1994;11:153–61.
61. University of Missouri. A celebration of 100 years of agricultural research. Columbia, MO: University of Missouri-Columbia, 1989.
62. Vitousek PM. Beyond global warming: ecology and global change. Ecology 1994;75(50): 1861–76.
63. Rapport D. The use of indicators to assess the state of health of ecosystems: an historical overview. Notebook of the international symposium on ecological indicators. Washington, DC: EMAP, U.S. EPA, 1990.
64. Crosson PR. Is U.S. agriculture sustainable? Resources for the Future 1994; No. 117:10–18.
65. Faeth P, Repetto R, Kroll K, Dai Q, Helmers G. Paying the farm bill: U.S. agricultural policy and the transition to sustainable agriculture. Washington, DC: World Resources Institute, 1991.
66. Blaikie P. Environment and access to resources in Africa. Africa 1989;59:18–40.
67. Henderson C. Famines, droughts, and the "norm" in arid western rajasthan: problems of modeling environmental variability. Res Econ Anthropol 1987;9:251–280.
68. ICIMOD. Plastic film technology. Newsletter of the International Centre for Integrated Mountain Development. Spring 1993. Kathmandu, Nepal.

69. Holling CS. Resilience and stability of ecological systems. Ann Rev Ecol System 1973;4:1–23.

70. Rambo A. Poverty, population, resources, and environment as constraints on vietnam's development. Honolulu: East-West Center, 1994.

71. Poore D. No timber without trees: Sustainability in the tropical forest. London: Earthscan, 1989.

72. Botkin D, Talbot L. Biological diversity and forests. In: Sharma NP, ed. Managing the world's forests. Dubuque: Kendall/Hunt, 1992.

73. Johnson N, Cabarle B. Surviving the cut: natural forest management in the humid tropics. Washington, DC: World Resources Institute, 1993.

74. Franklin JF. Sustainability of managed temperate forest ecosystems. In Ref. 2.

75. Lipske M. A new gold rush packs the woods in central Oregon. Smithsonian, 1994ₐn:35.

76. Richey J. Tropical waste resources management. In Ref. 2.

77. Pickup G, Morton SR. Restoration of arid land. In Ref. 2.

78. Kondolf G. Livestock grazing and habitat for a threatened species: landuse decisions under scientific uncertainty in the White Mountains, California, USA. Environ Manage 1994; 18(4):501–509.

79. Lusigi WJ. Measuring sustainability in tropical rangelands: a case study from northern Kenya. In Ref. 2.

80. U.S. National Research Council, Committtee on Rangeland Classification. Rangeland health: new methods to classify, inventory, and monitor rangelands. Washington, DC: National Academy Press, 1994.

81. MacKay KT. (International Center for Living Aquatic Resources Management). Global warming, fisheries and policy for sustainable development. Presented at the International Conference on Global Warming and Sustainable Development, Bangkok, June 10–12, 1991.

82. Sherman K. The definition and measurement of sustainability: the biophysical foundations of large marine ecosystems and fisheries. In Ref. 2.

83. Regier HA, Bocking AS. Sustainability with temperate zone fisheries: biophysical foundations for its definition and measurement. In Ref. 2.

84. Schmidt K. Scientists count a rising tide of whales in the seas. Science 263(7):25–26.

85. NOAA. Discussion paper for the development of a draft EIS and sanctuary management plan. Honolulu, 1993.

86. Ray G, McCormick-Ray G. Marine and estuarine protected areas. Canberrra: Australian National Parks and Wildlife Service, 1992.

87. GESAMP—Joint Group of Experts on the Scientific Aspects of Marine Pollution. The state of the marine environment. UNEP Regional Seas Reports and Studies No. 115. Nairobi, Kenya: United Nations Environment Programme, 1990.

88. Holling CS. The resilience of terrestrial ecosystems: local surprises and global change. In: Clark WC, Munn RE, eds. Sustainable development of the biosphere. Cambridge: Cambridge University Press, 1986.

89. Hilborn R. Can fisheries agencies learn from experience? Fisheries 1992;17(4):6–14.

90. Marker D, Ryaboy S. The quality of environmental databases. In Ref. 26.

91. Costanza R, ed. Ecological economics: the science and management of sustainability. New York: Columbia University Press, 1991.

92. Schaeffer DJ, Herricks E, Kerster H. 1988. Ecosystem health. I: measuring ecosystem health. Environ Manage 1988;12(4):445–55.

93. O'Neill R, et al. Landscape monitoring and assessment research plan. 620/R-94/009. Las Vegas: U.S. EPA, 1994.

94. O'Neill RV, Hunsaker CT, et al. Sustainability at landscape and regional scales. In Ref. 2.

95. U.S. EPA. Biological criteria: Technical guidance for streams and small rivers. Washington, DC: U.S. EPA/OST, 1993.

96. Izac A-M, Swift M. On agricultural sustainability and its measurement in small-scale farming in sub-Saharan Africa. Ecol Econ 1994;11:105–25.

97. Cole M. Soil quality as a component of environmental quality. In: U.S. National Research Council. Issues in risk assessment. Washington, DC: National Academy Press, 1993.

98. Risser P. Indicators of grassland sustainability: a first approximation. In: Munasinghe M, Shearer W, eds. The definition and measurement of sustainability: the biophysical foundations. Washington, DC: Joint publication of the United Nations University and The World Bank, 1995.

Scientific Uncertainty and Environmental Policy: Four Pollution Case Studies

Judith S. Weis

The Nature of Science and of Policy

In theory, it should be relatively easy for environmental scientists to recommend policy options for improving the quality of the environment. At first glance, it seems relatively straightforward to conduct assessments of the extent of adverse conditions in an area, to identify the causes of the adverse effects, and to evaluate options for minimizing the environmental impacts. However, things are not so simple for a number of reasons. The first is the nature of science and the role of statistics and probability. Science relies on statistics for its conclusions, and practitioners of environmental sciences generally start with a null hypothesis of "no effect" and require 95% confidence before concluding that some environmental factor or perturbation has had an effect on the organisms or communities studied. Thus, if the data show an effect but with only 90% or 70% confidence, the null hypothesis is accepted—that is, that no effect has been demonstrated. However, there is a tendency to assume not only that there is not enough evidence to reject the null hypotheses, but that there *was really no effect*. Policy makers may then believe that there was no impact, when, in fact the test was too weak or the data were too variable or too close for an effect to be demonstrated even if there had been one (a type II error). The power of a test is a function of the magnitude of the effect to be tested, the sample variance, and the number of replicates, and the significance value (usually 0.05). In policy making, on the other hand, weight of the evidence rather than 95% confidence is generally used. Policy decisions, unlike science, are not probabilistic, but are usually discrete choices among specific alternatives. In the light of the urgent need for informed decisions in environmental policy, scientists should become comfortable with recommending actions based on the weight of evidence, which is the standard in governmental practice.

Second, scientists appreciate the complexity of the ecosystems they study, and are likely to acknowledge the uncertainties and complexities of the situation, to discuss the need for further research, and to focus on complications,

160

disagreements, and parts of the problem that are not yet understood. Thus, when scientists advise policy makers, they may present many possible options and be hesitant to recommend a course of action unless they are 95% confident of its success. This is a major source of confusion and frustration for policy makers and often results in inaction regarding protection of the environment.

Current policies, which include cost-benefit considerations, put increasing pressure on the environment, since it is often difficult to demonstrate deleterious effects in unequivocal terms. Most examples of pollution permitting use an approach in which the environmental degradation is permitted if the resulting damage is uncertain but the costs of preventing degradation are known. The burden of proof is placed on those who must demonstrate harm before regulatory action can be taken. Monitoring approaches to detect this harm can seldom detect subtle changes until these effects are no longer subtle. It is difficult for field studies (monitoring) to link in situ effects with specific causes, and thus it is difficult to conclude that a certain kind of pollution or activity is causing damage. On the other hand, laboratory studies demonstrating adverse effects of toxic chemicals can be faulted for being environmentally unrealistic.

Our knowledge of ecosystems is insufficient and their variability is too great to explain many of the large-scale changes that may be noted. In the absence of our ability to link ecologic changes to specific causes, those who would continue polluting or otherwise changing the environment are allowed to do so. There are interests who profit greatly by the current status of who bears the burden of proof, and, since there is generally some scientific uncertainty, they argue against regulation. There can be considerable disagreements over how much uncertainty is acceptable and how much research needs to be done. For these reasons, governmental restrictions on toxic materials are often late in coming.

The Precautionary Principle

What if the burden of proof was on the other side? If chemical manufacturers or developers had to "prove" with 95% certainty that a particular substance or activity did *not* cause an adverse effect on the environment, then in all those cases that indicated an effect but with less than 95% confidence, regulators would conclude that the standard of "no effect" had not been proved, and the environment would be protected. Environmental policy would be more protective if it replaced the "wait-and-see" principle with a precautionary principle (1) that would be to refrain from actions with potential negative impacts, even in the absence of clear proof of their harmfulness.

Case studies of both successes and failures of science to influence environmental policy are described here. This chapter focuses on the scientific input and the policy actions regarding ocean dumping of sewage sludge, acid precipitation, the antifouling paint additive tributyltin (TBT), and estuarine eutrophication. The first two cases took place under a spotlight of press attention and considerable concern by the general public and environmental groups, while the second two have had considerably less attention from the press, the general public, and environmental advocates. All of these issues involved both the federal government and states, although only the first three have been incorporated into federal legislation. Only one of these four cases, TBT, may be considered a clear success.

Ocean Dumping of Sewage Sludge

Since sewage treatment plants collect effluent from factories and street runoff as well as domestic sinks and toilets, the exact constituents of sewage are highly variable. Sewage contains a large proportion of organic matter and nutrients, microorganisms, some of which may be pathogens, and toxic components, such as oils and metals, especially when discharges from industry are mixed with domestic waste. Through primary and secondary wastewater treatment, sewage is separated into two phases, a liquid and a solid (sludge); these may be disposed of separately. Sewage sludge generally consists of 1–5% dry solids. The disposal of sewage sludge can be a serious problem. It may be placed in landfills, spread on agricultural land, incinerated, or disposed of at sea. While risks are associated with all possible disposal options, dumping at sea is the cheapest and most convenient option for coastal communities (2). When sludge is dumped at sea, the effects on the environment depend on the rate and volume of the dumping, the nature of the sludge, and the physical characteristics of the disposal area. In areas where the material can accumulate, major impacts can be seen on the bottom. In severe cases, deposition of organic material leads to anoxic conditions, and the benthic fauna may be reduced to a few resistant species. If toxic chemicals are not a major component, the organic component of the sewage is largely degradable, and affected areas can recover after the input ceases. If sites are selected properly, the fertilizing nature of the nutrients in the sewage may be regarded as more important than its potential toxicity (3).

The 12-Mile Site

Sludge from sewage treatment plants in New York City and many northern New Jersey communities was dumped since 1924 in the ocean about 12 miles off shore in the New York Bight (Fig. 5.1). The amount of sludge in later years from New York City alone was about 236 dry tons of sludge per day (4). In 1973, the amount of sewage sludge was 150 million ft^3 (5). The dumped sewage sludge was comprised of roughly 5% solids, comprising two major fractions: one composed of heavier solids which sinks to the bottom near and downstream of the dump site; the second composed of lighter materials, which remain in the water column for varying periods of time after dumping, depending on its composition and the water circulation in the area. This second fraction includes

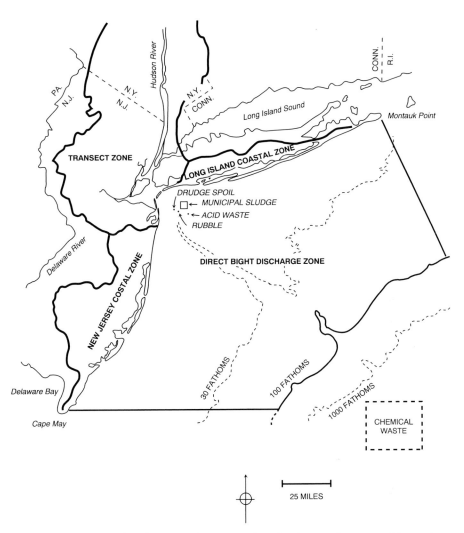

Figure 5.1 Geographical zones in the New York Bight, showing locations of dump sites. The site labeled "Chemical waste" became the 106-mile sludge dump site. Reprinted from NOAA, 1976. Contaminant Inputs to the New York Bight, NOAA Technical Memorandum ERL MESA-6. National Oceanic and Atmospheric Administration, Boulder CO.

both dissolved and suspended solids. Both fractions can contain toxic heavy metals and pathogenic materials. Sludge from major metropolitan regions such as New York City typically contains significant concentrations of toxic materials such as heavy metals and polychlorinated biphenyls (PCBs). This site accumulated rather than dispersed materials, so that persistent contaminants (metals, PCBs) would be localized and their effects would be detectable. However, since anaerobic conditions are likely to occur in accumulating sites, degradation of sludge would also be slower, and release of metals greater (6) than in more dispersive sites.

Scientific studies in the 1970s revealed a "dead sea" in the dump site area, due to the deposition of large quantities of sludge that contained toxic materials. The sewage sludge settling on the bottom had major impacts on benthic biota. Samples collected from the dumping area were frequently devoid of normal benthic fauna; when organisms were found, the diversity of species was greatly reduced (7). Organisms from the dump area were found to have elevated levels of disease. Fin erosion disease (8), noted in several fish species, is a progressive erosion of the fin rays and overlying epidermis, with erosion starting at the outer edges of the fins and progressing to the base. Its precise cause is unknown, but it is found in greater frequency in benthic fish living in polluted waters. Shell erosion disease, (9) a progressive necrosis and lysis of the exoskeleton of crustaceans, was found in crabs, shrimps, and lobsters from the New York Bight. It could be produced in healthy organisms maintained in the lab with sediments contaminated with sewage sludge.

Due to bacteriologic contamination, the area around the sludge dump site had been long closed to shellfishing. There was also concern that Long Island beaches could be at risk from the bacteriologic contamination from the sludge. Federal agencies began studies to relocate the site, while environmental organizations spurred a movement to ban the ocean dumping of sludge. The disposal of sludge represented the most serious environmental problem confronting the New York City region and the New York Bight, since cessation of dumping sludge requires that there be an alternative practical means of disposal.

The Court Case

In 1977, amendments to the Ocean Dumping Act (Marine Protection, Research and Sanctuaries Act) banned ocean disposal starting at the end of 1981, and the Environmental Protection Agency (EPA) took steps to enforce these provisions. This ban was contested in court by New York City, whose officials foresaw high expenses for building incinerators, which they concluded was the only feasible alternative for handling the dewatered sludge. The city had considered application of sludge as a soil conditioner at city parks, but realized it would be practical for only a limited period of time and would entail risks associated with pathogens in the sludge and the possible leaching of metals into surface and groundwater. In the long term, incineration was determined to be the only feasible alternative. The ash from the incinerators was anticipated to be categorized as hazardous waste, and its disposal would require transport to a secure landfill.

Furthermore, the city had been informed by the National Oceanic and Atmospheric Administration (NOAA) that cessation of sludge dumping would not measurably improve the water quality of the New York Bight, since the massive inputs of polluted water from the Hudson/Raritan estuary would have overriding effects on the Bight's water quality. Mueller et al (10) found that the major contaminant inputs to the New York Bight originate from the northern New Jersey and Hudson River drainage basins. Wastewater, runoff, and barge discharges are the major sources from this zone, and heavy metals comprised the most significant anthropogenic inputs.

New York City anticipated significant public opposition to its proposed plans for incinerating the sewage sludge, due to the potential for severe environmental impact (including air emissions of metals from the incinerators), and excessive costs (4). In their lawsuit they argued that the evaluation of environmental effects of ocean dumping should include a comparison with the anticipated impacts of the alternative land-based options of landfilling and incineration. Federal Judge Sofaer in "City of New York v. EPA" supported the city's position, since EPA had not determined whether the land-based alternatives would actually be environmentally preferable. The city's position was supported by a report from the federally appointed National Advisory Committee on Oceans and Atmospheres (NACOA) (11), which also advocated a multimedia approach to waste management.

The 106-Mile Site

Pressure to close the 12-mile site continued, and many congressional hearings were devoted to the subject of ocean dumping. In 1987, EPA mandated the closing of the New York Bight 12-mile dump site, but did not ban ocean dumping. Instead, EPA required the sludge-dumping municipalities to use another site 106 miles from shore, a more costly operation. Since the cessation of dumping at the 12-mile site, significant improvements have been noted in the environment of the area, including reduced total organic carbon and metal contamination, increased dissolved oxygen, and changes in abundance of benthic infaunal species and diet of some fish (12).

The new site was selected to minimize the chances of contaminants reaching shorelines and beaches, to minimize adverse effects on living marine resources, and to avoid areas most intensively fished (13). In the new site, the sludge was placed far away from coastal communities in a much deeper and more dispersive environment. Sludge is dispersed by tidal currents, thermoclines, and wind. Dilution varies depending on the speed of the discharge vessel, rate of discharge, and local hydrographic conditions (14). The projected consequences of dumping sewage sludge at this site were that water column contamination would be less than levels that would affect planktonic organisms or fish, and that seafloor contamination would be due primarily to sludge particles (15). These authors felt that the major sediment contaminants (PCBs and polynuclear aromatic hydrocarbons (PAHs)) could reach concentrations at the sediment surface of 0.2 and 0.3 μg/gm, respectively, after 100 years of sludge dumping. Those concentrations were not considered to be a threat to benthic life.

There have been many studies of effects of dumping sewage sludge in various environments. Studies have included measurements of water quality, sediments, metal levels in sediments and biota, distribution of benthic invertebrates, fish diseases, and fecal bacteria. In field surveys of various dump sites, some degree of enrichment of the seabed occurs. Field studies in dispersive sites generally indicated that the possible toxic effects might be counteracted by the nutrient enrichment (reviewed by Costello and Read, 16). Organic enrichment of the seabed is associated with high numbers of species and individuals at certain sites. In most of the dispersive sites reviewed, there was not a significant bioaccumulation of toxics in resident biota, nor was there a clear indication of deleterious effects on animal health. This may have been due to concentrations below toxic levels, acclimation of animals to the stress, or effects occurring but not being detected.

Despite the move to a more distant and dispersive site, pressure to end the dumping of sewage sludge continued. Some environmental and fishing groups claimed that the sludge dumping was causing death and disease of marine life at the 106-mile site. The site had previously been used for disposal of industrial waste and had therefore been studied extensively long before the sewage sludge began to be dumped there. These earlier studies generally found negligible adverse effects at the dumpsite or the surrounding area (17). Studies reviewed by NOAA in 1981 were very inconclusive (18, 19), despite the likelihood that these industrial wastes would have a greater deleterious effect than sewage sludge.

A panel of the National Academy of Sciences (20) concluded that when water depth and circulation are sufficient, sewage sludge can be disposed with few if any localized detrimental effects. They felt that suitable disposal sites existed over wide areas of the continental shelf and slope. The panel did express concern about trace contaminants such as metals and xenobiotic organic compounds, and felt that these chemicals should be managed by preventing their entry into the sewer system either by source control or prohibition from urban use. The Congressional Office of Technology Assessment (21) produced a report with similar findings—that is, that pollutants going into estuaries and coastal waters cause obvious degradation, but that the open ocean exhibits few adverse effects from waste disposal, partly because it can disperse most wastes widely. They agreed that more research was needed to determine long-term effects of toxic pollutants.

After the initiation of sewage disposal at the 106-mile site, one study using stable isotopes (22) indicated that some nutrients from the sludge were arriving at the seafloor, contrary to what was previously thought. The sewage-derived organic material was found to be utilized by certain deep-sea deposit-feeding animals, species of sea urchin and sea cucumber, and thus could enter the deep-sea food web. Although this indicates that disposal of sewage sludge in the open ocean does impact the benthic ecosystem and may alter benthic communities in favor of species that can readily utilize the enrichment, there was little evidence for environmental degradation at the deep-water site. However, additional chemical changes to the benthic environment, not anticipated by regulatory agencies, have also been identified.

Floatables on the New Jersey Beaches

In the summer of 1987 there were numerous incidents of coastal pollution—medical waste and other floating materials washed up on the beaches of New Jersey—deterring tourism and causing economic losses to coastal communities. A huge public outcry resulted. The public's response was out of proportion to the real level of risk. Sewage sludge does not normally contain floatables in it. The materials washing up on the beaches were not derived from the sludge dump site 106 miles away, but had come from the Fresh Kills landfill on Staten Island and from combined sewer overflow from storm and household sewers in the metropolitan area. The public's response to the waste on the beaches was not based on scientific assessments and may have obscured more serious pollution problems. Nevertheless, the appearance of the waste on the beaches and the public outcry prompted New Jersey Governor Thomas Kean, various environmental groups, and those with interests in the economy of the New Jersey shore to call upon Congress to prohibit sludge dumping totally, and Congress soon passed such a ban, led by the New Jersey delegation.

Public Misperception Drives the Policy

The sludge dumping issue is one in which science did not clearly indicate harm at the deep-water site, yet the issue received a great deal of public attention and press coverage. The bill that was passed banning ocean dumping was a result of pollution events having nothing to do with scientific assessment of the 106-mile sewage sludge dump site. Instead, the policy was a result of public misperception, which outweighed any scientific assessment. However, since then, a study (23) analyzing bottom trawl data showed that abundances of silver and red hakes, summer flounder, goosefish, black sea bass declined significantly in the vicinity of the 106-mile dumpsite during the years of use of the site for sludge dumping. The author felt it was likely that the sludge dumping could have been a factor contributing to this decline in abundance, although cause and effect could not be established. The body burdens of chlorinated pesticides, metals, and PCBs in fish in the area of the dumpsite were elevated.

The law banning sludge dumping may not be the final chapter on this issue. As land-based alternatives prove to be expensive and unpleasant to the public, some are expressing a desire to reopen the issue of ocean dumping.

Acid Precipitation

By the early 1980s it had become clear that the air pollutants sulfur dioxide (SO_2) and nitrogen oxides (NO_x), emitted from burning of fossil fuels and released from tall smokestacks could be transported long distances. When these gases were returned from the atmosphere to the Earth's surface, in the form of rain, snow, fog, or dry deposition, the acidity they produced could pose a risk to living resources and accelerate the weathering of various materials (Fig. 5.2). The best documented effects were those on aquatic ecosystems. High levels of acidity posed a threat to thousands of lakes and streams in the eastern United States and Canada. The sensitivity of a lake to acidic deposition depends mostly on the nature of the adjacent soil or bedrock. When there is little neutralizing capacity, the body of water is at risk. When the water's pH drops to 5–6, many species of aquatic organisms die and the ecosystem is greatly modified. These changes are due to the acidity and/or to the release of metals (especially aluminum) that occurs under acidic conditions.

There was some evidence that declining forest productivity was attributable to acid precipitation, although this was controversial and not proven. Forested areas subjected to acid deposition, ozone, or both were noted to have declining productivity and dying trees, but it was uncertain how much of this was attributable to air pollution or to other causes. Acid deposition could harm trees either directly by removing nutrients from leaves or indirectly by altering the

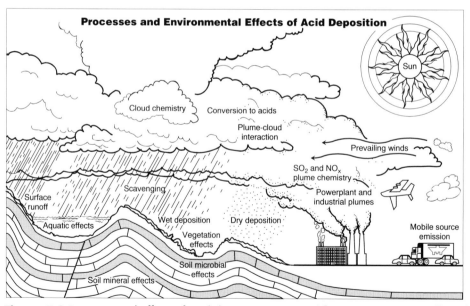

Figure 5.2 Processes and effects of acid deposition. Reprinted from U.S. Congress Office of Technology Assessment, 1976. Acid Rain and Transported Air Pollutants. Implications for Public Policy. OTA-O-204.

mineral composition of the soil by stripping essential nutrients such as calcium and magnesium from it. Trees at higher elevations were considered to be at greater risk. Agricultural productivity could also be affected adversely by acid rain, but experimental data were equivocal. Acid precipitation was damaging to many materials, such as stone, iron, and steel. Air pollutants could also impair visibility, and ambient levels of sulfates were suspected to contribute to human health problems, but the magnitude of the health risk was uncertain and not established.

To Act or to Study the Problem?

In addition to the scientific uncertainty, there were a number of challenging issues for policy makers. There were geographic issues since the activities generating the emissions were primarily in the Middle West, and the adverse effects were occurring mostly in the eastern states. There were international issues as well, since much of the pollution generated by the United States was deposited on Canada. Controls would be costly. Another set of disagreements in the policy debate were dealing with how to promote economic development while protecting the environment. The dilemma was whether to act quickly to control acid deposition or to wait for more scientific results to clarify the magnitude and nature of the environmental effects. There were uncertainties about how many lakes and streams had been damaged or were at risk, or what quantity of crops had been damaged, and whether damages were cumulative and/or irreversible. Knowledge of pollution sources and transport was not precise, so that scientists could only estimate the amount of sulfur pollution transported from one region to another. They could not predict to what degree reduction of emissions in one region of the country would reduce deposition in another or how much time it would take to correct the problem. Control programs also would carry risks of an economic variety. Electric bills would rise; if there would be a switch to low-sulfur coal, jobs in high-sulfur coal regions would be displaced. Likewise, there was no guarantee that a research program would provide data that would make policy choices easier.

Establishment of the National Acid Precipitation Assessment Program

In 1980, the U.S. Congress established a multimillion dollar research program to be organized by an Interagency Task Force on Acid Precipitation, the National Acid Precipitation Assessment Program (NAPAP). This was established to carry out a 10-year research and assessment program. The Interagency Task Force consisted of representatives of 12 agencies, led by the NOAA, the EPA, the Department of Energy (DOE), the Department of Agriculture (USDA), the Department of the Interior (DOI), and the Council on Environmental Quality (CEQ). NAPAP was mandated to develop programs to (1) identify sources of atmospheric emissions contributing to acid precipitation, (2) establish and operate a nationwide monitoring network to measure acidity of precipitation, (3) study atmospheric physics and chemistry in order to understand the processes by which acid precipitation is formed, (4) develop and apply models to predict long-range transport of the substances causing acid precipitation, (5) define geographic areas of impact through deposition monitoring and identification of sensitive areas, (6) broaden impact data bases by collecting existing data on water and soil chemistry and through trend analysis, (7) develop dose-response functions for soils, aquatic organisms, crop plants, and forest plants, (8) establish studies on plant physiology, aquatic ecosystems, soil chemistry, soil microbiology, and forest ecosystems, and (9) perform economic assessments of environmental impacts caused by acidic precipitation and of alternative technologies to remedy or ameliorate the harmful effects.

Research under NAPAP focused on atmospheric chemistry, atmospheric transport and modeling, atmospheric deposition and air quality modeling, terrestrial effects, aquatic effects, and effects on materials and cultural resources. They were to develop many comprehensive state-of-the-science reports and an integrated assessment that was to be a structured compilation of policy-relevant information to assist policy makers to evaluate the key questions regarding acidic precipitation:

1. What are the effects of concern, and what is the relationship between acidic deposition/air pollutant concentrations and these effects?
2. What is the relationship between acidic deposition, air quality, and emissions?
3. What is the sensitivity of ecosystems to change?
4. What are the estimates of future conditions (emissions, costs, deposition and effects) with and without additional emissions reductions?
5. What differences emerge from comparative evaluations of future scenarios?

After 2 years of planning, NAPAP presented Congress with a plan, which stated that an integrated policy-related assessment would be completed by 1987 and updated in 1989. A number of members of Congress and public interest groups felt this time frame was too slow.

The OTA Report

A 1984 report of the Congressional Office of Technology Assessment (OTA) (24) laid out four policy options for congressional action:

1. Mandating emissions reductions to control the sources of the pollutants. This would involve decisions on which pollutants to reduce (sulfur alone or nitrogen as well), how much to reduce them, from which

regions of the country, and over what time period. Congress would also have to choose specific mechanisms to implement the reductions, allocate their costs, and deal with undesired side effects of emissions reductions.

2. Liming lakes and streams to mitigate some of the effects of acid deposition. This had been done in Scandinavian countries and Canada to counter acidification in surface waters. The liming would need to be repeated every few years to prevent reacidification.

3. Modifying the federal acid deposition research program to provide more timely guidance for congressional decisions. This was in response to the NAPAP plan of having an assessment done by 1987.

4. Modifying existing sections of the Clean Air Act to enable the USEPA, states, and countries to more effectively address transported air pollutants other than acid deposition. The existing Clean Air Act legislation did not have provisions dealing with deposition and transport, only with emissions.

NAPAP's Interim Report

NAPAP released its interim report in 1987 (25), which presented its assessment of the causes and effects of acid rain. The report concluded that acid deposition contributed to the acidification of some lakes and streams in sensitive regions, including the Adirondack Mountains, the Pocono Mountains, parts of New England, parts of Florida, and the Upper Peninsula of Michigan, but that there were many regions of the country in which no harmful effects could be discerned. The report indicated that the sensitive northeast region was at a steady state with regard to acidification, and that it was unlikely that there would be an increase in the number of acidic lakes under the current levels of deposition. The Southern Blue Ridge area, in contrast, was considered to be at risk for increasing acidity if deposition continued at current levels. After examining eight cases of regional decline of forests in the United States, NAPAP concluded that in only two cases (eastern white pine in the eastern United States and Canada, and ponderosa pine in southern California) was air pollution (ozone) the cause of the damage. For the other six cases, definitive conclusions could not be drawn implicating acid rain or other air pollution. NAPAP stated that current levels of deposition did not pose a threat to healthy forests or to crops. The report concluded that acid rain can accelerate the weathering of materials, but the degree of acceleration was uncertain. In short, it concluded that acid rain had produced only minor effects on agriculture, forests, aquatic life, or materials. Since few effects had been demonstrated, NAPAP stated that economic analysis was premature.

Criticism of the Report

According to Schindler (26), NAPAP was subject to strong political pressures from the Reagan administration, which preferred to support more research rather than develop regulations to curb emissions that contributed to acid rain. Delaying action would allow emissions to remain high with the risk of further ecologic damage. The Reagan administration, supported by the coal and utility industries, preferred to wait for results of the research program, while many in Congress, supported by environmental groups, preferred to act sooner to prevent further damage. The political pressures on NAPAP caused many excellent scientists to disassociate themselves from it. The 1987 interim report which stated that the acid precipitation problem was small and had been exaggerated, was not at all the concensus of scientists working in the field. Schindler discusses the statement in the report that lakes whose pH was greater than 5 had not been damaged by acid rain. Yet investigators had shown that biologic damage starts as the pH drops below 6. Contrary to most evidence, the report also concluded that rapid reductions in emissions would have little effects on lakes. This report had little credibility within the scientific community, and in 1987 a new director of NAPAP took over, who attempted to restore the credibility of the program. While Schindler considers NAPAP's final report to be more credible, he feels that it also understates the seriousness of the issue. He states that some good research came out of the program, but that NAPAP did not get a good handle on nitrogen emissions and deposition, episodic acidification, and ecosystem-scale and long-term studies, despite the expenditure of $570,000,000. Meanwhile, emissions continued unabated through the 1980s.

The component of NAPAP involving research on forests did not begin until many years after the other studies, so investigators in that area could not gather data for more than 3 years before the final NAPAP report was due. Contrary to the views of many of those investigators, the assessment of effects on forests in the final report hinged on a very narrow definition of "damage" (tree death) and some of the scientists involved, such as Loucks, took exception to the report's statement of "no evidence" of forest damage, since they had clearly noted changes in forest condition and nutrient cycling in the soils (27). Loucks felt that the scientific questions for the forest component of the study had been appropriately laid out, but that the writers of the summary reports using "selected highlights" did not adequately explain the situation to the general public.

NAPAP has also been criticized for focusing too much on the pure science, and not enough on the process of assessment and the policy implications of the findings. An assessment process should analyze and interpret the scientific findings in order to make predictions. It should analyze the degree of uncertainty involved. It should discuss the benefits, costs, and risks that would be expected with different alternative courses of action. This policy analysis was not done adequately (28). It is interesting to note that many years earlier, OTA had seemingly anticipated such a problem (24) and had suggested that Congress consider the option of establishing a "two-track" research program within NAPAP, which would have included a separate policy assessment. This would have provided Congress with timely policy guidance without jeopardizing the long-term research program. The report envisioned an assessment that would evaluate a series of control alternatives in terms of the costs of each, secondary effects of each, expected deposition reductions, other air-quality benefits, and resource benefits.

The 1990 Clean Air Act

The Clean Air Act was reauthorized in 1990 with provisions to reduce SO_2 emissions from power plants considerably and reduce NO_x emissions to a smaller degree. The legislation mandates an annual 10-million-ton reduction of sulfur dioxide emissions below 1980 levels and targets electric utilities. The control program allows a market-based banking and trading system of emissions allowances and gives utilities flexibility in the choice of technology for reducing emissions. The law mandates a reduction in the emissions of nitrogen oxides of about 2 million tons from the 1980 baseline. This provision affects primarily utilities and other industries. Overall, the emissions of sulfur dioxide will be reduced by about 40%, and nitrogen oxides by about 10%. The 1990 Clean Air Act also included specific mandates for NAPAP to continue and to focus more on policy-relevant issues than it did in its first 10 years.

The final NAPAP report, issued after the passage of that legislation, had greater emphasis on NO_x and on mobile sources of emissions (automobiles), which were not dealt with in the law. The acidity of streams due to nitrates continues to be a problem and may need additional attention in the future. The "new" NAPAP has issued its 1992 report to Congress (29), which noted downward trends in sulfate and nitrate concentrations, and acknowledged that acidic deposition continues to affect sensitive forest, soil, and aquatic ecosystems. It firmly implicated acid deposition as a causal factor in the decline of high-elevation red spruce and found evidence that changes in soil chemistry and nutrient leaching in soils south of the Great Lakes have been increased by acidic deposition. It concludes that some watersheds receiving high nitrogen inputs have become nitrogen saturated since the inputs exceed the ability of the watershed to absorb nitrogen. Thus, the official government assessment has finally caught up to the conclusions of the independent scientists many years earlier. The report expressed the expectation that with greater emissions reductions, greater improvements in ecosystem health will result.

NAPAP's deficiencies were due to failure to adequately perform the assessment function and the holding of science hostage to a political agenda, a phenomenon more apt to happen in a multimillion dollar megaproject and where corporate profits are at risk than in less costly scientific endeavors. As seen in this case, the uncertainty of scientific research, especially on ecosystem studies, can be used as an excuse for delaying action.

Antifouling Paints

Antifouling paints are designed to keep "fouling" organisms, such as barnacles, from settling on boat bottoms, since their presence impairs the speed and efficiency of the boats. Through most of this century and before, copper-based paints, which require annual application, have been used for antifouling. In the 1970s, paints containing the chemical tributyltin (TBT) became extremely popular because TBT is an extremely effective and long-lasting antifouling agent (up to 7 years for one coat of paint).

Indications of Environmental Effects

As the use of tin-based paints expanded, suspicions began to arise that organotin leaching from the paint might harm the aquatic environment. Laboratory studies in the early 1980s (30, 31, 32) suggested that a variety of species could be affected by TBT at levels below 1 part per billion (ppb). These were concentrations below those detectable by standard analytical chemical techniques. During the same period of time, French scientists (33) noted abnormalities in shell growth of oysters (*Crassostrea gigas*) living near boatyards and marinas. The shells grew unusually thick as the oysters kept secreting additional layers of shell with a series of cavities, so that the shells became rounded rather than growing in the normal way. Inside the thick shell, the animals were stunted. Such oysters were unmarketable. The deformed oysters were found to contain high levels of tin, suggesting that organotins, rather than other aspects of boating activity, were linked to the observed effects. The number of oyster larvae decreased as well. Combining laboratory and field work, English investigators demonstrated that this abnormality was due to TBT (at extremely low levels). The shell-thickening phenomenon could be produced by concentrations as low as 150 parts per trillion (ppt) (32).

Other researchers studying snails found that another anatomical abnormality, a masculinization of females, called "imposex," was associated with TBT. Female snails affected by extremely low levels of the chemical grow a penis; when the condition is severe, they are effectively sterilized, and populations decrease due to reproductive failure (34). This malformation could be produced in the laboratory at levels as low as 20 ppt.

Laboratory studies on other species showed that TBT could have deleterious effects on a large number of marine organisms at very low concentrations. Because of its low solubility in water, much of the TBT in an estuary would concentrate in the sediments rather than in the water (35, 36). While this would reduce the TBT exposure of fish, organisms that reside on the bottom would remain exposed. TBT also accumulates in the surface microlayer of the water (37) to which intertidal animals are subjected as the tide moves in and out.

The special characteristic of this chemical is that it produces unique anatomical abnormalities in the oysters and snails, so that organisms in the field can be readily identified as having been affected by TBT. This is a highly unusual situation, since most "biomarkers" of toxic chemical exposure are more nonspecific and much less indicative of the particular chemical or stressor responsible. For example, many different chemicals could be responsible for tumors in fish, so that field observations of these diseases cannot easily be traced to a particular toxicant (38). In this way, TBT is unique. The typical uncertainty regarding lab versus field effects, and the customary lack of proof of cause and effect is replaced by a considerable amount of scientific certainty.

Restrictions on TBT

France passed restrictions on the use of the chemical in 1982. It was banned from use on vessels under 25 m (82 ft) long. The rationale for the size provision was that large commercial vessels spend more time at sea, while pleasure craft are more frequently in harbors, where impact from TBT would be greater. Since the ban, French scientists observed a decrease in the abnormalities in the oysters (39). Great Britain imposed restrictions on TBT paints in 1986.

When in 1985, the U.S. Navy proposed to paint the entire fleet with TBT, and stated that an environmental impact statement would not be necessary, U.S. marine scientists were energized to stimulate activity in this country to restrict the use of TBT. Dr. Robert Huggett of the Virginia Institute of Marine Sciences (currently head of EPA's Office of Research and Development), who had been aware of the research and policy actions in Europe, was able to interest the Virginia State Legislature and Congressional delegation in the issue because Virginia is home to much of the naval fleet, and there is concern over the environment in Chesapeake Bay. The issue was brought to more of the public by a series of articles in the Newport News, Virginia newspaper, the *Daily Press/Times-Herald*. These articles came at a time of increasing concern about pollution in the Chesapeake Bay in general. TBT seemed to be yet another insult to an already stressed ecosystem, and state officials responded quickly. Hearings were held before the Virginia Water Control Board and the Board of Agriculture and Consumer Affairs. Environmental groups such as the Environmental Defense Fund and the Chesapeake Bay Foundation participated, and urged the establishment of water-quality standards for TBT. The Virginia congressional delegation took federal leadership on the issue, and Senator Paul Trible attached a rider to the military appropriations bill in December 1985 that prohibited the Navy from using TBT until the EPA had completed a review of the chemical (40).

EPA's "Special Review"

In January 1986, the EPA began a "special review" of the chemical under the Federal Insecticide Fungicide and Rodenticide Act (FIFRA), under which TBT is regulated. FIFRA requires that EPA weigh the risks against the benefits of a pesticide to determine whether it causes unacceptable adverse effects on humans or the environment. The special review process is lengthy, and many are concerned that the burden of proof during a special review, unlike when a company is trying to get approval of a new product, is again on those who would remove or restrict a pesticide rather than on those who would prove it environmentally safe. Furthermore, when the "special review" process is triggered by some aspect of the chemical's effects, the special review must focus only on those effects and not on any others. For example, in the case of the insecticide carbaryl, a special review was triggered by the finding of teratologic effects in beagle dogs (41). Therefore, the special review was only on mammalian teratogenicity and could not consider effects on marine organisms (including teratogenicity). Thus, in weighing risk versus benefits, only a subset of risks are considered in the special review process.

Congressional and State Activity

In the meantime, Representative Herbert Bateman, whose district included Newport News, scheduled the first congressional hearing on the issue, before a subcommittee of the Merchant Marine and Fisheries Committee. A total of three congressional hearings were held in 1986 and 1987. In all three, independent scientists and environmentalists supported legislation to restrict the use of TBT, while spokesmen for EPA, the navy, and manufacturers of TBT and TBT paints opposed such restrictions. The paint and chemical companies claimed that restrictions would cause economic hardship and urged that further studies be done. The navy contended that TBT paints were safe and that banning them would be expensive. They estimated that painting the naval fleet with TBT would save $100 million a year as a result of more efficient fuel use by vessels without fouling organisms and by not having to repaint the ships as often. The navy further claimed that the use of TBT enhanced national security by allowing its ships to travel faster. The EPA also opposed restrictive legislation because it would circumvent EPA's own regulatory process under FIFRA, in which decisions are made after carefully weighing risks and benefits. EPA wanted to let the FIFRA process work, since FIFRA is a framework for regulating any pesticide. EPA opposed the concept of individual laws for every chemical that could pose a problem in the environment (42).

During 1987, several states, impatient with the EPA process, passed their own restrictive legislation. Virginia and Maryland (home to naval operations and to oysters in Chesapeake Bay), followed by New York, California, and

Oregon, banned the use of TBT on boats smaller than 82 ft in length. The size provision follows the same rationale used by France. The state legislation further required large vessels that are still permitted to use TBT to use a formulation with a very low release rate. By this time, limits or prohibitions had been enacted in a number of other countries, including France, England, Canada, Japan, Switzerland, and West Germany. In late 1987, the U.S. Congress, also frustrated with the length of time needed by the EPA to complete its regulatory review process, and led by members from Maryland and Virginia, also passed similar federal legislation. In addition to the 82-ft size restriction, a release rate maximum of 5 μg cm^{-2} day^{-1} was required.

When EPA's "Preliminary Determination" on TBT came out, 4 months after the last congressional hearing, it recommended similar provisions, but emphasized that studies were still in progress and there was no need to implement restrictions until the studies were completed, which could take several more years. Its risk-benefit analysis showed that increasing cost would result from the restrictions on the chemical, but that the financial costs would not be as severe as those predicted by industry (43). Due to the passage of the federal law, the restrictions were implemented much sooner than would have occurred had the FIFRA process continued.

Uniqueness of TBT

This is the major success story among these four case studies, but it is sufficiently unusual that it may not offer a precedent for other substances. TBT is the first chemical to be regulated solely on the basis of its environmental—rather than potential human health—impacts. But TBT has been described as "the most toxic compound ever deliberately introduced by societies into natural waters" (44), so it is unique in its toxicity at extremely low concentrations. Since it produced unusual anatomic malformations (in an organism of economic importance), it could be readily identified in field studies as the cause of these effects. The events regarding the restrictions on TBT took place with relatively little fanfare, public awareness, or press coverage. The movement to restrict it was spearheaded by marine scientists rather than environmental groups, although they did support the restrictions. The conclusion in this case is unique because of the unique nature of the chemical. Since the scientific uncertainty was very low, quick action could be taken to regulate it.

Second, reasonable alternatives to TBT are available. Although they are more costly because more frequent painting is required, copper copolymer paints are effective antifoulants. Although copper is also not without deleterious effects in the aquatic environment, the magnitude of its toxicity is small compared with TBT. Since the regulatory actions were relatively easy to implement and did not cause major societal dislocations or severe economic consequences, legislative activity occurred a few short years after scientists became aware of the problem. Barely 3 years elapsed between awareness in the United States of a possible problem and the enactment of restrictive legislation. Since copper is also toxic to marine organisms, research into the development of nontoxic anti-fouling alternatives is under way.

Estuarine Eutrophication

While nutrients are necessary for all life, excessive amounts of nutrients enter-ing aquatic ecosystems can cause negative changes in the ecosystem. With increases in the human population, in farms that use fertilizers, in cities with paved streets and parking lots, in suburbs with fertilized lawns, there are more land-based nutrients generated and at the same time there are fewer forested areas and natural buffers to absorb them. Therefore, greater amounts run off into aquatic systems, spurring the growth of algae. These microscopic plants can grow in such great densities that they can block the penetration of light needed by the submerged aquatic vegetation. The submerged vegetation, which pro-vides both food and habitat for many organisms, declines. The algae photosyn-thesize during the day, but use up much dissolved oxygen at night, leading to lowered oxygen levels. Furthermore, when the algae die, they sink to the bot-tom and decompose, a process that uses up much of the dissolved oxygen in the water, which in turn causes stress and mortality for the bottom-dwelling species. These are all signs of eutrophication. Low levels of dissolved oxygen can be found in large areas of the deeper portions of many of our estuaries, especially in the summer. Symptoms of coastal eutrophication include algal blooms, nutrient enrichment, low dissolved oxygen, changes in the food chain, and changes in diversity of organisms. However, there remain considerable sci-entific uncertainties about the relationship between nutrient reductions and improved oxygen concentrations in a given estuary.

The major nutrients responsible for eutrophication are phosphorus and nitrogen, found in human and animal wastes, fertilizers, and plant material. Both can enter the aquatic environment through point sources (e.g., pipes from sewage treatment plants) and nonpoint sources (runoff from farms and city streets, seepage from septic systems). Nitrogen can also enter from the atmo-sphere. Eutrophication has been studied to a greater extent in freshwater sys-tems, in which it was clear that phosphate was the major cause. Phosphorus control, through bans on phosphate detergents and upgrading sewage treatment plants in the 1970s and 1980s was fairly effective in improving the environ-mental condition in many lakes and rivers. When symptoms of overenrichment appeared in many of the nation's estuaries—algal blooms, oxygen depletion, fish kills, and the decline of submerged aquatic vegetation—there was a ten-dency to assume that again phosphorus was the major cause. It was assumed that inputs from sewage treatment plants, particularly phosphates, were responsible.

The Role of Nitrogen from Agricultural Nonpoint Sources

As research in estuaries progressed, evidence accumulated that nitrogen, rather than phosphorus, was the limiting nutrient. Additional evidence in the Chesapeake Bay watershed implied that much of the nitrogen input came not from sewage treatment plants but from agricultural runoff, due to the extensive use of fertilizers and manure. Scientific workshops in the early 1980s concluded that these nonpoint inputs were the most significant source of nitrogen and the most important cause of eutrophication. However, management was refractory to this evidence and continued to focus on phosphorus control from sewage treatment plants, since point sources could more easily be regulated and the technology for removing phosphorus was cost-effective (45). Controlling nitrogen discharges from sewage treatment plants is more technologically difficult and more expensive than controlling phosphorus. There are a variety of biologic and physicochemical processes for removing nitrogen. It should be noted that such processes, which will reduce nutrient emissions, will lead to creation of greater amounts of nitrogen-rich sludge, which will have to be disposed of.

The Chesapeake Bay Program gave compelling evidence of the primary role of nonpoint nitrogen inputs and recommended that controls be implemented. Nitrogen is more difficult to control than phosphorus in runoff. It is water soluble and more prone to leaching into groundwater. Nonpoint sources were estimated to be responsible for more than half of the nitrogen inputs to the bay. Farms with chickens and cows produce large amounts of manure, the largest source of agricultural inputs. In the past, manure had been used for fertilizer on agricultural fields, but now, because of the abundance of chemical fertilizers, there is more manure than can be spread on land for fertilizer. Therefore, both the manure and chemical fertilizers run off into aquatic systems. In addition, atmospheric deposition can occur. Nitrate, from atmospheric emissions of NO and NO_2, can be deposited in estuaries, where it can contribute to eutrophication rather than acidification (46).

When they finally acknowledged the role of nonpoint sources, Chesapeake Bay managers focused on ''best management practices'' (BMPs) for farmlands, which involve reducing soil loss (45). Since phosphorus adheres to soil particles, these practices do reduce phosphorus inputs; however, nitrogen is more soluble and is more likely to enter the groundwater and, from there, the estuary. Yet, the management community resisted this information until the studies became unquestionable. To reduce nitrogen in runoff, BMPs must be supplemented by nutrient management plans. Such plans determine the optimum amount of fertilizer to be used based on factors such as soil conditions and crop rotation, and maximize the benefits of fertilizers while minimizing impacts on water quality. The 1987 Chesapeake Bay agreement calls for a 40% reduction in nitrogen and phosphorus inputs by the year 2000.

Nonpoint Source Provisions in the Clean Water Act

The 1987 reauthorization of the Clean Water Act included provisions for nonpoint source pollution, relying on voluntary controls by agriculture. The time lag between the scientific findings and the management action was due to politics and economics. Unlike TBT, there is no quick and easy solution to nutrient enrichment. The problem is large, and actions needed to reduce nonpoint source inputs are complicated and difficult, involving sensitive issues of land use, which have never been issues for the federal government, but rather for local authorities. Whether voluntary nutrient management will be effective or whether mandatory policies will be needed remains to be seen.

Nitrogen from Other Nonpoint Sources

Estuarine eutrophication in other areas, including parts of New England, Florida, and the West Coast, is due primarily to nitrogen leached through the groundwater from septic systems (47, 48). In Hillsborough Bay, FL, the construction of a wastewater treatment plant led to rapid decreases of phytoplankton and improvements in water quality (49). In rural areas such as Cape Cod, however, the construction of major wastewater treatment facilities is very expensive and could encourage further development; therefore advanced on-site wastewater treatment devices are viewed as more desirable. A number of technologies are available to remove nitrogen and prevent it from reaching the groundwater (50). Since such systems would have to be installed by individual homeowners, issues of who should pay and whether installation should be voluntary versus mandatory have arisen and are being debated.

Estuarine eutrophication is another case of delaying action until the scientific data are unquestionable; delaying because of the political and economic difficulties involved in implementing regulations or other changes, particularly since it entails land use and actions by individual landowners. The 1990 amendments to the Coastal Zone Management Act (CZARA) deal with nonpoint pollution in the coastal zone and are expected to have positive impacts.

Alternative Means to Alleviate Eutrophication

Some scientists and policy makers are discussing alternative means to alleviate eutrophication. It is possible that once the nutrients have entered the water, they can be channeled into productive directions. One possibility is seaweed farms that would utilize the nutrients to grow commercially important products. It is clear that macroalgae can take up dissolved nutrients from the water, and they have been used for this purpose in some experimental systems (51, 52). To give them a competitive advantage over the microalgae, their biomass should be near the surface, shading the water column and limiting growth of the microalgae. Another suggestion is to enhance the standing crop of grazers who would consume the abundant phytoplankton. There is growing evidence that bivalve suspension feeders, such as oysters, can exert a top-down control on phytoplankton. It has been suggested that the eutrophication in Chesapeake Bay may be partly due to the striking decreases in the populations of oysters (53, 54). Therefore, bivalve aquaculture might promote increases in water quality in eutrophic estuaries.

The Brown Tide

It is ironic that the acceptance, by the scientific and management communities, of nitrogen as the primary cause of estuarine eutrophication, has led, in at least one case, to anomalous management recommendations when nitrogen had been shown to be unrelated to the particular problem. In the late 1980s an unusual algal bloom appeared in various bays in Long Island. This bloom, dubbed the "brown tide" was particularly severe in the Peconic Estuary system, where it attained 1 million cells per millimeter, eradicated important bay scallop populations, and decimated eel grass beds. The eel grass beds, which were a critical nursery and spawning area for shellfish and finfish, were affected, probably due to reduced light penetration caused by the dense bloom. The bloom was unpredictable in terms of its onset, duration, and cessation. The organism responsible for the bloom is a previously unknown, very tiny form, *Aureococcus anophagefferens*. In response to this bloom, a task force, the Brown Tide Comprehensive Assessment and Management Program (BTCAMP) was set up, spearheaded by Suffolk County, to investigate the causes and effects of the bloom, investigate conventional water quality problems in local areas, and recommend management actions. The investigation focused on nitrogen from sewage treatment plants as the most likely cause of the bloom. The sewage treatment plant in the town of Riverhead was the most significant one in terms of nitrogen loading to the estuary. Other major nitrogen sources in the area, such as duck farms, had decreased in the years prior to the appearance of the brown tide, such that total nitrogen loadings had decreased during the previous decade. However, sediment flux contribution of nitrogen was greater than all other sources of this nutrient. After considerable study, the causal factors for the brown tide remained unknown. It was found that conventional nutrient inputs such as nitrogen and phosphorus were clearly not the triggers for the bloom. Research implicated micronutrients such as organic nutrients, chelators such as citric acid, and trace metals such as iron and selenium. It was found that the brown tide organism produced acrylic acid and dimethyl sulfide, which might be toxic to zooplankton that would otherwise graze on and limit the organism. Certain meteorologic patterns, including wind and rainfall, also appeared to be associated with the bloom. There was some evidence that viruses might play a role in the subsidence of the bloom. These findings were tentative and uncertain.

The report produced by BTCAMP (55) had recommendations for further research on the chemicals and meteorologic patterns implicated in triggering the bloom, the sediment flux of nitrogen, the role of viruses in the decline of the brown tide, and the relationship between zooplankton grazing and acrylic and dimethyl sulfide. It also recommended future research on the mechanisms of the brown tide's adverse effects on shellfish: was it poor nutrition or toxic to shellfish, or did it inhibit shellfish reproduction through other mechanisms?

The policy recommendations of the report, however, seemed to be relatively unrelated to the scientific findings; they focused on reducing further nitrogen inputs. Specifically, they recommended no net increase in nitrogen inputs to the surface waters from the sewage treatment plants, advocated upzoning of developable land in the groundwater contributing area, other land-use management techniques such as open space, setbacks from the water, and natural landscaping to minimize fertilizer use. They further recommended storm water runoff remediation, additional pumpout stations and holding tanks for boats. These recommendations are generally applicable to reducing estuarine eutrophication from point and nonpoint sources and will probably contribute to better water quality in the area. However, regarding the brown tide organism, they ignore the major scientific finding of the study, namely that nitrogen was *not* the cause of this particular bloom. This surprising scientific information could apparently not be integrated into the policy recommendations. Since there was uncertainty as to the real cause of the bloom, the policy recommendations focused on controlling the "usual suspect," although it had already been shown to be "innocent" in this particular case. Given the uncertainty as to the true cause of the algal bloom and the inability to control weather patterns, it is likely that the brown tide will return some season in the future despite the management actions that will be taken. (It in fact returned in the summer of 1995.)

Discussion and Conclusions

The Limits of Science

Monitoring, even by the best-designed programs, is unlikely to detect subtle changes before considerable damage has been done to an ecosystem. Furthermore, in most cases the agent(s) responsible for damage is (are) difficult to discern (TBT being a notable exception to the rule). Given these limitations of science, as well as the conflict between environmental protection and economics, it is likely that increasing pressure will be put on the environment and continued degradation will result. One way to get around this problem is to intensify monitoring programs and obtain more long-term data. Challenges for the scientific community are to identify early indicators of stress. There is great need to find early warning indicators of subtle rather than gross effects (56), to detect and quantify long-term changes at the ecosystem level by means of holistic environmental assessments, and to develop probability-based ecologic risk assessment methodologies to identify which threats are greatest and which control actions would be the most effective.

However, during the time that these studies are being performed, the environment is likely to continue to be impacted. A more protective approach in the case of scientific uncertainty would be to use a "precautionary concept" even in the absence of clear proof, as proposed by Dethlefsen (1). According to this argument, scientific uncertainty should be followed by prudent measures to prevent degradation rather than continued degradation. Peterman (57) similarly suggested that the burden of proof be placed on industry or developers to demonstrate that their actions will not have deleterious environmental effects. Environmental degradation would then be prevented rather than dealt with only after it has reached unacceptable proportions. Acceptance of this principle would contribute to sustainability of natural ecosystems and our ability to provide a quality environment for future generations. However, there are significant economic costs associated with protective measures such as controlling point and nonpoint sources of pollution, restricting development, and others. The precautionary principle has been criticized on the grounds that it is unscientific in that it accepts suspicion of effects rather than the scientific evidence of effects as sufficient to trigger action (58). Given the economic impacts, there will be great resistance to changing the burden of proof, and it is impractical. Nevertheless, when human health is involved there may be enough public pressure for the precautionary principle to be used. Such is the case with bans on smoking in public places prior to scientific proof that second-hand smoke is harmful to one's health.

The growing use of risk assessments for human health and ecologic effects may be able to resolve the issue of burden of proof. These processes for determining the magnitude, range, and likelihood of adverse effects include explicit considerations of uncertainty and produce probability-based estimates of risk. Thus, best judgment and weight of the evidence are appropriate in recommending policy decisions.

In environmental issues, the scientific uncertainties may lie in the area of how to prevent or mitigate effects. However, even with increased scientific understanding, predictive ability, and the ability to devise effective management action to prevent or control the adverse effects, there will remain divisions within the scientific community and a gap in the ability of government to manage resources sustainably.

One issue and potential source of conflict is the ability to distinguish normal from abnormal in complex and naturally diverse ecosystems. Distinguishing anthropogenic stress from the normal variability in systems such as estuaries can be quite difficult. In addition, there is often a communications gap between the scientific community and the general public and policy makers in their understanding of a problem. Public perception of risk may disagree with scientific assessments. Furthermore, the public is critical of science because it has often been unable to predict the adverse consequences of various activities. This may be a reason for the limited influence of the scientific community in the public-policy arena.

Roles of Science and Values in Policy Making

Scientists must be more active in communicating effectively the results of their analyses to policy makers, managers, and the public. (As advocates on both sides of environmental issues are well aware, the timely communication of information to policy makers may be more important than the quality of that information.) However, it is naive and unrealistic to expect that good science alone will drive the development of public policy. While scientific consensus can influence the debate, policy makers are influenced by many sources of information and will base their decisions on economic, social, political, and psychological factors in addition to science. Furthermore, policy makers have few ways to resolve opposing conclusions which have been arrived at by their various sources of information.

Differing values will inevitably lead various groups to draw different conclusions from the same body of scientific information. The central and most divisive issue is *how do we distinguish acceptable and unacceptable change?* This distinction is the crux of the issue and will be made by society as a whole, not by scientists alone, although scientists should certainly feel free to participate in the debate. The existence of scientific uncertainty will focus the debate initially on whether the magnitude of damage and the effectiveness of control efforts are sufficiently understood and then on the issue of how much damage must occur in order to justify regulatory or legislative remedies; that is, how much degradation is "acceptable"?

The challenges facing the scientific community are to promote the incorporation of scientific principles and information into environmental policy and management, to familiarize themselves with the statutes and regulations under which the policy is made, and to ensure public support for the long-term research and assessments upon which the management should be based. The central question—how much environmental damage is acceptable?—is a question of values, not science, and is one that will be long debated in policy circles throughout the world.

Acknowledgments

I appreciate the reviews and comments on the chapter by Marjorie Holland, Maurice Zeeman, Leonard Cole, Glenn Suter, and Robert DeLuca.

References

1. Dethlefsen V. Marine pollution mismanagement: towards the precautionary concept. Mar Poll Bull 1986;17:54–57.
2. Guarino CF, Nelson MD, Townsend SA, Wilson TE, Ballotti EF. Land and sea solids management alternatives in Philadelphia. J Wat Pollut Control Fed 1975;47:2551–2564.
3. GESAMP (IMCO/FAO/UNESCO/WMO/WHO/IAEA/UN/UNEP Joint Group of Experts on the Scientific Aspects of Marine Pollution). The Health of the Oceans. UNEP Regional Seas Reports and Studies No. 16, 1982.
4. McGough J. Ocean dumping of sewage sludge: a perception from New York City. In: Ocean waste management: policy and strategies. International Ocean Disposal Symposium Series, abstract: 1983:54–55.
5. National Oceanic and Atmospheric Administration, Ocean Dumping in the New York Bight. NOAA Tech. Report ERL 321-MESA 2. Boulder, CO, 1975.
6. Grunseich GS, Duedall IW. The decomposition of sewage sludge in seawater. Water Res 1978;12:535–545.
7. Pearce J. Benthic fauna. MESA Atlas Monograph #14. Albany, NY: New York Sea Grant Inst., 1980.
8. Ziskowski J, Murchelano R. Fin erosion in winter flounder. Mar Poll Bull 1975;6:26–29.
9. Young JS, Pearce JB. Shell disease in crabs and lobsters from New York Bight. Mar Poll Bull 1975;6:101–105.
10. Mueller JA, Jeris JS, Anderson AR, and Hughes CF. Contaminant inputs to the New York Bight. NOAA Technical Memorandum ERL-MESA-6. Boulder, CO, 1976.
11. National Advisory Committee on Oceans and Atmosphere. The role of the ocean in a waste management strategy. Special Report to the President and Congress. Washington DC, 1981.
12. Steimle FW. Sewage sludge disposal and winter flounder, red hake, and American lobster feeding in the New York Bight. Mar Environ Res 1994;37:233–256.
13. National Oceanic and Atmospheric Administration. Evaluation of proposed sewage sludge dumpsite areas in the New York Bight. NOAA Technical Memorandum ERL MESA-11. Boulder CO, 1976.
14. Lewis RE, Riddle AM. Sea disposal: modelling studies of waste field dilution. Mar Pollut Bull 1989;20:124–129.
15. O'Connor TP, Okuba A, Champ M, Park PK. Projected consequences of dumping sewage sludge at a deep ocean site near New York Bight. Can J Fish Aquat Sci 1983;40 Suppl 2: 228–241.
16. Costello MJ, Read P. Toxicity of sewage sludge to marine organisms: a review. Mar Environ Res 1994;37:23–46.
17. Csanady G, Flierl G, Karl D, Kester D, Connor TO, Ortner P, Philpot W. Deepwater dumpsite 106. In: Goldberg ED, ed. Assimilative capacity of U.S. coastal waters for pollutants, Boulder, CO: U.S. Dept of Commerce, 1979:123–147.
18. Longwell AC. Cytological examination of fish eggs collected at and near 106-mile site. In: Assessment report on the effects of waste dumping in 106-mile ocean waste disposal site. Boulder, CO: NOAA, Dept. of Commerce, 1981:257–276.
19. Pathobiology Division. Histopathology of organisms sampled at and near 106 mile site. In: Assessment report on the effects of waste dumping in 106-mile ocean waste disposal site. Boulder, CO: NOAA, Dept. of Commerce, 1981:277–293.
20. National Academy of Sciences. Ocean disposal systems for sewage sludge and effluent. Marine Board Committee on Ocean Waste Transportation. Washington, DC: National Academy Press, 1984.
21. U.S. Congress, Office of Technology Assessment. 1987. Wastes in marine environments. OTA-O-334. Washington, DC, 1987.
22. Van Dover C, Grassle JF, Fry B, Garritt R, Starczak V. Stable isotope evidence for entry of sewage-derived organic material into a deep-sea food web. Nature 1992;360:153–156.
23. Chang S. Analysis of fishery resources: potential risk from sewage sludge dumping at the deepwater dumpsite off New Jersey. Fish Bull 1993;91:594–610.

24. U.S. Congress, Office of Technology Assessment. Acid rain and transported air pollutants: implications for public policy. OTA-O-204. Washington, DC, 1984.

25. National Acid Precipitation Assessment Program. Interim assessment: the causes and effects of acidic deposition. Vol. 1. Executive Summary, 1987.

26. Schindler DW. A view of NAPAP from north of the border. Ecol Appl 1982;2:124–130.

27. Loucks OL. Forest response in NAPAP: Potentially successful linkage of policy and science. Ecol Appl 1992;2:117–123.

28. Cowling EB. The performance and legacy of NAPAP. Ecol Appl 1992;2:111–116.

29. National Acid Precipitation Assessment Program. Report to Congress, 1992.

30. Smith BS. Male characteristics on female mud snails caused by antifouling bottom paints. J Appl Toxicol 1980;1:22–25.

31. Thain JE. The acute toxicity of bis (tributyltin) oxide to the adults and larvae of some marine organisms. Internat. Council for the Explor. of the Seas Coop. Res. Rept. CM 1983/ E:13, 1983.

32. Waldock MJ, Thain J. Shell thickening in *Crassostrea gigas:* organotin antifouling or sediment induced? Mar Pollut Bull 1983;14:411–415.

33. Alzieu C, Heral M. Ecotoxicological effects of organotin compounds on oyster culture. In: Persoone G, Jaspers E, Claus C, eds. Ecotoxicological testing for the marine environment. Vol. 1. Belgium: The State University of Ghent, 1984:187–196.

34. Bryan GW, Gibbs PE, Hummerstone LG, Burt GR. The decline of the gastropod, *Nucella lapillus* around south-west England: evidence for the effect of tributyltin from antifouling paints. J Mar Biol Assoc UK 1986;66:611–640.

35. Maguire RJ, Tkacz R. Degradation of tri-*n*-butyltin species in water and sediment from Toronto Harbor (Canada). J Agr Food Chem 1985;33:947–953.

36. Salazar MH. Environmental significance and interpretation of organotin bioassays. In: Organotin symposium. Proc. Oceans 86 Conf. Sept. 23–25, 1986. Washington, DC, 1986: 1240–1245.

37. Cleary JJ, Stebbing ARD. Organotin in the surface microlayer and subsurface waters of Southwest England. Mar Pollut Bull 1987;18:238–246.

38. Hinton DE, Lauren DJ. Liver structural alterations accompanying chronic toxicity in fishes: potential biomarkers of exposure. In: McCarthy JF, Shugart LR, eds. Biomarkers of environmental contamination. Boca Raton, FL: Lewis, 1990:17–57.

39. Alzieu C. TBT detrimental effects on oyster culture in France—evolution since antifouling paint regulation. IEEE Oceans '86 Conference Proceedings, 1986:1130–1134.

40. Weis JS, Cole L. Tributyltin and public policy. Environ Impact Assess Rev 1989;9:33–47.

41. U.S. Environmental Protection Agency. Aspects of the Pesticidal uses of Carbaryl (Sevin) on Man and the Environment. Carbaryl Decision Document, 1975.

42. U.S. Congress. Antifouling paints. Hearing before the Subcommittee on Fisheries and Wildlife Conservation and the Environment and the Subcommittee on Oceanography of the Committee on Merchant Marine and Fisheries. House of Representatives, July 8, 1987.

43. U.S. EPA. Office of Pesticides and Toxic Substances. Preliminary Determination to cancel certain registrations of tributyltin products used as antifoulants (mimeographed), 1987.

44. Goldberg, ED. 1986. TBT: An environmental dilemma. Environment 22:17–20, 42–44.

45. Malone TC, Boynton W, Horton T, Stevenson C. Nutrient loadings to surface waters: Chesapeake Bay case study. In: Uman M. ed. Keeping pace with science and engineering. Washington, DC: National Academy Press, 1993:8–38.

46. Paerl HW. Enhancement of marine primary production by nitrogen-enriched acid rain. Nature 1985;316:747–749.

47. Valiela I, Costa J. Eutrophication of Buttermilk Bay, a Cape Cod coastal embayment: concentration of nutrients and watershed nutrient budgets. Environ Manage 1988;12:539–551.

48. Valiela I, Costa J, Foreman K, Teal J, Howes B, Aubrey D. Transport of groundwater-borne nutrients from watersheds and their effects on coastal waters. Biogeochemistry 1990;10: 177–197.

49. Johansson JO, Treat SS, Clark PA. Long term trends of nitrogen loading, water quality, and biological indicators in Hillsborough Bay, FL. Proceedings Tampa Bay Area Scientific Inform. Symp. 2. Feb. 27–Mar. 1 1991. Tampa, FL, 1992.

50. Waquoit Bay National Estuarine Research Reserve. Nitrogen removal on-site wastewater treatment systems: technologies and regulatory strategies. Position paper from a conference Feb. 27–28, Waquoit, MA, 1992.

51. Ryther JH, DeBoer JA, Lapointe BE. Cultivation of seaweeds for hydrocolloids, waste treatment and biomass for energy conversion. Proc Int Seaweed Symp 1979;9:1–16.

52. Harlin MM, Thorne-Miller B, Thursby GB. Ammonium uptake by *Gracilaria* sp (Florideophyceae) and *Ulva lactuca* (Chlorophyceae) in closed system fish culture. Proc Int Seaweed Symp 1978;9:285–292.

53. Newell RIE. Ecological changes in Chesapeake Bay: are they the result of overharvesting the American oyster *Crassostrea virginica?* In: Understanding the estuary: advances in Chesapeake Bay research. CRC Publ 1988;129:536–546.

54. Newell RIE. Top-down control on phytoplankton populations. Presented at National Research Council Marine Board Workshop on Coastal Eutrophication. Stony Brook, NY, 1993.

55. Suffolk County Department of Health Services. Brown Tide Comprehensive Assessment and Management Program, Summary, 1992.

56. Bewers JM, Wells PG. Challenges for improved marine environmental protection. Mar Poll Bull 1992;25:112–117.

57. Peterman RM. Statistical power can improve fisheries research and management. Can J Fish Aquat Sci 1990;47:2–15.

58. Gray JS, Calamari D, Duce R, Portmann JE, Wells PG, Windom HL. Scientifically based strategies for marine environmental protection and management. Mar Poll Bull 1991;22: 432–440.

6 Uncertainties Associated with Extrapolating from Toxicologic Responses in Laboratory Systems to the Responses of Natural Systems

John Cairns, Jr. and Eric P. Smith

Introduction

The purpose of risk assessment is to assess or measure the stress on an ecosystem that can be attributed to a typically presumed harmful source. The process of risk assessment involves discovering and investigating the relationship between the stressor and the ecosystem and developing models of that relationship. Because of the complex nature of ecosystems, risk assessments involve assumptions and choices.

Approaches to ecologic risk assessment may be viewed as either bottom-up or top-down. Top-down refers to attempts to describe the system being studied and how a stressor affects that system, typically using a computer simulation model or a measurable surrogate of the system chosen to indicate the system's "health." All approaches to risk assessment involve uncertainty, and the level of uncertainty can limit the ability to make sound decisions.

In this chapter, uncertainty in ecotoxicologic investigations is discussed, along with sources and implications of uncertainty and some common methods for dealing with it. The importance of assessing the validity or usefulness of a model is also presented. Finally, some general approaches for reducing uncertainty are described.

Sources of Uncertainty

One view of uncertainty is in terms of effects. From a statistical perspective, the appropriate effects to consider are variance and bias. Variance of a response is a measure of how different the response is expected to be in, for example, a repeated measurement or at a different site. Variance in risk is important because it may be used to provide a range of expected risk. Bias refers to how different the response is from its expectation. Bias is especially important for long-term prediction of a model of risk since small errors may lead to a large bias over time.

There are many ways to view uncertainty in risk assessment. Another view is based on the sources of uncertainty (1). Thus, uncertainty is related to any factor that causes results to vary from their true values. A list of some of the sources common to ecologic risk assessment and their importance is given in Table 6.1.

Rather large uncertainties are often associated with general lack of knowledge of the system under study and the scale of measurement. Although the problem of global warming is generally accepted, uncertainty is great as to the consequences. Little information exists on the amount of increase in temperature (lack of data), and there is disagreement on the consequences of global warming due to the complexity of the response. Hence, the disparity is great between predicted effects and timing of effects. Forman and Godron (2) assert that measurements of ecologic structure and function must be taken at a scale appropriate to the process being observed. Many global and smaller-scale changes that occur in the environment will be cast in the framework of landscape ecology (3). What is not resolved is how ecotoxicologists will adjust to this expanding synthesis. If no new methods are developed appropriate to larger spatial and temporal scales, how will present ecotoxicologic methods be used in this new landscape context?

Individuals aspiring to be *ecotoxicologists* should be professionals who use ecologic parameters to assess toxicity or stress. This would require merging the present fields of ecology and environmental toxicology. Most important, the attributes or end points used to determine response thresholds should have some ecologic significance at higher levels of biologic organization than species or populations. Landscape ecology focuses on the following attributes: (1) landscape structure, or the spatial arrangement of ecosystems within landscapes; (2) landscape function, the interaction among the component ecosystems as determined by the flow of energy, materials such as nutrients, and organisms; and (3) any alterations of these attributes through stresses caused by human society as well as natural stresses (2, 4).

Persuasive evidence shows that toxicants are either emitted or spread soon after release over immense areas. Kern (5) speculates that the haze in the Arctic results from industrial smog spreading from Europe. Hirao and Patterson (6) note that damage from car exhausts originating in the coastal California cities was detectable in the Sierra Nevada Mountains located in the eastern part of the state. Spencer and Cliath (7) demonstrated that over 90% of a number of toxic pesticides had volatilized into the atmosphere within a week, presumably able then to affect areas quite distant from their area of application. As an example of this, Kurtz and Atlas (8) note that volatilization and atmospheric transport of hexachlorocyclohexanes have resulted in the contamination of surface water throughout the Pacific Ocean. Hunsaker et al (9) and Suter (10) have commented that what appear to be small-scale environmental impacts can

Table 6.1 Some Uncertainties and Their Importance in Ecological Risk Assessment

SOURCE OF UNCERTAINTY	IMPORTANCE	MAGNITUDE OF EFFECT
Poor knowledge of system	Without any knowledge of the system, it is not possible to build a useful model	Many orders of magnitude
Extreme variation, incorrect scales	Great variation in weather, for example, may cause a large change in the importance of the stressor. Modeling large-scale phenomenon using a small-scale model may lead to great uncertainty	
Wrong model, end points, exposure routes	Measuring the wrong end point may lead to missed effects. Lack of knowledge of the exposure or model may lead to large errors	
Surprises and episodic events	Unexpected effects may occur caused either by important gaps in knowledge or by random effects. Despite low probability of occurrence, effects can have great consequences	
Data collection practices	Errors in data collection and entry may lead to mistakes in interpreting statistical analyses	Order(s) of magnitude
Design of laboratory experiments and quality control	Adherence to laboratory standards is necessary to avoid errors in measurement	

Table 6.1, continued

SOURCE OF UNCERTAINTY	IMPORTANCE	MAGNITUDE OF EFFECT
Variability in mesocosms or other ecosystem surrogates	Mesocosm studies have higher variability than laboratory studies and need to be carefully designed	
Extraneous variables	Physical conditions may have a strong effect on laboratory or field results	
Mistakes in statistical analysis	Outliers, wrong statistical model	
Interactions	Uncertainty may be introduced by failing to account for interactions among species or combined effects of chemicals or stressors	
Parameterization of computer model	Parameter estimates are taken from the literature, not from a fit to actual observations	
Mistakes in computer code of simulation models	Errors in code may lead to gross prediction errors	
Extrapolations across one species to another species to community laboratory to field spatial (local to regional)	Using a model developed for a simple end point may lead to errors when applied to estimate a more complex end point	
Spatial and temporal variability	It is difficult to predict with precision either over long time periods or over space	

Table 6.1, continued

SOURCE OF UNCERTAINTY	IMPORTANCE	MAGNITUDE OF EFFECT
Variability in laboratory test conditions	Variation in test organisms or concentrations of chemicals, for example, may cause under- or overestimation of effects	Up to one order of magnitude
Minor mistakes in choice of statistical model	Including variables that are not necessary in the model may lead to increased variance; missing variables may add a bias	
Statistical design of manipulative studies (choice of stressor levels, randomization, number of experimental units, number of units per treatment)	Proper statistical design is important in laboratory and field studies especially when sample sizes are limiting. Estimates of quantities, such as no-observed-effect levels, may be greatly affected by sample size and other factors	Potentially of great importance
Design of field study	Haphazard design of field studies may lead to incorrect decisions regarding effects	

Source: Adapted from Cothern (39) and Smith and Shugart (40).

become landscape stresses when similar impacts are repeated over a large area. Woodman and Cowling (11) present a persuasive example of ecologic causality analysis by considering the causal association between air pollutants and deleterious effects upon forests.

Unquestionably, expanding the spatial and temporal scales has added much realism to the determination of toxicologic responses in natural systems! However, increasing each adds to the uncertainty (variability) because (1) increases in temporal and geographic scales inevitably do so and (2) end points and parameters must be included in the tests if they are to be environmentally realistic. These were not present in the simple single-species laboratory toxicity testing that dominated the field just a few decades ago and still represents the major information base in regulatory agencies. While the U.S. Environmental Protection Agency (USEPA) EMAP Program (12) will substantially expand the information base of the agency on natural systems using end points not customarily used by that agency, it will also add a number of uncertainties not commonly associated with the historic data base of USEPA. One of the major advantages of the simple single-species laboratory toxicity tests was their high degree of replicability. This was achieved, in large part, by reducing the environmental realism. Thus, laboratory studies have lower variance at the expense of bias.

As spatial and temporal scales expand, replicability will arguably be virtually impossible. Ecosystems are the result of a *sequence* of climatic and biologic events influenced strongly by the geology of the region. Each of these sequences is likely to be unique and, therefore, by definition, difficult or impossible to replicate. Large spatial scales involving natural systems inevitably involve a patchy distribution of components with the relationship of these components regularly undergoing change. A patch that has been a source of a particular species exporting colonists to the surrounding region may ecologically change and become a sink for the same species. Mosaics are extremely difficult to replicate even for a short time span and, arguably, impossible over a longer time span. One of the ''buzz words'' commonly used these days is *biologic integrity* or *ecologic integrity*. Regrettably, although few would argue that integrity is not an important attribute, robust measurements over modest spatial and temporal scales are exceedingly difficult, although the situation is improving considerably despite the seeming intractability of this problem. Presumably, an ecological landscape with a high degree of integrity would be self-maintaining—that is, resilient following perturbations and exhibiting the characteristic structural and functional attributes despite successional and other dynamic changes. If long-term sustainable use without abuse of natural systems is considered desirable, then systems with high biologic integrity should be desirable because, presumably, these would incur no management costs, or at least modest costs, compared to highly managed systems.

Uncertainties Associated with Top-down and Bottom-up Extrapolations

Cairns and Cherry (13) have designed a table contrasting covarying gradients in the design and application of toxicity tests, which includes a bottom-up appraisal of risk and a top-down appraisal of risk. The main categories are design complexity from low to high; design scale, low to high; and difference in complexity and scale between tests and application, high to low (see also Cairns and Smith [14]). As Cairns and Cherry (13) note, it helps to place all toxicity tests along covarying gradients of complexity, spatial scale, time scale, and similarity to assessments of natural systems. It is a *sine qua non* that, in natural systems, complexity, size, and time scale increase together. In contrast, in designed systems, complexity and size can be individually manipulated so small, complex systems and large, simple systems are possible. There is a contrasting or countervailing gradient of complexity in the methods by which toxicity test data are customarily applied to prediction of environmental outcome. As a general rule, toxicity tests low in environmental realism require more accessory information and the most complex methods for translation into ecologically relevant effects, while the most environmentally realistic or complex test methods are most simply translated into site-specific effects. A significant number of publications exist on this subject, but their appearance could hardly be called overwhelming as yet. Two publications giving additional information have been written by Norton and colleagues (15, 16). This discussion does not explore these two approaches in depth but points out that, in both toxicology and ecology (17), the bottom-up approach based on simple components has been most commonly employed.

Uncertainties Associated with Sustainable Growth

The term *sustainable development* has received public attention in the Bruntland Report in 1987 and other publications since then (18). Sustainable development has a wide variety of definitions often applied to specialized subjects such as sustainable agriculture and sustainable forestry. The Bruntland Report focused on human society at a global level as well as environmental quality at a planetary level. The intent was to manage human populations and the global environment so that human society on the planet can continue to persist, presumably with a quality of life equal to or greater than the present while simultaneously maintaining ecologic integrity. It is regrettable that the term *sustainable development* was used because the verb *develop* means to expand through growth and other related terminology. Used in this context, sustainable growth is an oxymoron because prolonged growth of any population, including humans, is impossible because eventually some limiting factor is reached and growth ceases, sometimes with a precipitous population decline.

Uncertainty and the Bottom-up Approach

It is difficult (and in some cases impossible) to measure directly how a stressor affects an ecosystem. Just as society cannot allow a nuclear power plant to discharge radioactive material in order to see what happens, neither can it wait 20 years to observe the effect of low levels of hazardous pesticides on duck populations. Thus, given the need to assess effects and to model the response of stressors, surrogate studies and models must be used. For example, a model of the toxicity of a chemical may be developed based on the chemical structure and activity of the chemical. Similarly, the toxicity of a chemical to one species may be used to estimate the effect the chemical has on the growth of the species or on the growth of other species. When models are used in this manner, uncertainty arises that is attributable to the need to extrapolate across end points, species, and even systems (19, 20). For example: Results may be based on small-scale studies and applied over a larger scale. For instance, since a toxicity test often spans a short time period (e.g., 7 days), effects of low dose may be misleading because a long-term low dose may give a different effect than a short-term one. An ecologic end point of interest may be replaced by a measurable end point. For instance, laboratory studies often involve a mortality or growth in a single species, while the end point of interest may be a measure of ecosystem function. Uncertainties may occur as a result of the genetic composition of laboratory species relative to field species. Uncertainties may occur as a result of laboratory conditions not reflecting field conditions or may relate to factors not considered in laboratory studies (e.g., hardness of the water). Additional discussion of this problem is covered in Smith and Cairns (19), Suter (1), Forbes and Forbes (20), and Forbes and Forbes (21). There are approaches to estimating the uncertainty due to extrapolation, and these are discussed below.

A potentially important source of uncertainty in single and multispecies tests is uncertainty associated with experimentation. The sources of uncertainty include those associated with quality control, stochasticity, choice of experimental design, or choice of statistical model.

An important aspect of any research study is quality control. If a good assessment is to be produced, it must be based on good information. Hence, attention must be paid to the quality of the information and, when collecting data, measures should be taken to ensure that the data are of high quality. Basic principles of quality assurance include control of laboratory and field operations, documentation of procedures and the reporting of results of sample collection and analytical activities (22). These principles are designed to reduce the chance of errors in laboratory and field procedures, to decrease variation due to researcher effects, and to aid in correcting errors when they occur.

Another important source of uncertainty is the design used to collect information. The design of a study involves questions about the amount of information collected (sample size) and the manner in which data are collected (design). For example, in estimating the effects of a single toxicant on an end point such as mortality or fecundity, it is common to estimate a no-observable-effect level. The study to estimate this quantity requires choices, including what doses to use and how many organisms to test.

Two measures of statistical uncertainty are power and variance. In designed experiments, power is an important measure of the strength of a statistical test. In testing statistical hypotheses, null and alternative hypotheses are stated. The null hypothesis refers to the hypothesis of no effect. In the dose-response study above, the null hypothesis is that the dose is not different from the control. The alternative hypothesis is that the dose is different. As indicated in Table 6.2, there are two errors associated with this process: type I (rejecting the null when it is true) and type II (not rejecting the null when it is false). Power is a measure of how likely it is that the null is rejected when it is false.

A number of authors (23, 24) have criticized risk assessments for lack of power. Peterman (23) found that the sample sizes used in some impact assessment studies were inadequate to assess all but gross changes in ecologic end points. Peterman (23) also noted that power was also limited by lack of attention to reducing variation. Improper sampling, inadequate controls, and natural and unnatural influences are potential factors that may have a strong influence on ability to make inferences. For example, if the interest is in the effect of a chemical plant discharge on an aquatic system and upstream sites used as controls, then a flood is a natural influence that can greatly alter habitat in both upstream and downstream sites. Sampling upstream and downstream sites on different dates may increase variance due to temporal effects.

A third source of uncertainty arises from the use of a model to describe the phenomenon of interest to the risk assessment. For example, these models may describe the effect of the stressor on the ecosystem, the effect of a single toxicant on a species, the bioavailability of a toxicant, or the spatial distribution of a toxicant in soils. Uncertainty in modeling arises from incomplete information, from assumptions about the model form, and stochasticity. As models involve parameters, there is also uncertainty that arises from the use of estimated values for model parameters. Model uncertainty due to lack of information may be

Table 6.2 Errors Associated with Hypothesis Testing

		DECISION	
		DO NOT REJECT	REJECT H_0
Truth	H_0	OK	Type I
	H_1	Type II	OK

great. One problem is the choice of an end point. In some cases, it is clear what biologic characteristic is relevant. For example, in studies on power plants, the loss of fish in the plant intake is important. Thus, models of fish abundance are relevant. In studies on recovery of wetlands, what to model may not be clear. There are differences of opinion as to which species may be relevant, which taxonomic group is relevant to monitor, which physical and chemical variables are valuable to monitor, and how the system should "recover." Models developed around the wrong end point can have high uncertainty and be entirely misleading.

Another area of uncertainty is associated with stochasticity. Stochasticity is associated with the natural variability of the system of interest. It is generally perceived that ecosystems frequently have a high degree of stochasticity; for example, in studies of fisheries abundance, changes of 50% or more may be difficult to detect (25). Stochasticity is not necessarily bad. For example, taxonomic stochasticity ensures a high diversity of species.

Uncertainties due to Sequencing and Episodic Events

The media has lately covered episodic events such as volcanic eruptions (e.g., Mount St. Helen's and, more recently, New Guinea), floods, hurricanes, and the like. In addition to episodic events, each ecosystem results from the sequence of climatic and biologic events that preceded its present condition. These sequences over multiyear periods are highly unlikely to be repeated in precisely the same way. Episodic events may be statistically predictable, but, as the Mississippi River flood has shown, the precise timing generally eludes society until the event has actually occurred. As a consequence, while approximations of sequences can be made or naturalistic sequences based on a longer period of history can be developed for laboratory or other controlled experiments, it is unlikely that there will be a precise match with future sequences wherever they may occur. Inevitably, there is a certain uncertainty about how closely the actual sequence, climatic or biologic, will match the predicted sequence or historic sequence. Additionally, episodic events will also occur, but their timing relative to the hazard being assessed cannot be precisely determined. However, uncertainties of this type can be incorporated by including a safety factor based on the probability of a match between utilized sequence and deviation from this or the probability of a particular episodic event occurring during a crucial time period.

Methods for Dealing with Uncertainty

As indicated above, uncertainty is likely to be present in all risk assessments. Two important issues then arise: how is uncertainty to be incorporated into the assessment and what can be done to reduce the uncertainty in future assessments? Taking uncertainty into account depends on the type of model that is used and the amount of information available to estimate uncertainty.

Perhaps the simplest approach to incorporating uncertainty into a risk analysis of toxicants is through an uncertainty factor. This approach is most often applied when the result of the risk analysis is a single number, such as the amount of a pesticide or chemical that is allowed into a system. The approach may be quite simple; for example, the best available information is used to pick a number (related in some way to risk) and then the number is adjusted by a safety or extrapolation factor (e.g., 10, 100) that indicates the degree of magnitude of uncertainty in extrapolation from laboratory or test results to the environment. Perhaps the most common example of an uncertainty factor involves the quotient method (26), which computes the ratio of the environmental concentration that is expected by the test end-point concentration (e.g., LC_{50}). The ratio is then multiplied by a risk factor (or an assessment factor) and compared to 1.0. Values less than 1.0 indicate safety. Bioaccumulation factors and acute/chronic ratios also may be viewed as uncertainty factors.

A more recent approach developed to account for uncertainty in choosing a safe level of a stressor is called the extrapolation method. In this approach, the unit to extrapolate over is defined. For example, to set a level of safety for a chemical, tests on single species may be used with an interest in protecting all the species in an ecosystem. Thus, since single-species results are used to extrapolate across species, the unit is the species. Each unit is tested separately, and the test result (e.g., an LC_{50} value) is viewed as a number representing a sample from the population of units. By relating the population of units to a distribution of values, statistical theory can be used to estimate a number that is smaller than most of the values in the population (e.g., smaller than 95% of the values) for most sets of test units of the same sample size (e.g., 95%). Theoretically, the method is better than the simple approach of choosing an uncertainty factor, since the uncertainty factor applied is based on the amount of data used to assess the stressor effects and the variation in response. The method is being used currently to set safety levels for new chemicals in the European Community (27), and a similar model is being used by the USEPA (28). This approach does not include all potential sources of uncertainty, however, and may be subject to biases (for a more extensive summary along with references, see Suter [1]).

Uncertainty factors can be made more precise when data relating different scales or end points are available. For example, if a chemical model is used to estimate the toxicity of a new chemical, the relationship between other chemicals and toxicity can be used to predict the toxicity of the new compound. Uncertainty then is estimated based on the variance of the prediction (29). This method is useful for the evaluation of uncertainty in the toxicity of chemicals, requires a reasonably large data base, and works best for chemicals of similar characteristics (30). An example in which this approach is used to estimate uncertainty attributable to extrapolation across different levels of complexity is described by Barnthouse et al (31) and is called the analysis of extrapolation error (AEE). The method uses information on a well-studied end point (e.g., rainbow trout LC_{50}) to predict the distribution of a desired end point (e.g., production of juvenile brook trout), using data available on different stressors (usually chemicals). The prediction is usually carried out using regression models and may extend over several levels of extrapolation, based on available data. The analysis ends by estimating the probability of risk by computing the probability that exposure exceeds the effected end point. Although the approach usually assumes that the end points follow a normal or lognormal distribution, other statistical models may be used. Details and examples are given in Suter (1).

An assessment of the uncertainty in using single-species tests to make population risk assessments is described in a series of papers by Barnthouse et al (31–33). These authors address, for example, the following situation. If data are available on a single-species test, then are the uncertainties in using one type of single-species test greater than those of using a different type of test? If information is not available on one species and a test is used from another species, how is uncertainty affected? These authors compare uncertainties associated with low knowledge information from QSARs through life-cycle tests, which have much higher information content. They report that uncertainty in extrapolation depends greatly on the level of information provided in the test and on the degree of relatedness of the species in the test and the species of interest for the risk assessment. Results (31–33) based on a life-cycle test on the species of interest produced upper and lower estimates that were only two orders of magnitude apart. However, results with other information resulted in much greater uncertainty. The upper estimate of risk based on a life-cycle test on a different species was roughly 100 times larger than the lower estimate. With acute test data, the upper prediction of risk was 150 times that of the lower, while the ratio for QSAR estimates were around 300. Barnthouse et al (31–33) also indicate that QSAR results showed substantial bias compared to other types of assessments (due to a poor prediction of toxicity from solubility for the chemical they used in their investigation). Thus, uncertainty generated from the level of the test can greatly influence interpretation of the risk and the effect at the ecosystem level.

Uncertainty and Complex Models

Models of ecologic risk are sometimes more complex than simple single-species laboratory studies combined with a field assessment of the stressor. These complex models are often computer simulation models and attempt to link together the ecologic response and the stressor. Uncertainty in more complex models are often described through the use of Monte Carlo uncertainty analysis (34) and the assessment of statistical uncertainty. The method of uncertainty analysis is reasonably simple to describe but may be difficult to implement when models are complex. A model of a system may be viewed as having three components: model form, parameters, and noise. Each of these components may be viewed as having uncertainty. Consequently, varying these may lead to information about that uncertainty. For example, in an assessment of parameter uncertainty, a parameter may not be viewed as being fixed but as having uncertainty. This uncertainty may be modeled by assuming a distribution for the parameter. By sampling from this distribution, the uncertainty in the outcome due to the uncertainty in the parameter may be estimated. In the same manner, uncertainty due to stochasticity (error) may be evaluated.

Complexities can easily arise since models for ecosystems often involve hundreds of parameters. A distribution for each of these parameters would be required. Often there is little information to describe the variation in the parameter, for example, only upper and lower bounds may be known. A large number of parameters involves computation difficulties in selecting a random sample of parameters (35). While it is common to expect that there are interactions between processes in ecosystem studies, there is often an assumption of independence between parameters. Describing multivariate distributions is complicated, especially given the lack of information available. Ignoring the relationship between parameters can be important, especially when estimating small probabilities or values associated with small probabilities (e.g., EC10) (36).

Another source of uncertainty due to stochasticity can also be incorporated into complex models using simulation methods. Stochasticity may represent natural variation associated with site differences, weather effects, and other factors. These are generally factors that are included in the model but not as deterministic components of the model, or they may be factors that are not directly modeled but are included as general error terms.

Evaluating uncertainty due to stochasticity may be quite valuable. For example, the magnitude of stochasticity leads to a better understanding of the variation to be expected. Knowledge of variation adds information about the size of stress that can be detected. Understanding stochasticity leads to a better evaluation of the "where, when, and how" of the stressor effects.

Statistical Uncertainty

Statistical uncertainty is commonly used in top-down models based on surrogate measures or field studies. Uncertainties in statistical models are expressed as parameter or end-point variance, confidence, or power. Variance is a measure of the constancy of the estimate of an ecologic parameter. Associated with variance is the confidence interval for the parameter. Confidence intervals represent plausible values of the parameter, given the information in the data. Wider confidence intervals are associated with greater uncertainty.

In the design of statistical studies, an important decision is the choice of sample size. There are two approaches to addressing sample size, depending on whether interest is in a good estimate or in making a good decision. When interest is in a parameter, sample size is often based on the width of the confidence interval. If this width can be specified, then the sample size can be chosen to achieve this width with high probability.

When interest is in making sound decisions, power analysis is a useful approach. In a power analysis, the size of an effect that is ecologically relevant is chosen. Then, based on the test used to aid in making the decision, an estimate of variance, and the type I error rate, the sample size required to make the correct decision can be calculated. Power analysis can be quite valuable in ecologic assessments because it provides a method for evaluating the strength of the decision for a given sampling design.

Validation

An important component of any modeling exercise is the validation of the model. *Model validation* is usually the term applied to the process of assessing how well a model fits a particular set of data (verification) and how well it may predict a new set of values. As pointed out by a number of authors, most recently by Oreskes and colleagues (37), validation is a misnomer for the process commonly used. Philosophically, validation refers to assessing how true something is. Thus, the truth of the coding of a computer program (free of logical mistakes) can be assessed, but not whether the output is true. This problem leads to the use of the terms *goodness of fit* and *predictive capability* in lieu of the terms *verification* and *calibration*.

Another method of assessing the utility of an ecotoxicologic test is the assessment of error rates. As indicated in the discussion of statistics, there are two error rates of interest: those associated with false positives and false negatives. For a given method, field data can be used to assess how frequently the method indicates an ecologic effect when, in fact, there is not one (false positive) and how frequently the method finds no ecologic effect when, in fact, there is one.

An example of such an assessment is the study of Commencement Bay, Washington (38), in which the apparent effect threshold (AET) was estimated for a number of toxicants based on the values of three biologic measures. An AET is the lowest sediment concentration above which ecologic effects are expected. To determine an AET, sampling stations are generally arranged in order of concentration of the toxicant of interest and the biologic indicators at those stations are compared with reference stations. If the biologic indicators show significant statistical difference from the reference, then there is an effect.

The AET is the station with the largest concentration that does not yield a statistically significant result. An AET is used like a safe concentration, and safety is assessed with the quotient method. If the ratio of the field concentration to the AET exceeds 1, then the station indicates effects.

However, these studies are not without difficulties. For example, a random sample of sites should be used to make the assessment. Also, the sites selected for the comparison should be independent of one another. In addition, it may be difficult to assess if a site is actually affected. Thus, the error rate may depend on the degree of effect. It is important to note that it is the marginally affected cases that are the most important to investigate since these indicate the sensitivity of the method. Further, the classification of a site into the "affected" and "not affected" class should be made independently of the test.

Reducing Uncertainty

Since uncertainty is an element of all analyses, some thought must be given to reducing uncertainty. Although details about reducing uncertainty are typically specific to a study, some general comments are possible. General methods for controlling uncertainty are quality control, power analysis and variance control methods, iterative approaches to risk assessment, and multiple lines of evidence.

Quality control should be a component of any ecotoxicologic study. A quality assessment of a study should investigate the components of a study and the potential sources of error (22). Recommendations can then be made to control the errors and avoid other errors. These would include checks on data quality, investigation of odd observations, adherence to good laboratory standards and use of standard methods for the collection and identification of organisms, processing chemical samples, and collecting data.

A second approach to controlling uncertainty is the use of statistical procedures to control error. These include the use of power analysis to increase the sensitivity of a study and the use of methods for reducing variability. Power analysis is a method useful in the design of studies to evaluate hypotheses. For a given hypothesis, a decision is made on the amount of deviation from the hypothesis that requires detection. Based on this deviation, the size of the sample can be calculated, which will lead to detection of a change with high probability.

Statistical methods can also be valuable in designing field studies. Two approaches for controlling variability are the use of composite samples and co-variates. Compositing is a relatively inexpensive way to reduce variation by combining samples. For example, the concentration of a chemical varies over space. In a study, samples could be taken from several sites within a segment of a river. Rather than measuring each sample, the samples could be combined and one measurement made. By combining the samples, the variation within the segment is reduced and the cost is typically less.

Covariates are variables that are useful for controlling variation. For example, a study that involves sampling over a number of sites usually has spatial variability. Some of this variability can be accounted for by collecting information on another variability and adjusting for this information. For example, in a study on toxicity in aquatic systems, the hardness of the water may be influential. Consequently, hardness measurements may be used to adjust the sites so that the sites may be viewed as having equal hardness. If the variability explained by hardness is relatively large, making the adjustment leads to a more sensitive test.

Another method for reducing uncertainty is the use of an adaptive, rather than a strict, approach. For example, in assessing the groundwater below a hazardous waste site, a typical standard approach might be to compare the data from wells below the site with data from wells upstream. A difference would suggest contamination. However, some uncertainty is associated with this approach. For example, the signal indicating differences may be false. To reduce the uncertainty with the decision, a sensible approach would be to take additional samples following rejection.

Finally, a method helpful in reducing uncertainty is the use of multiple lines of evidence. For example, a field study to assess the effect of an industrial plant on the aquatic ecosystem might compare upstream taxa with sites below the plant. A difference may be due to the plant or to other factors, for example, habitat. A toxicity test using effluent from the plant may be useful in providing additional information to establish causal links.

Summary and Conclusions

Ecotoxicologic investigations involve uncertainties, regardless of top-down or bottom-up approaches. The importance of uncertainty depends on the nature of the questions posed in the study, the amount of information available, and the level of resolution needed in the investigation. Listing and quantifying the uncertainty in the study are valuable components of the investigation and should be elements in all ecotoxicologic investigations.

References

1. Suter GW, II. Ecological risk assessment. Chelsea, MI: Lewis, 1993.
2. Forman RTT, Godron M. Landscape ecology. New York: Wiley, 1986.
3. Cairns J, Jr. Will there ever be a field of landscape toxicology? Environ Tox Chem 1993; 12:609–610.
4. Risser PG. Landscape ecology: state of the art. In: Turner MG, ed. Landscape heterogeneity and disturbance. New York: Springer-Verlag, 1987:1–14.
5. Kern RA. Arctic haze actually industrial smog? Science 1979;205:290.
6. Hirao Y, Patterson CC. Lead aerosol pollution in the High Sierra overrides natural mechanism which exclude lead from food chains. Science 1974;184:989.
7. Spencer WF, Cliath MM. Movement of pesticides from the soil to the atmosphere. In: Kurtz DA, ed. Long-range transport of pesticides. Chelsea, MI: Lewis, 1990:1–16.

8. Kurtz DA, Atlas EL. Distribution of hexachlorocyclohexanes in the Pacific Ocean, air and water. In: Kurtz DA, ed. Long-range transport of pesticides. Chelsea, MI: Lewis, 1990:143–160.

9. Hunsaker CT, Graham RL, Suter GW, II, O'Neill RV, Barnthouse LW, Gardner RH. Assessing ecological risk on a regional scale. Environ Manage 1990;14:325–332.

10. Suter GW II. End points for regional ecological risk assessment. Environ Manage 1990;14:9–23.

11. Woodman JN, Cowling EB. Airborne chemicals and forest health. Environ Sci Technol 1987;21:120–126.

12. US Environmental Protection Agency. R-EMAP: Regional environmental monitoring and assessment program, EPA/625/R-93/012. Cincinnati: USEPA Office of Research and Development, 1993.

13. Cairns J, Jr, Cherry DS. Fresh-water multi-species test systems. In: Calow P, ed. Handbook of ecotoxicology. Oxford: Blackwell Scientific, 1993:101–116.

14. Cairns J, Jr, Smith EP. Developing a statistical support system for environmental hazard evaluation. Hydrobiologia 1989;184:143–151.

15. Norton S, McVey M, Colt J, Durda J, Hegner R. Review of ecological risk assessment methods, EPA/230-10-88-041. Springfield, VA: National Technical Information Service, 1988.

16. Norton SB, Rodier DJ, Gentile JH, van der Schalie WH, Wood WP, Slimak MW. A framework for ecological risk assessment at the EPA. Environ Toxicol Chem 1992;11:1663–1672.

17. Harte J, Torn M, Jensen D. The nature and consequences of indirect linkages between climate change and biological diversity. In: Peters R, Lovejoy T, eds. Global warming and biodiversity. New Haven, CN: Yale University Press, 1992;325–343.

18. Bruntland G. Our common future. New York: Oxford University Press, 1987.

19. Smith EP, Cairns J, Jr. Impact assessment using the before-after control-impact model: Comments and concerns. Can J Fish Aquat Sci 1993;50:627–637.

20. Forbes TL, Forbes VE. A critique of the use of distribution based extrapolation models in ecotoxicology. Funct Ecol 1993;7:249–254.

21. Forbes VE, Forbes TL. Ecotoxicology in theory and practice. London: Chapman and Hall, 1994.

22. Liabastre AA, Carlberg KA, Miller MS. Quality assurance for environmental assessment activities. In: Hewitt CN, ed. Methods of environmental data analysis. New York: Elsevier Applied Science, 1992:259–299.

23. Peterman RM. Statistical power analysis can improve fisheries research and management. Can J Fish Aquat Sci 1990;47:2–15.

24. Parkhurt DF. Statistical hypothesis tests and statistical power in pure and applied science. In: von Furstenberg GM, ed. Acting under uncertainty: Multidisciplinary conceptions. Boston, MA: Kluwer Academic Press, 1990:181–202.

25. Van Winkle W, Vaughan DS, Barnthouse LW, Kirk BL. Can J Fish Aquat Sci 1981;38:627–632.

26. Barnthouse LW, DeAngelis DL, Gardner RV, O'Neill RV, Suter GW, Vaughan DS. Methodology for environmental risk analysis, ORNL/TM-8167. Oak Ridge, TN: Oak Ridge National Laboratory, 1982.

27. van Leeuwen K. Ecotoxicological effects assessment in The Netherlands. Environ Manage 1990;14:770–792.

28. Stephan CE, Mount DI, Hanson DJ, Gentile JH, Chapman GA, Brungs WA. Guidelines for deriving numeric national water quality criteria for the protection of aquatic organisms and their uses, NTIS PB85-227049. Duluth, MN: U.S. Environmental Protection Agency, 1985.

29. Slooff W, van Oers JAM, de Zwart D. Margins of uncertainty in ecotoxicological hazard assessment. Environ Toxicol Chem 1986;5:841–852.

30. Lindgren F, Eriksson L, Hellberg S, Jonsson J, Sjostrom M, Wold S. A strategy for ranking environmentally occurring chemicals. Part 4: Development of a chemical model system

for characterization of halogenated aliphatic hydrocarbons. Quant Struct-Act Relat 1991; 10:36–42.

31. Barnthouse LW, Suter GW, II, Rosen AE. Risks of toxic contaminants to exploited fish populations: influence of life history, data uncertainty and exploitation uncertainty. Environ Toxicol Chem 1990;9:297–311.

32. Barnthouse LW, Suter GW, II, Rosen AE, Beauchamp JJ. Estimating responses of fish populations to toxic contaminants. Environ Tox Chem 1987;6:811–824.

33. Barnthouse LW, Suter GW, II, Rosen AE. Inferring population-level significance from individual-level effects: an extrapolation from fisheries science to ecotoxicology. In: Lewis M, Suter GW, II, eds. Aquatic toxicology and environmental fate, 11th volume, STP1007. Philadelphia, PA: American Society for Testing and Materials, 1989:289–300.

34. Gardner RH, O'Neill RV, Mankin JB, Carney JH. A comparison of sensitivity analysis and error analysis based on a stream ecosystem model. Ecol Model 1981;12:173–190.

35. Iman RL, Conover WJ. Small sample sensitivity analysis techniques for computer models, with an application to risk assessment. Commun Statist Theor Meth 1979:1749–1842.

36. Ferson S. Naive Monte Carlo methods yield dangerous underestimates of tail probabilities. Proceedings of the High Consequence Safety Symposium, Sandia National Laboratories, 1994.

37. Oreskes N, Shrader-Frechette K, Belitz K. Verification, validation, and confirmation of numerical models in earth sciences. Science 1994;263:641–646.

38. Barrick RC, Beller HR. Reliability of sediment quality assessment in Puget Sound. In: Proceedings of Ocean '89. New York: Institute of Electrical and Electronics Engineers, 1989: 421–426.

39. Cothern CR. Uncertainties in quantitative risk assessments—two examples: trichloroethylene and radon in drinking water. In: Cothren CR, Mehlam MA, Marcus WL, eds. Advances in modern environmental toxicology. Vol. 15: Risk assessment and risk management of industrial and environmental chemicals. Princeton, NJ: Princeton Scientific, 1988:159–180.

40. Smith EP, Shugart HH. Uncertainty in ecological risk assessment. Washington, DC: Risk Assessment Forum, USEPA, 1994.

The Conservation of Biodiversity: Scientific Uncertainty and the Burden of Proof

7

John Lemons

Biologic diversity is a term that generally refers to the variability among living organisms from all sources and the ecologic complexes of which they are a part; this includes diversity within species, between species and of ecosystems. Consequently, there are many hierarchical levels of biodiversity, ranging from the genetic level to the landscape level, with each subsequent level supporting the next. At the smallest scale is genetic diversity, which includes the variety of genes within a population or a species. Genetic variation is necessary within a population in order for it to maintain reproductive vitality, resistance to disease, and the ability to evolve or to adapt to changes in the environment. Populations contribute to species diversity, which involves not only the number of different species (the variety component) but also how the total number of species is divided up (the relative abundance component). Add to species diversity the interactions among these species to form the next level in the hierarchy, community diversity. Biodiversity also includes abiotic processes, and at this larger scale is then considered ecosystem diversity. Clusters of interacting ecosystems are often looked at on an even greater scale, the landscape level. These different levels of biodiversity include not only the structural diversity in each but also the variety of functional processes occurring at each scale.

Biodiversity is valued for a variety of reasons. Biologic resources are used by humans in products such as food, pharmaceuticals, and fiber. Biodiversity also supplies humans with what have been called "ecosystem services," which includes the maintenance of atmospheric gases, climate control, and nutrient cycling. Other values are not related to consumptive resource uses, such as aesthetic and recreational values. Some also argue that biodiversity has intrinsic value, separate from its current or potential uses by humans.

Because of the variety of values associated with biodiversity, worldwide concern has arisen over its loss, which has primarily been studied in terms of species extinction (1). Although the process of extinction is natural, the rate of extinction has increased since the arrival of humans on earth. It is difficult for scientists to determine to what extent the loss of species has been caused by human impacts. The natural rate of extinction, or extinction that would occur in the absence of human influence, is estimated from the fossil record. This background rate of extinction is best known for birds and mammals, where the current extinction rate has been calculated to be 100 to 1000 times greater than the natural extinction rate (2). The often-cited total number of species lost per year is 100,000 (3). The cause for this elevated rate of extinction is attributed to a variety of human influences although the contribution of each to species extinction or the exact manner in which it might increase the probability of extinction often are not known very well. These include (a) the introduction of exotic species, (b) destruction of habitat, (c) overexploitation of species due to hunting and deliberate extermination, (d) habitat fragmentation, (e) pollution, and (f) the spread of disease.

In addition to the uncertainty surrounding the rate of species extinction and the factors responsible for it, other uncertainties pose difficulties to the enhanced protection of biodiversity. Many of these have to do with knowing the spatial and temporal patterns of diversity, how to define diversity more precisely and measure it more accurately, and which theories of the mechanisms influencing biodiversity are more explanatory or robust. In addition, much uncertainty surrounds the question of how to accurately determine the minimum viable population (MVP) required to minimize the risks of extinction for a threatened or endangered species.

In general, science is called on to develop methods and technologies for the conservation of biodiversity and, more recently, the sustainable use of biologic resources. The recovery of endangered species and the restoration of damaged ecosystems also will demand the expertise of scientists. The conservation of biodiversity is not solely a scientific problem; it also involves public policy, economic, and ethical decisions whose outcome often depends on how questions regarding the scientific uncertainty pervading biodiversity issues are resolved. The purpose of this chapter is to discuss some of the public-policy issues and implications of scientific uncertainty to problems of conserving biodiversity. For purposes of describing some of the areas of scientific and other uncertainty pertaining to the conservation of biodiversity, examples are drawn from (a) the calculation of MVPs, (b) conservation efforts focused on community, ecosystem, or landscape levels, (c) the status of ecology to provide knowledge of impacts of human activities upon ecologic resources of which species depend, (d) concepts of ecologic integrity used to promote protection of biodiversity, (e) value-laden methods of ecology, and (f) value-laden conservation legislation. Following a discussion of these examples, I discuss in general terms (a) some possible roles for science in biodiversity decison making and (b) implications of scientific uncertainty in fulfilling burden-of-proof requirements of conservation legislation.

Some Sources of Uncertainty

Calculation of Minimum Viable Populations

Conservationists have attempted to counteract the effects of chance events and human activities on populations by calculating and managing for MVPs. A population's MVP is the smallest number of individuals it can contain and still continue to survive natural stochastic and human-induced events. However, in the calculation of MVPs, neither the length nor the probability of survival are known for most species. Estimates of the percent chance of survival and duration of existence for a population are determined arbitrarily by scientists before calculating an MVP. These estimates commonly are not agreed upon, and typically they vary from 95% chance of survival for 100 years (3) to 99% chance of survival for 1000 years (4).

Presently, conservation biologists attempt to calculate MVPs using what is known as a population viability analysis (PVA). This analysis attempts to take into consideration the chance events that may affect populations; however, efforts to design models that consider all four stochastic events have been slow to develop (5). So far, PVAs developed have dealt with only one or two of these factors (6). This is due partly to the fact that the interactions between the various factors (genetic, demographic, environmental) are not understood well (7). Consequently, current estimates of MVPs have been criticized because they are based on very few data and lack feedback among demographic and genetic processes. Estimates of MVPs also are criticized because they are based on the assumption that loss of genetic diversity within populations affects all species equally; however, few data are available that can be used to verify this assumption (8). Estimating MVP size also is difficult because there is no reliable way of predicting how severe future or human-induced environmental fluctuations will be or of assessing their consequences on population growth rates.

Some things may help biologists to be able to predict MVP sizes in the future better. More data on the genetic variation within a species are needed. Although genetic information is currently known about many captive populations, few wild populations have been studied in depth. One of the best-studied mammal populations is the lions of Ngorongora crater in Tanzania. Here, scientists have combined data from genetic surveys with genealogic data in order to estimate the rate of genetic change within a population (9). Biologists compared the genetics of this small lion population with that of a much larger population from the Serengeti; such a comparison may help them decide when inbreeding may become a problem for wild populations (7). If biologists know when a population is small enough to be in danger of inbreeding, they may then be able to manage the population by establishing corridors between existing populations or by transporting individuals between populations in an attempt to prevent the problems of inbreeding. However, knowing when a population has reached the size where inbreeding is a problem has been difficult to predict. Extensive studies of the genetics and responses to inbreeding of the population of concern may be necessary before estimates of MVPs can be made (8).

The existence of metapopulations also may be helpful in terms of the long-term survival of threatened species, because gene flow may occur between sub-populations. However, estimating MVP size for metapopulations is much more complex than for single populations (6). More information is also needed about life histories of species as well as about the temporal and spatial distribution of resources (5). Detailed demographic studies of populations will be necessary in order to have data adequate to predict MVP sizes.

It is important to understand that estimates of MVPs must still be performed on a case-by-case basis and that models should be built for particular species of concern (6, 10). No single MVP model is applicable to all species, and scientists do not see this as a possibility in the future (5, 10). The genetic, demographic, environmental, and human-induced factors that influence the survival of one species are not the same for another species, so developing a universally accept-able protocol for determining MVP size is unlikely. "Comprehensive, realistic, and reliable methods for applications to all situations or precise prescriptions that can be applied uncritically will take a long time to develop, if they are even possible" (6).

Despite these problems, it is important to note that even with limited knowledge, biologists have been able to estimate MVPs for several populations, including Florida panthers (*Felis concolor coryi*), Bali starlings (*Leucospar roths-childi*), and the Sumatran rhinoceros (*Didermocerus sumatrensis*) (3). What has become evident from this work is that in order to ensure survival of some of these species, especially mammals, populations must contain a substantial num-ber of individuals. Using both genetic and demographic extinction models, sci-entists have estimated the MVP size for the Tana River crested mangebey, an endangered forest primate, to be 8000 individuals. It is hoped that this will ensure a 95% chance of survival of these primates for 100 years (11). In order to ensure similar survival success for the grizzly bear (*Ursus arctos*), researchers estimate that a population of 50 to 90 individuals must be maintained. This MVP was calculated using models that incorporated estimates of environmental and demographic stochasticity (12). Although the MVP is smaller for grizzlies than for mangebeys, conservation biologists also must consider the amount of habitat needed by MVPs. Large mammals such as grizzlies are often wide-rang-ing and have a large habitat requirement per individual. It has become appar-ent, therefore, that most national parks and other protected areas are not of sufficient size to maintain MVPs of these types of species (13). In situ conser-vation will require that these threatened species have habitats extensive enough for their continued survival (14).

To summarize, scientists can do certain things in order to get a better idea of the MVP size of a population, including the gathering of more genetic and demographic data for species of concern. The use of models to predict MVPs will continue to be difficult due to the inherent unknown factors affecting a population's survival, such as the occurrence of natural environmental fluctuations and human activities. Also, the chance that a single model will be applicable to all populations is highly unlikely, and studies therefore will have to be performed on a case-by-case basis. However, the work of conservation biologists in this area has pointed out three important issues: (a) the effect of various chance events on a population's continued survival, (b) the time frame to use in conservation planning, and (c) the degree of security desired for populations of concern (15). The second and third issues relate to the criteria that should be used in defining MVPs. Should we plan for continued survival of a species for 100 or 1000 years or longer? Do we plan for an 80% chance of survival or a 99% chance? Answering these questions will involve disciplines not traditionally considered a part of science, such as ethics, economics, and politics.

Conserving Biodiversity at Larger Scales

In addition to the variety of species, biodiversity also includes diversity at levels such as the community, ecosystem, and landscape. Although conservation efforts traditionally have been focused at the population or species level, scientists are now assessing the conservation of biodiversity at larger scales (15–17). Proponents of this wider view claim that the traditional reductionist approach of science wherein biologic systems are reduced to their component parts in order to try to understand them does not necessarily foster successful conservation of biodiversity. By employing a more holistic approach to the study of biodiversity and by enlarging the scale of reference, scientists can obtain a more accurate assessment of biodiversity, because each level has unique properties that can only be understood by studying that level. For example, if healthy ecosystems depend on specific interactions among organisms functioning within food webs and responding to the abiotic environment, then storing species ex situ (in zoos or botanical gardens) and later reintroducing them to the wild will not recreate a functional ecosystem (17). This view is reflected by the fact that studies at the ecosystem level emphasize the processes essential to healthy ecosystem functioning rather than end products (18, 19). Ultimately, however, the question of what scale is appropriate will be determined by the particular research question being asked.

Conserving biodiversity at the ecosystem level means striving to protect ecosystems' basic trophic structure and the energy flow and nutrient cycling patterns that result from that structure. If an ecosystem is protected adequately, then the assumption is that all of its resident species also are protected. Whereas traditionally species that have economic or instrumental value have been the targets of conservation, here other species, such as bacteria, fungi, invertebrates, and even those species that are not yet known are protected as well. These species may in fact be more important to the healthy functioning of an ecosystem than large, charismatic species (17). Further, by protecting large areas, species that move between habitats or live where two habitats meet will have a better chance of survival. Another important objective of conservation at this scale is to protect a representative sample of ecosystems on a worldwide basis. Most of the large areas that have some sort of conservation status today are in national parks, protected reserves, or forests. However, these areas were not originally set aside with the goal of protecting biodiversity, and their boundaries do not make sense ecologically. Grumbine points out the inability of U.S. national parks to provide habitats of adequate size for large vertebrates and some long-lived plant species (20).

Although the importance of the concept of protecting ecosystem diversity is now agreed upon by most conservation biologists, the question of how best to achieve conservation at this level is still being debated. To begin with, the classification of ecosystems into a manageable system has been a major problem. The range of classification systems is great, with some systems classifying terrestrial ecosystems, for example, according to their plant communities and other systems taking a more general approach based on an area's physical characteristics and appearance. Part of the problem is that these systems are based upon the assumption that ecosystems are discrete units that can be delineated and distinguished from each other instead of a series of interacting parts of a greater and highly variable continuum. It is extremely difficult to determine exact areas for ecosystems and even more difficult to estimate rates of habitat loss.

Some research on spatial and temporal scales of ecosystems has focused on the biogeographic consequences of fragmentation. Ecosystem fragmentation causes large changes in the physical environment as well as biogeographic changes, resulting in landscapes that consist of remnant areas of native flora and fauna surrounded by land modified by human influences. Fluxes of physical and biotic inputs across ecosystem boundaries are altered, thereby affecting native species in natural remnant areas. Biologic consequences of the isolation of protected areas due to the modification of adjacent lands is significant and varies as a result of the time since isolation and distance from other natural remnant areas. Research also has indicated that the consequences of fragmentation are influenced by size, shape, and distance of remnant areas from each other. Controversy exists regarding whether one large reserve will protect more species than several smaller areas with a total area equivalent to that of the larger reserve. Generally, larger reserves are more buffered from adverse consequences of fragmentation and therefore are thought to be better than smaller reserves. Unfortunately, most research on size and shape of reserves has provided little of practical value to resource managers for the reason that managers of protected lands are dealing with ecologic conditions that are a fait accompli.

With few exceptions, protection of biodiversity must occur on lands already set aside for particular management goals but whose boundaries were based upon political or cultural as opposed to biotic considerations. Consequently, a critical need exists to develop integrated approaches to land management that place conservation of biodiversity in the context of factors that influence overall use of landscapes.

One strategy being employed by conservation biologists to help determine what areas are in need of protection is gap analysis. Although the term "gap analysis" is relatively new, the process has been used for many years (21). The idea is to compare the locations of habitats or ecosystems with those of existing reserves in order to find the gaps in the system (22). In the area to be considered, the biodiversity is identified and classified, often in several different ways (e.g., by ecosystems, vegetation types, habitat types, species). Existing and proposed protected areas are then identified, and by comparing these with the biodiversity data, protection of biodiversity can be enhanced (21). Gap analysis can be done on a large international scale or at more local levels. At the largest scale, biogeographers designated eight terrestrial biogeographic regions worldwide which were then further subdivided into 227 provinces and have evaluated whether these regions are protected in existing reserves, national parks, national monuments, wildlife reserves, or protected landscapes. These provinces are also associated with 14 biomes. This information can be used to identify high-priority ecosystems and to guide recommendations for the establishment of future parks and reserves (2, 21). For example, after assessing what percent of each of the 14 biomes is currently protected, it was discovered that least protected are temperate grasslands (0.78%) and lake systems (1.28%), whereas the biomes protected most are subtropical/temperate rain forests/woodlands (9.32%) and mixed mountain systems (7.71%) (3).

Gap analyses are also being used in all U.S. states to determine if existing preserves, parks, and refuges are protecting biodiversity adequately. This effort is primarily being orchestrated by the Nature Conservancy, a private nonprofit organization, along with Defenders of Wildlife and the U.S. Fish and Wildlife Service (22). Even though there is no agreed-upon system of vegetation classification, the Nature Conservancy, working with state government agencies, has established Natural Heritage Data Centers in most states (21). Using several vegetation classifications as well as available data on species distribution, data needed for conservation efforts are gathered and organized. One application of this information is in preserve selection, so that limited resources can be better used to protect biodiversity in priority areas (23). Scientists have used gap analysis in several states by overlaying maps displaying current land ownership patterns, the location of threatened and endangered plants and animals, and vegetation types in order to look for gaps in the protected area network. Because data were unavailable for locations of all species, vegetation maps were used to predict where animals might be found, and these areas were then field-checked to test these predictions (22).

Much of the recent work in gap analysis has been aided by a technology known as Geographical Information Systems, a computer-based process that allows maps containing diverse data such as vegetation types, climate, soils, species distribution, and current land ownership to be overlaid (16). After converting traditional maps to digital (computer-compatible) maps, scientists can combine these data with information from other sources, such as Landsat Thematic Mapper Imagery, to create overview maps that provide useful information for specific conservation questions. Geographical Information Systems are becoming a powerful tool in the conservation of biodiversity, but it should be noted that some sources of error are associated with this technology. Errors can result from mistakes in data input (including using data from inappropriate or out-of-date sources), processing and analyzing data, and the output and presentation of data (24).

In recent years, efforts to conserve biodiversity at the ecosystem level have yielded new information. However, some areas of uncertainty and several limitations in this approach must be recognized. As mentioned previously, defining, classifying, and delimiting an ecosystem is still problematic. In addition, many ecologists traditionally have thought that ecosystems and their communities would eventually reach a steady state, and many management activities in conservation areas have attempted to stabilize the ecosystems being managed (19). Accordingly, concepts of ecosystem stability have been a focus of study for many conservation biologists. However, the concept of stability has stimulated considerable debate among scientists. For example, it is not known whether stability is due to species diversity or the cause of it. Further, concepts of stability can variously emphasize the resistance to disturbance of an ecosystem, the time an ecosystem requires to recover from damage, the zone from which an ecosystem will return to a stable state, the degree to which the pattern of secondary succession is not an exact reversal of the retrogression following environmental impact, and the degree to which a stable ecosystem established after disturbance differs from the original steady state. More problematically, some ecologists question whether concepts of stability have any real ecologic meaning (25). Consequently, management decisions based on concepts of stability must recognize the uncertainty surrounding the different concepts and the practical implications of managing for one concept as opposed to another.

Further, the goal of managing ecosystems in relatively stable or equilibrium states may conflict with the goal of maintaining high levels of biodiversity, because some systems need to be unstable in order for natural biodiversity to remain high (19, 26). Management of ecosystems for stable states also conflicts with recent views on the nonequilibrium nature of ecosystems. Pickett et al describe the classical paradigm in ecology as the "equilibrium paradigm," and the new paradigm as the "nonequilibrium paradigm" (18). The equilibrium paradigm emphasized the stable end points of ecologic systems and the idea that ecosystems were functionally and structurally complete and self-regulating. This implies that ecosystems, once set aside in parks or preserves, will maintain themselves as they were at the time of protection and that if disturbed they will return to their original state. In contrast, the contemporary nonequilibrium paradigm includes the following ideas: (a) natural systems are open; (b) processes rather than end points are emphasized; (c) a variety of scales are considered; and (d) episodic disturbances are recognized. In order to practice conservation under the modern paradigm, conservationists must focus on the processes of communities and ecosystems and work to maintain the dynamics of the system while recognizing that change and disturbance are important to the continued health or integrity of ecosystems (27). These processes might include the effect of herbivores on vegetation, fire, rainfall, and other natural disturbances. The nonequilibrium nature of ecosystems has important consequences for decisions about sustainable development and the conservation of biodiversity. If decisions are made to attempt to maintain static ecosystems so that they may provide specific biologic resources or services to humans on a sustainable basis, then significant and perhaps unrealistic levels of intervention in biologic and ecologic systems may be required of resource managers. Because of these concerns, Angermeier and Karr have proposed that concepts of ecologic integrity rather than elements of biologic diversity be used as a basis for policies to protect biologic resources because the former emphasize the organizational processes of ecosystems that generate and maintain all of the elements of biodiversity instead of only the presence or absence of particular elements (28).

By practicing conservation at larger scales, scientists have the opportunity to see how various processes operate at the ecosystem level, and it then becomes important for those involved in the management of protected areas to enable these dynamic processes to occur. This approach has been adopted by those in the field of landscape ecology who realize that ecosystems are not isolated units but are interacting systems that exchange nutrients, energy, and species. However, despite the apparent advantages of managing biodiversity at the landscape level, there is a concern that such a scale is so large that not all species will be able to be conserved. Communities that are small but rich in species (such as streams, wetlands, and coastal habitats) may not be apparent at larger or coarser scales (26). For example, much of the diversity of African ecosystems is found in areas that are too patchy or narrow to be picked up by large-scale approaches using vegetation maps. Endangered or threatened species in these smaller areas may not be adequately protected by ecosystem-level strategies, and endemic species are often not identified at such large scales. Consequently, the use of finer-scale resolution may be necessary to discern additional areas of conservation concern. Further, maps displaying vegetation types are often used to predict the locations of animal species in landscape analysis because of the assumption that if all vegetation types are included in protected areas, then all animal species also will be protected (21). However, it is not known whether this assumption should be accepted.

The conservation of biodiversity requires not only scientific information regarding numbers of species, extinction rates, and MVPs but also the successful management of biologic resources and ecosystems so that they are protected from adverse impacts of human activities. In this sense, conservation of biodiversity requires knowledge of the impacts of human activities on biologic and ecosystem attributes.

Ecologic Integrity

Recently, concepts of ecologic integrity have been proposed to facilitate enhanced protection of biodiversity against the threat of human activities because ecosystems that encompass facets of integrity would be protected better against activities that cause ecologic change or impairment (29, 30).

Most definitions of ecologic integrity focus on the ability of ecosystems to cope with stress and maintain their self-organizational capacities; however, the concept is not defined precisely (31, 32). Concepts of ecologic integrity have been derived from studies of ecosystems based upon complex systems theories. Such systems are described as nonlinear whose properties or behavior cannot be explained or predicted by knowledge of lower levels of hierarchical organization within them. In addition, complex systems have multiple organizational states and processes based upon nonequilibrium paradigms that include the following notions: (a) ecosystems are open; (b) processes rather than end points are emphasized; (c) a variety of temporal and spatial scales are emphasized; and (d) episodic disturbances are recognized.

Presumably, the meaning of ecologic integrity must be understood and accepted by decision makers if it is to be used as a basis for public policy. Angermeier and Karr propose that ecologic integrity refers most appropriately to ecosystems whose operations and evolution have been minimally influenced by human interventions (28). Kay has proposed that ecologic integrity encompasses three facets of ecosystems: (a) the ability to maintain optimum operations under normal conditions; (b) the ability to cope with changes in environmental conditions (i.e., stress); and (c) the ability to continue the process of self-organization on an ongoing basis (i.e., the ability to continue to evolve, develop, and proceed with the birth, death, and renewal cycle) (31). By optimum operations, Kay means the situation where the external environmental fluctuations that tend to disorganize ecosystems (i.e., make them less effective at dissipating solar energy) and the organizing thermodynamic forces that make ecosystems more effective at dissipating solar energy are balanced. It is not clear what Kay means by "normal environmental conditions" and, hence, to what extent ecologic integrity can or should refer to human interventions in ecosystems. Westra proposes a definition slightly different from Kay's, wherein she says that ecosystems can be said to have ecologic integrity when they have the ability to maintain operations under conditions as free as possible from human intervention, the ability to withstand anthropocentric changes in environmental conditions (i.e., stress), and the ability to continue the process of self-organization on an ongoing basis (30). She argues that concepts inherent in ecologic integrity emerge from continuing scientific, legal, and ethical analysis and that while they correspond in her mind to more or less "pristine nature," they cannot be described or predicted precisely because ecosystems are constantly changing and evolving.

Regier provides an abstract definition stating that ecologic integrity exists when an ecosystem is perceived to be in a state of well-being (33). In part, a more precise definition of ecologic integrity is dependent on peoples' perspectives of what constitutes complete ecosystems. In addition to reflecting the concerns and values of scientists, definitions of ecologic integrity also must reflect various social and ethical values relevant for public-policy decisions regarding protection of ecosystems. One reason for inclusion of these various values is because there is no a priori scientific definition of ecologic integrity, and therefore the concept encompasses perspectives or ways of viewing the world that inevitably reflect value-laden judgments. The ambiguity of ecologic integrity is a recognition that its definition, like many ecologic concepts, is determined, in part, on the basis of value-laden judgments and not solely on so-called value-free or precisely defined scientific criteria.

According to Schneider and Kay, ecosystems can respond to environmental changes in five qualitatively different ways: (a) after undergoing some initial structural/functional changes, they can operate in the same manner prior to the changes; (b) they can operate with an increase or decrease in the same structures they had prior to the changes; (c) they can operate with the emergence of new structures that replace or augment existing structures; (d) different ecosystems with significantly different structures can emerge; and (e) they can collapse with little or no regeneration (32). Although ecosystems can respond to environmental changes in one of these five ways, there is no inherent or predetermined state to which they will return. Further, none of these ways indicates a priori whether a loss of integrity has occurred.

As Kay points out, any change that permanently alters the normal operations of an ecosystem could be said to affect its integrity (31). Accordingly, the last four types of ecosystem responses would constitute a loss of integrity. One problem with this view is that it seems to reflect a commitment to preserve ecosystem attributes as they exist and does not recognize sufficiently the fact that ecosystems are dynamic and evolve; attempts to maintain them in existing states often require intensive management interventions that can have adverse ecologic consequences. Further, there is no scientific reason why a changed ecosystem necessarily has less ability to maintain optimum operations under normal environmental conditions, cope with changes in environmental conditions less effectively, or be limited in its ability to continue the process of self-organization on an ongoing basis. Alternatively, it could be said that any ecosystem that can maintain itself without collapsing has integrity. Accordingly, the first four types of ecosystem responses would constitute integrity. One problem with this view is that it would not be very helpful to agency decision makers because it accepts all ecosystem responses to change as constituting integrity, with the exception of total collapse, which occurs rarely and is clearly undesirable. Consequently, Kay concludes that between these two extremes is the option that some ecosystem changes might represent a loss of integrity, while others might not. However, while in theory science might inform decision makers about the responses of ecosystems to environmental change, it cannot provide a scientific or so-called objective basis for deciding whether one change is more desirable than another. In other words, the selection of criteria to use in such a decision must be based on human judgment regarding the acceptability of a particular change.

If concepts of ecologic integrity are to be used as a basis for enhanced protection of biodiversity, judgments about whether particular ecologic or species' responses to environmental change are consistent with the concepts will have to be made and will have to be reconciled with uncertainties stemming from two other sources: (a) value-laden judgments inherent in the methods and tools of ecology, and (b) value-laden judgments stemming from ambiguous normative words and phrases contained in the legislation designed to protect biodiversity. When these types of uncertainties are not recognized, then scientific conclusions and public-policy decisions appear to be more quantitative, less speculative, and more value-neutral than is warranted.

Capabilities of Ecologic Methods and Tools

Cognizant of the aforementioned theoretical and practical problems that constrain the scientific understanding of biodiversity as well as the ability to assess the ecologic impacts of human activities, a number of researchers have critically analyzed the extent to which the methods and techniques of ecology are capable of yielding reasonably certain information appropriate to serve as a basis for the successful protection of biodiversity (34, 35). Westman concludes that many current policies to conserve biodiversity are based on concepts no longer accepted in the ecologic community, including notions that all species in a community are interdependent (25). He suggests that policies should recognize the individualistic distribution of species over a landscape and that research should be directed toward better understanding of the relative abundance of coevolved relationships in different biomes and whether and to what extent one species can substitute functionally for another in an ecosystem. Lemons has analyzed the different meanings of "stress" as applied to species and ecosystems and has concluded that the theoretical differences between the different meanings are so great that when combined with informational uncertainty concerning the assessment and evaluation of the causes of stress and their effects, little basis for reasonably certain predictions exists (36). Lubchenco et al have identified numerous scientific uncertainties regarding sustainable development and protection of biodiversity and have proposed a research agenda to obtain more information about biodiversity (37). In their report, they acknowledge the limited role scientists can play in making reasonably accurate predictions about the effects of human interventions in ecologic systems. Cairns and Niederlehner note that in theory, both structural and functional attributes of ecosystems can be used as a basis for ecologic predictions but that practically speaking, there is a significant lack of knowledge about them (38). Based upon a review of the role of science in the protection of biodiversity in national parks, Lemons and Junker concluded that scientific information ideally required as a basis for management decisions typically is not adequate to serve as a basis for firm predictions (39). Huston has provided an in-depth assessment of the theoretical and informational knowledge concerning biodiversity, including the major areas of uncertainty (40). Finally, the Committee for the National Institute for the Environment has developed a comprehensive proposal to reform environmental research in the United States in order to provide more adequate information for decision making (41).

Value-Laden Scientific Methods

Many so-called scientific methods and tools upon which scientific information is derived also are embedded with the subjective values of scientists which serves to increase uncertainty. Shrader-Frechette and McCoy (35), Mayo and Hollander (42), and Westra and Lemons (43) have presented critical analyses of how and why numerous value-laden judgments, evaluations, assumptions, and inferences are embedded in scientific methods of ecosystem and human health risk identification, assessment, evaluation, and management, as well as in more basic research methods of ecology. For example, scientists often have to make judgments about which species or ecosystem attributes to study without having a firm scientific knowledge base to inform their choice. Often, ecologists use simplified models with many built-in assumptions that cannot be validated or verified. Interdisciplinary studies used in ecology require the synthesis of information and methods from different disciplines, which introduces subjectivity into the studies. Many studies are by necessity limited to small spatial and temporal scales, yet scientists often make long-term predictions extrapolated from them, even though such predictions cannot be verified or validated. Scientists also have to make decisions about whether to minimize type I or type II errors in their evaluation of acceptance or rejection of testable hypotheses. In other words, they must decide whether it is better to have a higher probability of accepting false positive or false negative results. Type I error errs on the side of protection of biodiversity, while type II error errs on the side of activities that potentially threaten biodiversity. Accepted scientific norms favor the minimization of type I error in order to reduce speculative thinking. However, this decision is value-laden and is not a scientific type of decision. Finally, scientists often have personal interests in certain attributes of biodiversity and attach their own values to them. Consequently, they often acquire knowledge and define scientific problems based, in part, on their interests and values. Each of these types of decisions and judgments reflects a combination of the professional expertise of scientists as well as some of their values.

The mention of the fact that scientific methods and tools are value-laden is not a criticism of science. Rather, I raise this point because a failure to recognize the existence of the value-laden dimensions of science can cast doubts about scientific and technical studies used to inform decisions about the conservation of biodiversity. In other words, unless the value-laden dimensions of scientific and technical studies and information are disclosed, the positions of decision makers will appear to be justified on what appear to be more quantitative, objective, or value-neutral scientific reasoning, when in fact they will be based, in part, on often controversial and conflicting values of scientists and decision makers themselves.

Value-Laden Decision Making

In part, actions taken to conserve biodiversity inevitably will conflict with other economic or developmental interests for which agencies have responsibilities or with which they have to deal. Consequently, even though many of the laws that might be applied in support of conserving biologic diversity contain normative language to promote or encourage biodiversity values, such language generally does not offer firm legal or public-policy prescriptions regarding how or where an administrator should balance conservation of biodiversity with conflicting uses of resources. Consequently, many decisions about the use of natural resources and biodiversity reflect the values of the decision makers themselves (44). Unless or until the value-laden dimensions of such decisions are made more explicit, most people will assume erroneously that decisions about biodiversity are made on more sound scientific information and more prescriptive legislative mandates used to guide agency decisions than is the case.

The U.S. Endangered Species Act can be used to demonstrate the fact that many decisions about biodiversity are value-laden in a manner that exacerbates problems of uncertainty. Specifically, the law permits the secretary of the interior to list a plant or animal as endangered for any one of five reasons: (a) present or threatened destruction of habitat; (b) overutilization for commercial, recreational, scientific, or educational purposes; (c) losses due to disease or predation; (d) inadequacy of existing laws and regulations to protect the organism in question; and (e) other natural or human factors affecting the continued existence of a species (including subspecies and populations). The law also mandates that listing decisions be based on the best scientific and commercial data available.

Briefly, decisions reflect the values of decision makers to a large extent in several key areas. First, significant uncertainty exists about the ecologic status of most species and their habitat requirements. Decision makers must make the value-laden decision of how conservative to be given conditions of scientific uncertainty. Second, although the Endangered Species Act directs federal departments and agencies to utilize their authorities in furtherance of the purposes of the act, most agencies have the authority to use administrative discretion so long as it is not arbitrary or capricious. Consequently, many agency decision makers attempt to balance decisions about listing of endangered species with the economic and social costs associated with protecting such a species, as well as with the goals of their own particular agencies. Neither the Endangered Species Act nor the legislative mandates for federal agencies prescribe how or where concerns about protection versus concerns about social and economic costs should be balanced. Third, although the Endangered Species Act permits the listing of subspecies and populations, it does not mandate it. Hence, decision makers have discretion concerning the taxonomic basis for listings. Further, there is no firm scientific or practical definition of a subspecies or population. Consequently, the judgments of scientists themselves regarding how a particular group of organisms should be classified are not based on scientific information solely. Fourth, the Endangered Species Act requires that conservation measures include all methods and procedures necessary to bring any endangered or threatened species to the point at which the measures pursuant to the act are no longer necessary. Because of the scientific uncertainties regarding what constitutes an MVP, decision makers have considerable discretion and can therefore base their decisions on numbers that are so low that prospects for recovery are low. This is a likely outcome if the decision makers hold the social and economic costs of listing to be higher than its benefits. Fifth, there often is bias toward such taxonomic groups such as mammals and birds compared with others. Sixth, lists of endangered species may carry their own bias which is not recognized by decision makers. For instance, rare or restricted species are not necessarily the most endangered but may receive more attention. Seventh, the Endangered Species Act is biased toward the protection of recognized species but cannot effectively target unrecognized species or taxa for action. Eighth, cultural and political bias exist insofar as nations are more willing to protect species within their own boundaries. Similar types of value-laden issues are inherent in other laws that attempt to protect species and ecosystems (45, 46).

The Use of Science in Decision Making

Given that we know some things about biodiversity but not others, how should science be used in resolving conservation biology problems? Some researchers recommend that the use of science in decision making be based upon classical scientific methods and techniques that they believe will yield robust scientific knowledge and that they believe will minimize speculation (47, 48). Peters recommends that the primary way to improve the utility of the ecologic sciences in decision making is to judge every ecologic theory on the basis of its ability to predict (49). These recommendations are based on assumptions that (a) there should be a strong emphasis on quantitative data acquisition, (b) more and better data will solve problems, and (c) the scientific method as commonly understood has great ability to discern facts about the natural world. Consequently, the necessary goal of science in protection of natural resources is said to include (a) formulating hypotheses and conducting observations to test them, (b) acquiring quantitative data, (c) developing an understanding of processes and linkages among variables, and (d) developing reasonably certain predictions (50). According to this line of reasoning, scientific methods and techniques should be based ideally on hypothesis-deduction methods or other methods yielding "strong inferences" in order to increase the likelihood of accurate results and predictions. Further, when interpreting the results of ecologic or environmental studies, many researchers maintain that type I rather than type II error be minimized because it is the most conservative course of action in situations of uncertainty and because minimization of type I error conforms to traditional forms of scientific rationality (51). This means that scientists would apply tests of statistical significance that reject an experiment's or study's results if the probability of their being due to chance is greater than, say, 5%. By minimizing type I error, it is said that science can play a central role in informing debates about problems of protecting biologic resources because the likelihood that decisions will be based on speculative thinking will be reduced (52).

Of course, not all ecologic studies involve the use of hypothesis-deduction methods; other forms of scientific explanation are not inductive and do not test for inferences, but are deductive. Shrader-Frechette and McCoy have critically analyzed numerous methods and techniques of ecology, including hypothesis-deduction methods but also various other approaches used in site-specific studies, statistical models, mechanistic models, comparative studies, theoretical modeling, historic studies, and case studies (35). Their conclusion is that the methods and techniques of ecology are of limited utility in providing decision makers with descriptive scientific conclusions for conservation issues because they have not yielded sufficient information for the development of general theories capable of serving as a basis for precise or reasonably certain predictions useful for conservation decisions. All of the methods and techniques are limited by both informational and theoretical uncertainty, and they all involve value-laden judgments, assumptions, inferences, and evaluations that serve to increase uncertainty. Consequently, Shrader-Frechette and McCoy conclude that ecology is more of a science of case studies than a science of generalizable laws and that its primary value is with its heuristic rather than predictive capabilities. Their conclusions are consistent with those of other researchers who have critically analyzed the extent to which the methods and techniques of science are capable of yielding reasonably certain information appropriate to serve as a basis for management and public-policy decisions.

For example, Sagoff argues that the role of ecology should be to identify ecologic indicators that might allow scientists to diagnose perturbations in species or ecosystems early enough so that mitigation measures could be implemented (53). This type of diagnosis does not depend on knowing generalizable laws and basing predictions upon them; rather it involves the integration of diverse information to make a general argument for one rather than another interpretation of the causes or consequences of ecologic impacts. Recently, Bella et al have argued that the role of the ecologic sciences in problems of environmental change ought to be in the identification of indicators of ecologic change rather than in the prediction of the consequences of human activities with reasonable certainty (50).

Other researchers have recommended that the role of science in problems such as conserving biodiversity should reflect the limited predictive capabilities of ecology (35, 52). Consequently, so-called holistic or "postnormal" scientific approaches have been proposed for use in natural resources decision making (34). Proponents of postnormal approaches reject the belief that (a) problem solving should reflect an emphasis on data acquisition per se; (b) more data will necessarily solve problems; and (c) the scientific method is objective and value-neutral. Consequently, the postnormal science approach emphasizes (a) adequate formulation of problems so that data will contribute to public-policy goals; (b) that most results from scientific studies will not yield reasonably certain predictions about future consequences of human activities and that many problems of protecting biodiversity therefore should be considered to be "transscience" problems requiring research directed toward useful indicators of change rather than precise predictions; and (c) the need to evaluate and interpret the logical assumptions underlying the empirical beliefs of scientists with a view toward ascertaining more fully the validity of scientific claims and their implications. While postnormal science is not easy to characterize, it seeks a broad and integrated view of problems and places more emphasis on professional judgment and intuition and is less bound by analytically derived empirical facts. Proponents of the application of postnormal science to problems such as the conservation of biodiversity maintain that its claims are more amenable for practical public-policy purposes than the claims of predictive science approaches (50, 52). Postnormal approaches are based more on retroduction and conceptual

analysis than are hypothesis-deduction methods, and by necessity they emphasize explanation and heuristic understanding of the complexities of nature
rather than predictions.

Clearly, there is a need to improve the capabilities of science and its use in
order to enhance the conservation of biodiversity. Although there is a debate
about the role science should play in decision making, there is no singular
scientific approach or methodology that can be applied to their understanding
or resolution, because there are different types of problems (e.g., ranging from
determining the minimum viable population for a threatened or endangered
species to determining the role exotic species play in affecting other species)
that need to be understood and resolved. However, as suggested in the remaining sections of this chapter, efforts to enhance the protection of biodiversity not
only require improving the capabilities of science but also require an examination of its role in fulfilling the prescriptive mandates of legislation affecting
the biodiversity. For example, the pervasive uncertainty inherent in the understanding of biodiversity combined with science's emphasis on minimizing type
I error (thereby increasing the chances of accepting a false conclusion that no
harm will be done to biologic resources) means that those promoting the
enhanced protection of biodiversity will have difficulty in meeting burden of
proof requirements imposed by science and law. Allowing uncertainty to delay
decisions to protect biodiversity is to make a tacit decision to allow and thereby
promote the status quo; no decision is in fact still a decision. A precautionary
approach to protection of biodiversity might very will be to shift the burden of
proof to those seeking to undertake activities that potentially threaten biodiversity to demonstrate that their activities will not cause harm. Although the
conservation of biodiversity is a global problem, the following discussion is confined to the general implications of uncertainty to decision making under U.S.
laws.

Legislation Affecting the Conservation of Biodiversity

Nations' laws prescribe various goals and levels of protection for biodiversity.
Federal laws and policies in the United States that can be used to conserve
biodiversity include (a) wildlife statutes, (b) agencies' legislative mandates, (c)
laws that create liability mechanisms, (d) laws that guide federal actions, (e)
laws and policies to compel or encourage conservation by private parties, (f)
specific laws to conserve natural resources, and (g) federal executive orders and
policies.

In the United States, examples of wildlife statutes include (a) the Lacy Act
(16 U.S.C. 701), which promotes the restoration of game and wild birds to areas
where they have become extinct; (b) the Federal Aid in Wildlife Restoration
Act (16 U.S.C. 669e[a][1]), whose purposes include land acquisition for wildlife

rehabilitation and wildlife habitat enhancement for mammals and birds; and (c) the Federal Aid in Fish Restoration Act (16 U.S.C. 777c[a][1]), which promotes restoration and enhancement of fishery resources for sport or recreation. Examples of agencies' legislative mandates include the National Park Service Organic Act (16 U.S.C. et seq.) and the Forest Service Organic Act (16 U.S.C. et seq.), which established the overall management goals of those agencies. An example of a law that creates liability mechanisms is the Comprehensive Environmental Response, Compensation, and Liability Act (42 U.S.C. et seq.), which contains provisions for the recovery of monetary damages by government organizations for natural resource injuries resulting from the release of hazardous substances. Laws that guide federal planning include (a) the National Environmental Policy Act (42 U.S.C. et seq.), which is intended to promote harmony between humans and the environment, to prevent or eliminate environmental damage, and to enhance the quality of renewable resources; and (b) the National Forest Management Practices Act (16 U.S.C.A. 1600–1614), which requires the U.S. Forest Service to assess renewable resources and to develop a National Renewable Resources Program, including plans for integrated and interdisciplinary management of national forests. An example of a law that directs private parties to conserve natural resources is the Surface Mining Control and Reclamation Act (30 U.S.C. 1201 et seq.), which established a nationwide program to mitigate or prevent the adverse environmental effects of surface mining for coal. Examples of laws to protect and conserve biologic resources include (a) the Endangered Species Act (16 U.S.C. 1531 et seq.), which is designed to conserve threatened and endangered fish, wildlife, and plants; (b) the Marine Mammal Protection Act (16 U.S.C. 1361 et seq.), which authorizes actions to conserve and replenish those stocks and species that are endangered or severely depleted; (c) the Fish and Wildlife Conservation Act (16 U.S.C. 2901 et seq.), whose purpose is to provide states with financial and technical assistance to conserve nongame wildlife; (d) the Fish and Wildlife Coordination Act (16 U.S.C. 661–667e), which requires a federal agency or permitee to consult with the Fish and Wildlife Service prior to controlling or modifying a body of water in order to prevent or lessen potential damages to natural resources; (e) the Soil and Water Resources Conservation Act (16 U.S.C. 2001 et seq.), whose purpose is to conserve soil, water, and related resources; (f) the North American Wetland Conservation Act (16 U.S.C. 4401–4413), which is designed to conserve and restore wetland ecosystems and maintain healthy populations of migratory birds in North America; and (g) the Clean Air Act (42 U.S.C. 7401 et seq.) and the Clean Water Act (33 U.S.C. 1251 et seq.), whose purposes are to prevent and control pollution in order to enhance and protect human and ecosystem health. Federal executive orders and policies include (a) Executive Orders 11514 and 11991, which commit the federal government to provide leadership in protecting and enhancing the quality of the nation's environment to sustain and enrich human life, including resource restoration; (b) Executive Order 11988, which was designed to avoid adverse impacts of flooding associated with floodplain development.

The above laws and policies affect forests, mined lands, wetlands, plants and wildlife, and ecosystems. Berger posits that in combination with the many other laws affecting natural resources, a sound legal foundation exists for the conservation of biodiversity (54). On the other hand, while the types of laws mentioned serve as examples of measures that can be used to promote the conservation of biodiversity, they have limited capabilities to do so. For example, the Endangered Species Act has been criticized because it (a) primarily protects high-profile individual species rather than overall biodiversity; (b) lacks clearly defined thresholds to delineate endangered, threatened, and recovered species; (c) does not protect metapopulations adequately; (d) does not adequately document many biologic determinations and therefore prevents meaningful scrutiny and participation by the public and scientific communities in decisions about protection of species; (e) does not protect habitat reserves sufficiently to sustain recovered populations; and (f) allows or fosters the discounting of uncertain or nonimmediate factors in the decision-making process about species' protection (55). The treatment of biodiversity in environmental impact assessments conducted under the National Environmental Policy Act has been viewed as inadequate on two accounts: (a) biodiversity often is not considered in the assessment process even when there are reasons to do so, and (b) when biodiversity is considered in impact assessments, its treatment is inadequate (56). Further, most other legislation designed to protect species or ecosystems does not establish criteria for protection of attributes of ecosystem health or integrity.

These types of problems are compounded because legislative mandates guiding the goals and policies of federal agencies usually provide for administrative discretion in the balance chosen between use of resources and activities that potentially impairs biodiversity or other natural resource values and their protection, and administrative decisions therefore have favored short-term consumptive use of resources over long-term protection of biodiversity and other natural resources.

This has been shown to be true for agencies like the U.S. National Park Service, which has relatively strong legislative mandates to preserve biotic resources in their natural state, as well as for agencies such as the U.S. Forest Service (45). In other words, it must be recognized that while agency administrators may have the force of law or administrative discretion to conserve biodiversity, in the absence of stronger legislative directives to deal with scientific uncertainty by erring on the side of enhanced protection of biodiversity, they will not abandon their own existing statutory authority and responsibility for their individual agencies.

Prescriptive Legislation and the Burden of Proof

As a matter of public policy, we commonly understand legislation or administrative decisions designed to enhance the protection of biodiversity to be prescriptive. Typically, prescription is understood to mean that certain acts or behavior is being recommended, suggested, or required. If behavior or acts are prescribed in a law, then they are understood to be required under the threat of sanction. In this sense, prescription is understood dialectically in that it is opposed to proscription, which is to suggest against or to forbid certain behavior or acts. Despite the fact that we usually might think that legislation and administrative decisions designed to protect biodiversity might be prescriptive, the explicit language of ostensibly prescriptive legislation and administrative decisions at the practice level functions proscriptively. Consequently, any prescriptions generally lie outside the explicit language of statutes, and they serve to place a burden of proof on those seeking to enhance the conservation of biodiversity. Importantly, in the face of scientific uncertainty this burden of proof often cannot be met with the result that activities and behaviors that threaten biodiversity continue unless and until more certain information about the threats is obtained.

When a statute explicitly forbids activities, it functions proscriptively. Decision makers who wish to act legally take the statutes as a prohibition of explicitly identified activities and assume that other activities are either prescribed or allowed unless they are known to constitute an abuse of discretionary power. Evidence of that proscribed behavior thus operates to review conduct under a legislative mandate; attempts to prove the negative are not. However, the fact that many statutes and administrative decisions designed to conserve or to allow the conservation of biodiversity function proscriptively in practice has implications for the role of science in decision making which stem from (a) the discretion that decision makers often have under the law(s), (b) the ambiguity of legislation and conflicting interpretations of it, and (c) scientific uncertainty.

Were scientists to be certain of their scientific information, we might ideally expect decision makers to apply that information to biodiversity problems and to generate a definitive and "correct" solution to the problems. Practically speaking, this expectation is not met often because science is seldom capable of providing reasonably certain information or descriptions to immediate biodiversity problems. In theory, waiting for more scientific data may increase the level of certainty (but it may not). However, while waiting for more data the status quo (i.e., the immediate problems) continues. Waiting for more data (called more certainty) is a decision to allow current problematic or unacceptable threats to biodiversity to continue. Decision makers would like to avoid making a decision without sufficient data so as to avoid the appearance of being wrong or subjective. However, waiting for more data is tantamount to an assumption that deteriorating resource conditions will go on hold until the next quanta of data arrive. In part, few activities have been explicitly identified either by the courts or by decision makers as being prohibited because of (a) the pervasive scientific uncertainty that exists about the threats to biodiversity, (b) the conflicting interpretations of statutes pertaining to the conservation of biodiversity, and (c) the administrative discretion generally granted to agency decision makers.

Because in practice many statutes that can be used to enhance the protection of biodiversity function in a proscriptive manner, scientific uncertainty tends to work against the identification of explicitly prohibited activities. Regardless of when during the accumulations of data a manager makes a decision, the legal check on that decision is the same as the abuse-of-discretion scope of review. This review considers when a manager made his or her decision. If knowing a particular scientific fact can be shown to require a manager to decide differently than he or she did, the manager has not abused his or her discretion if the particular fact was only known or knowable after the discretionary decision was made. In other words, decision makers are not responsible for knowing all that is knowable at any given time; they are held to the same standard as that which exists for judges in courts of general jurisdiction in that they need only to make decisions based on the facts presented or available (46). This situation acts in concert with other proscriptive manifestations of science and law that make it difficult to fulfill burden-of-proof arguments to conserve biodiversity.

In order to obtain more accurate and credible results, scientists often will define problems in a narrow manner by isolating and studying selected variables under controlled conditions. This approach leads to formulating environmental problems in particular ways, but it presents what is almost an intractable problem due to the fact that it attempts to understand complex systems by isolating a few variables so that they can be studied under narrowly controlled and simplified conditions. This approach becomes problematic because when dealing with more complex ecologic systems the understanding of the interactions among variables that determine the way in which individual variables express themselves is not able to be discerned. Consequently, because of the pervasive scientific uncertainty inherent in complex environmental problems, establishing cause-and-effect relationships between the activities and impacts at, say, the 95% confidence level is precluded in many instances. Further, by simplifying and reducing parts of complex environmental problems to a more manageable scale, scientists often end up studying a scientific problem that may be very different from the more complex environmental problem from which it stems. For example, in attempting to provide answers to problems of aerial spraying of pesticides in New Brunswick forests in Canada, researchers utilized strict scientific research guidelines and norms in order to attempt to discern whether there was a cause-and-effect relationship between pesticide exposure from spraying and human health effects. While this problem theoretically is amenable to scientific investigation, the complexities involved in understanding ecologic phenomena as well as human responses to the pesticides precluded a definitive answer to the problem. More importantly, the research approach served to focus attention away from more fundamental questions concerning the misdirection and redesign of resource policy (52).

The act of a scientist positing a hypothesis and deducing and testing principles therefrom or favoring minimization of type I rather than type II error is analogous to the setting of the burden of proof in law. So long as logically deduced principles do not contradict each other, a scientist's hypothesis is considered valid. The inherent and implicit power or right that a scientist has to posit a hypothesis is not a scientific process, but it is analogous to placing the burden of proof in law because it sets the stage for identifying and describing certain facts about nature but not others. Those who may wish to contest the validity of principles logically deduced from a tested hypothesis generally are assumed to have the burden to prove a competing hypothesis. Further, when scientists tend to favor minimization of type I error as being the most conservative course of action because it reduces speculation, they are establishing rules of conduct; namely, that an experiment's results cannot be assumed to be valid if the probability of their being due to chance is, say, greater than 5%. With respect to determining whether a harm exists to a particular species due to human activity, the effect of such a rule is to act as if it is better to minimize type I error (rejecting a false null hypothesis that there is no harm to a species) than it is to increase the chances of concluding a finding of harm to a species when there is none. Such a rule increases the chance that a scientist will make credible scientific conclusions, but it also increases the chance of a conclusion that there is no harm to a species when, in fact, there is.

In the United States, the notion of stating the law, stating environmental problems for public policy purposes, and assigning the burden of proof under the law is done with great emphasis upon personal liberties and individual freedoms. This emphasis is consistent with most environmental statutes because generally they allow individuals and corporations to proceed with whatever activities they wish unless or until someone else proves that such activities are a harm to a resource covered under a relevant statute. When such harm is proven, one response is for Congress to enact legislation to mitigate against it; a second is for a manager to take mitigation actions under his or her agency's statutes; and a third is litigation under existing statutes. However, as we have seen, the demonstration of harm with reasonable certainty is difficult under conditions of scientific uncertainty. Moreover, the scientific framing of conservation biology problems as established by scientists determines the possible field of responses from which Congress or governmental agencies can choose when making decisions about activities that might cause harm to the resources.

Conclusion

Biodiversity is being threatened by a variety of human activities. Some people seeking to enhance the conservation of biodiversity have made recommendations to improve scientific capabilities in order to provide more information to decision makers. While these are important recommendations, in my opinion they need to be supplemented by an examination of the implications of the placement of the burden of proof to demonstrate threats to biodiversity and a reassignment of the burden to those who seek to conduct activities that might cause harm to prove that their activities will not do so.

Various types of burden-of-proof requirements are entrenched in prevailing practices of science and law. Given the various levels of legal protection to biodiversity from environmental statutes, the amount of discretion which administrators have to make decisions, the pervasive scientific uncertainty surrounding the understanding of biodiversity, and science's emphasis on minimizing type I error, it seems that those promoting the enhanced protection of biodiversity will continue to have difficulty in meeting burden-of-proof requirements imposed by both generally accepted practices of science and law. While recommendations to improve the scientific capabilities to enhance the conservation of biodiversity have merit, simply calling for more support for research or for more data will not eliminate uncertainty about the effects of human activities on biodiversity, because uncertainty is inherent in the nature of scientific practices and, in particular, in the field of ecology. Consequently, legal burden-of-proof requirements are not likely to be met by those seeking to demonstrate harm to biodiversity. Allowing uncertainty to delay decisions to conserve biodiversity is to make a tacit decision to allow and thereby promote the status quo; no decision is in fact still a decision.

In my opinion, the question of how uncertainty regarding the conservation of biodiversity should be considered or dealt with in decision making should be addressed through legislative mandates. One suggestion would be to establish legislation that clearly adopts precautionary principles to deal with uncertainty better. In other words, the burden of proof could be reassigned whereby certain activities that might cause harm to biodiversity could not continue or go forward until those advocating it can prove, with the same degree of scientific certainty now required of those who would seek curtailment, that the activities will not cause harm. The recommendation to adopt such a precautionary principle and shift the burden of proof to those undertaking activities that might cause harm to biodiversity to prove that the activities will not cause harm is consistent with views that it is better to minimize type II error in conservation decisions. In other words, from an ethical perspective, it is more prudent to accept a higher risk of an erroneous conclusion that activities will cause harm than it is to accept a lower risk of a false null hypothesis that no harm will result from activities that potentially threaten biodiversity (35).

References

1. Lemons J, Morgan P. Conservation of biodiversity and sustainable development. In: Lemons J, Brown DA, eds. Sustainable development: science, ethics, and policy. Dordrecht, The Netherlands: Kluwer Academic, 1995:77–109.
2. Primack RB. Essentials of conservation biology. Sunderland, MA: Sinauer Associates, 1993.
3. Groombridge B. Global biodiversity, status of the earth's living resources. London: Chapman and Hall, 1992.
4. Shaffer ML. Minimum population sizes for species conservation. Bioscience 1981;31:131–134.
5. Gilpin ME, Soulé ME. Minimum viable populations: processes of species extinction. In: Soulé ME, ed. Conservation biology: the science of scarcity and diversity. Sunderland, MA: Sinauer Associates, 1986:13–18.
6. Shaffer ML. Population viability analysis. In: Decker DJ, Krasny ME, Goff GR, Smith CR, Gross DW, eds. Challenges in the conservation of biological resources: a practitioner's guide. Boulder, CO: Westview Press, 1991:107–118.
7. Woodruff DS. The problems of conserving genes and species. In: Western D, Pearl M, eds. Conservation for the twenty-first century, Oxford: Oxford University Press, 1989:76–88.
8. Lacy RC. The effects of inbreeding of isolated populations: are minimum viable population sizes predictable? In: Fiedler PL, Subodh KJ, eds. Conservation biology: the theory and practice of nature conservation, preservation and management. New York: Routledge, Chapman and Hall, 1992:277–296.
9. Packer C, Pusey AE, Rowley H, Gilbert DA, Matenson J, O'Brien SJ. Case study of a population bottleneck: lions of the Ngorongoro Crater. Conserv Biol 1991;5:219–230.
10. Soulé ME. Where do we go from here? In: Soulé ME, ed. Viable populations for conservation. Cambridge: Cambridge University Press, 1987:175–183.
11. Kinnaird MF, O'Brien TG. Viable populations for an endangered forest primate, the Tana River Mangabey (*Cercocebus galeritus galeritus*). Conserv Biol 1991;5:203–213.
12. Shaffer ML, Samson FB. Population size and extinction: a note on determining critical population size. Am Natur 1985;125:144–152.
13. Newmark WD. Legal and biotic boundaries of western North American national parks: a problem of congruence. Biol Conserv 1985;33:197–208.
14. Soulé ME, Simberloff D. What do genetics and ecology tell us about the design of nature reserves? Biol Conserv 1986;35:19–40.
15. Shaffer M. Minimum viable populations: coping with uncertainty. In: Soulé ME, ed. Viable populations for conservation. Cambridge: Cambridge University Press, 1987:69–86.
16. Scott JM, Csuti B, Jacobi JD, Estes JE. Species richness: a geographic approach to protecting future biodiversity. BioScience 1987;37:782–788.
17. McNaughton SJ. Ecosystems and conservation in the twenty-first century. In: Western D, Pearl M, eds. Conservation for the twenty-first century. Oxford: Oxford University Press, 1989:109–120.
18. Pickett STA, Parker VT, Fiedler PL. The new paradigm in ecology: Implications for conservation biology above the species level. In: Fiedler PL, Subodh KJ, eds. Conservation biology: the theory and practice of nature conservation, preservation and management. New York: Routledge, Chapman and Hall, 1992:65–90.
19. Walker B. Diversity and stability in ecosystem conservation. In: Western D, Pearl M, eds. Conservation for the twenty-first century. Oxford: Oxford University Press, 1989:121–132.
20. Grumbine E. Viable populations, reserve size, and federal lands management: a critique. Conserv Biol 1990;4:127–134.
21. Burley FW. Monitoring biological diversity for setting priorities in conservation. In: Wilson EO, ed. Biodiversity. Washington, DC: National Academy Press, 1988:227–230.
22. Allen L. Ecology forum: plugging the gaps. Nature Conserv 1992; Sept/Oct:8–9.
23. Jenkins RE. Information management for the conservation of biodiversity. In: Wilson EO, ed. Biodiversity. Washington, DC: National Academy Press, 1988:231–239.

24. DeGloria SD. Elements of geographic information systems for resource conservation. In: Decker DJ, Krasny ME, Goff GR, Smith CR, Gross DW, eds. Challenges in the conservation of biological resources: a practitioner's guide. Boulder, CO: Westview Press, 1991:153–166.

25. Westman, WE. Managing for biodiversity. BioScience, 1990;40:26–33.

26. Huntley BJ. Conserving and monitoring biotic diversity: some African examples. In: Wilson EO, ed. Biodiversity. Washington, DC: National Academy Press, 1988:248–262.

27. Karr JR. Ecological integrity: protecting earth's life support systems. In: Costanza R, Norton BG, Haskell BD, eds. Ecosystem health. Washington, DC: Island Press, 1992:223–238.

28. Angermeier PL, Karr JR. Biological integrity versus biological diversity as policy directives. BioScience 1994;44:690–697.

29. Johnson SP, ed. The earth summit. The United Nations conference on environment and development (UNCED). London: Graham & Trotman/Martinus Nijhoff, 1993.

30. Westra L. An environmental proposal for ethics: the principle of integrity. Lanham, MD: Rowman and Littlefield, 1994.

31. Kay J. A non-equilibrium thermodynamics framework for discussing ecosystem integrity. Environ Manage 1992;15:483–495.

32. Schneider ED, Kay JJ. Order from disorder: the thermodynamics of complexity in biology. In: Murphy MP, Luke A, O'Neill J, eds. What is life: the next fifty years. Reflections on the future of biology. Cambridge: Cambridge University Press, in press.

33. Regier HA. The notion of natural and cultural integrity. In: Woodley S, Francis J, Kay J, eds. Ecological integrity and the management of ecosystems. Delray Beach, FL: St. Lucie Press, 1993:3–18.

34. Lemons J, Brown DA. The role of science in sustainable development and environmental protection. In: Lemons J, Brown DA, eds. Sustainable development: science, ethics, and policy. Dordrecht, The Netherlands: Kluwer Academic, 1995:11–38.

35. Shrader-Frechette K, McCoy E. Method in ecology. Cambridge: Cambridge University Press, 1993.

36. Lemons J. Ecological stress phenomena and holistic environmental ethics: a viewpoint. Int J Environ Stud 1986;27:9–30.

37. Lubchenco J, Olson AM, Brubaker LB, et al. The sustainable biosphere initiative: an ecological research agenda. Ecology 1991;72:371–412.

38. Cairns Jr J, Niederlehner BR. Ecological function and resilience: neglected criteria for environmental impact assessment and ecological risk analysis. Environ Prof 1993;15:116–124.

39. Lemons J, Junker K. The role of science and law in the protection of national park resources. In: Wright RG, ed. National parks and protected areas: their role in environmental protection. Cambridge, MA: Blackwell Science, in press.

40. Huston MA. Biological diversity. Cambridge: Cambridge University Press, 1994.

41. Committee for the National Institute for the Environment. A proposal to create a national institute for the environment (NIE). Environ Prof 1994;16:93–191.

42. Mayo DG, Hollander RD, eds. Acceptable evidence: science and values in risk management. Oxford: Oxford University Press, 1991.

43. Westra L, Lemons J, eds. Perspectives on ecological integrity. Dordrecht, The Netherlands: Kluwer Academic, 1995.

44. Lemons J. Conservation biology: the role of science, values, and ethics. In: Majumdar SK, Miller EW, Baker DE, Brown EK, Pratt JR, Schmalz RF, eds. Conservation and resource management. Easton, PA: The Pennsylvania Academy of Sciences, 1993:333–348.

45. Lemons J. United States' national park management: values, policy, and possible hints for others. Environ Conserv 1987;14:328, 329–340.

46. Brown DA. The role of law in sustainable development and environmental protection. In: Lemons J, Brown DA, eds. Sustainable development: science, ethics, and policy. Dordrecht, The Netherlands: Kluwer Academic, 1995:64–76.

47. Murphy D. Conservation biology and scientific method. Conserv Biol 1990;4:203–204.

48. Drew GS. The scientific method revisited. Conserv Biol 1994;8:596–597.

49. Peters RH. A critique for ecology. Cambridge: Cambridge University Press, 1991.

50. Bella DA, Jacobs R, Hiram L. Ecological indicators of global climate change: a research framework. Environ Manage 1994;18:489–500.
51. Simberloff D. Simplification, danger, and ethics in conservation biology. Bull Ecol Soc Am 1987;68:156–157.
52. Miller A. The role of analytical science in natural resource decision making. Environ Manage 1993;17:563–574.
53. Sagoff M. Ethics, ecology, and the environment: integrating science and law. Tenn Law Rev 1988;56:78–229.
54. Berger J. The federal mandate to restore: laws and policies on environmental restoration. Environ Prof 1991;13:195–206.
55. Rohlf DJ. Six biological reasons why the endangered species act doesn't work—and what to do about it. Conserv Biol 1991;5:273–282.
56. Hirsch A. Improving consideration of biodiversity in NEPA assessments. Environ Prof 1993; 15:103–115.

Can We Resolve Uncertainty in Marine Fisheries Management?

Charles F. Cole

Introduction

Despite a virtually unanimously held opinion that the world's marine fisheries resources show every sign of being overfished, fisheries biologists and fishermen are normally deeply divided over proposed solutions. Fisheries biologists generally have seen their role to be advocates of long-term wise use of resources and usually begin any dialogue by recommending harvest limitations well in advance of a serious decline in the resource. These risk-adverse recommendations are rarely acceptable to the fishing industry.

On the other hand, fishermen in near daily contact with the resource remain convinced that fisheries biologists are either ignorant of reality or, because they are government employees, cannot speak the unfettered truth. Furthermore, biologists, as advocates of the resource, are often seen as impractical because their recommendations would keep trawlers tied up to a dock and thus adversely impact peoples' lives. An inability to resolve this dilemma is a serious problem being repeated worldwide, and the longer it remains unresolved the farther off is any resource rebound that might lead to an acceptable yield. Far worse is the biologic view held by some that, because of overharvesting, large oceanic ecosystems have been permanently changed and that no rebound is possible.

This resource-use dilemma has been well described by Graham (1, 2), was placed in general economic terms by Gordon (3), and has subsequently been reviewed by many others. Hardin's (4) paper entitled "Tragedy of the Commons" put the underlying concept of common property resources into layman's terms. Hardin argued that a resource without valid ownership and thus not under an owner's prudent stewardship will inevitably be harvested at levels exceeding the resource's ability to sustain anything but a declining level of yield. Further, in fisheries, this problem will persist as long as profit exists for anyone engaged in the fishery. Individual fishermen who have already committed family resources to buying a boat see little choice but to fish the boat as hard as possible. Once at sea, they must compete with their increasingly numerous and capable fellow fishermen, and inevitably each, joined by yet others, will do so as long as expenses can be made. Under such increasing and unregulated pressure, fish stocks inevitably decline. First, fishermen deny the problem exists, then biologists explain the problem in complicated mathematical terms, banks compound the problem by funding bigger, better-equipped boats, and political entities finesse the problem by stepping back from hard management decisions. In turn, each constituent element has accelerated the decline. Given the conditions described, the unresolved dilemma is how might we protect ourselves from ourselves.

Others have argued that fisheries resources are more equitably divided at least among inshore users (5, 6, among others). They suggest that inshore fishermen are essentially part-time, even artisanal, fishermen and that self-regulation in the best long-term interest does occur within the local fishing community. However, such local regulating, they would concede, is not being practiced by larger, offshore trawlers. Further, Hanna and Smith (7) believe that trawler captains have a variety of attitudes about work, resource use, and fishery management that are not accurately captured by Hardin's model. They contend the premise that commercial fishermen do not care about the state of the resource is false but that they are instead trapped by a system that allows free access and thus unlimited entry.

The biologist's problem, irrespective of the open access, common property nature of the resource, is to determine the size and dynamics of a mobile and also invisible resource. If one wishes to design a rational management for this resource, one must certainly know how big the resource is and how it behaves biologically. Presuming that the first step to be an ecologic assessment, the biologist is drawn into an assessment of a resource that cannot be seen. By contrast, if fish grew on trees in forests, the assessment problem would be relatively simple. Trees in a forest are owned by someone, private or public, and thus are not common property. Furthermore, each tree grows in plain sight of its owner and the owner may, at any time, count the trees, measure their growth and bulk, and then decide whether or not to harvest based on the owner's best economic interest. Unfortunately, fish populations under water cannot be similarly assessed.

At best, fishery stock assessments are indirect and the techniques used typically complex and often mathematically obscure. The resulting estimates of population sizes, the impact of harvest, and the dynamics of population responses inevitably have wide confidence limits and, when played out, frequently seem not to mirror reality seen by commercial fishermen as they fish the resource being evaluated. Regardless, the prudent fisheries biologist has little choice but to be conservative when recommending management aimed at keeping fish populations at an optimal and sustainable level. Inevitably the biologist chooses a risk-averse harvesting recommendation at which point uncertainty returns to enter the discussion. Because there is a presumption of overfishing, how much can be removed now, what will the stock look like next year, in 10 years, and later, become integral steps in the assessment process. Reducing fishing pressure is presumed to lead to increased spawning, increased juveniles, and finally a large fish stock in the future. Some would contend the biologic evidence for this presumption is very limited (8). Although these questions have been on the table for nearly 100 years, there is a history behind the present state of marine fisheries management that can provide guidance to this process. To understand the cause of the present uncertainty, one needs to review the past.

Historical Perspective on U.S. Practices

Marine resource management off the coast of northeastern United States can be divided into five periods, and many of the world fisheries have traced similar paths.

1. An age of plenty: pre-Colonial times to 1945
2. Growing international problems: 1945–1965
3. Success of the 200-mile-limit theory: 1976–1982
4. We meet the enemy—us: 1982 to present
5. The unresolvable future

An Age of Plenty: Pre-Colonial Times to 1945

Commercial fishing by European fishermen was already underway on the oceanic banks off the North American continent before colonial times. The Maritime Provinces of Canada and much of New England were settled, at least in part, and sustained by an abundance of fish easily available to early colonists. Likewise, fish and forestry products from the New World were important to the colonizing nations of Europe. The struggle for colonies and the resulting commerce inevitably led to maritime wars.

Although the maritime wars of the sixteenth and seventeenth century often began as the dominant nations attempted to control and limit foreign shipping, by the end of the eighteenth century the international community had begun to agree that the cost of restricting effectively the oceanic commerce of others far exceeded any benefits. Accordingly, when it became apparent that control of oceanic commerce was impracticable, unrestrained commercial shipping, and inferentially fishing, in international waters beyond territorial waters became acceptable international practice. Except for boundary disputes between nations, such as the United States and the British colonies of what is now Canada, control of foreign fishing rarely became part of the problem.

Accordingly, an untrammeled use of oceans included fishing and became a customary practice among nations, a sufficient definition of international law. However, such intercourse between nations was inevitably modified by a national concept of sovereignty—something that ends at the limit of a nation state's territory. Further, except for the states' extended control over its citizens and vessels, all constitutional law also ended at the outer limit of territoriality, piracy, and a few other exceptions notwithstanding (9, 10).

Sovereignty over land was gradually extended seaward, first over a nation's internal waters and then over the nearshore or territorial waters. In 1793, President Thomas Jefferson established the first U.S. territorial limits when he proclaimed them to extend 3 nautical miles from land, a distance essentially that of shore battery range. A plethora of similar settings soon followed from other nations, not all of whom settled on 3 miles, and whether the width would ever be standardized became an increasingly contentious topic. In 1930, a conference of 48 nations was convened in the Hague to address the issue but, predictably, failed to agree on a set value. This issue would arise again following World War II, but appeared to remain one of sovereignty and commerce. Its role in fisheries management was yet to emerge.

Contentions over watery boundaries that defined territorial waters were troublesome and often had fisheries connotations. Once the United States and Canada resolved their own boundary differences, scientists from both countries began to cooperate on common, transboundary, resource problems.

Cooperation between Russia, Japan, Canada, and the United States resulted in the Fur Seal Convention of 1911, which served as an early model for allocating a common property resource (11). Coordinated research by Canadian and U.S. scientists and managers on the halibut fishery off the west coasts of both countries preceded a 1923 bipartite Convention on North Pacific Halibut, which established the Pacific Halibut Commission (12).

In 1902, the first international arrangement—the International Convention for the Exploration of the Sea—was created for coordinating research and management in the North Sea (13). Fisheries scientists from nations bordering the North Sea faced, among their many other tasks, the daunting problem of developing valid techniques for estimating the size and condition of specific fish stocks and to coordinate their own research programs. Important concepts in the population dynamics and stock assessment of marine fish resources were developed from such international cooperation. Of crucial importance was Beverton and Holt's classic study (14), which provided a scientific basis for groundfish (cod, haddock, and flounders) stock assessment. Their techniques, greatly enhanced by the subsequent work of many others, are still in use.

Foreign fleets returned to the east coast of North America after World War II. Initially they were not considered a threat to fish stocks and until 1965, U.S. and Canadian fishermen essentially behaved as if their respective nations owned the offshore waters and that their governments were only permitting limited foreign fishing. But from 1793 forward, neither nation had claimed sovereignty over waters of the Northwest Atlantic beyond 3 miles, and thus no one had the legalized right to manage and control use of the resource. Legal authority for regulating the harvest of fish stocks inside the 3-mile limit belonged to the states, and no authority existed to regulate fish harvest beyond 3 miles. Beyond 3 miles there were neither rules nor anyone to enforce them.

Unfortunately, waters off the three main coasts of the United States beyond 3 miles included much of the highly productive shallow waters of the continental shelf in which most commercial fishing would later occur. Yet these waters remained international and the fisheries resources under no nation's jurisdiction. As early as 1930, Canadian and U.S. fisheries workers were becoming concerned over the status of several species whose distributions were not only transboundary but also extended onto the Grand and Georges Banks off North America. Nonetheless, both state and federal jurisdiction remained at 3 miles until shortly after the end of World War II.

Growing International Problems: 1945–1965

In an effort to control smuggling during Prohibition, the United States unilaterally determined that waters out to 12 nautical miles were of special importance and represented a zone within which smugglers could legally be stopped and searched for contraband. These additional 9 miles would become known as the contiguous zone. Although 12 nautical miles had been a distance often selected by other nations as the width of their own territorial waters, the United States continued to recognize no territorial limits greater than her own 3 miles. However, the U.S. position appeared to change in 1945 when President Truman issued two presidential proclamations, the first of which extended U.S. "jurisdiction and control" over the continental shelf bottom to the edge of the outer continental shelf (15). The second proclamation defined U.S. capacity to enter into fisheries agreements that reached beyond territorial limits. This concept had been discussed in an earlier international meeting and the proclamation was used to validate U.S. commitment to salmon and halibut agreements then in force with Canada. These two unilateral actions, when uncritically linked together, became known as the Truman Doctrine and served as an element in justification for nations to extend partial jurisdiction or their territorial limits (16).

Inevitably, erroneous interpretations of the Truman Proclamations would also be used by nations to restrict foreign fleets from fishing continental shelf stocks and even to restrict free passage of commerce. The Icelandic Cod Wars with Great Britain in 1955 exemplified the former (17). Chile interpreted the Proclamations very stringently and in 1970 seized U.S. tuna boats fishing in her exclusive zone. Not only were fisheries rights being adversely affected, but also being affected were the traditional maritime rights such as the innocent passage of men-of-war and that of commercial steamers attempting to pass through previously international waters.

Upon becoming independent of the Netherlands, Indonesia unilaterally proclaimed all waters within its 10,000 islands to be her internal waters and thus not subject to the right of innocent passage, a "right" restricted by conventional practice to territorial waters. Any vessel passing through her internal waters would require prior permission which, in effect, would close a major shipping route from Australia to Singapore. To make U.S. disapproval clear, in 1957 the cruiser USS *Bremerton* without consent cruised unchallenged along the normal shipping route between the islands of Indonesia. The United States was reflecting a position long held by most maritime nations that freedom of the seas could not be unilaterally abrogated by any nation simply by extending its territory capriciously beyond the same limit then being claimed by the United States. To do otherwise could gradually shut down the innocent passage of merchant, as well as military, vessels throughout the world, to say nothing of passage of fishing vessels (9).

The motive behind the first Truman Proclamation had been to claim offshore oil reserves in the Gulf of Mexico for the United States and to prevent their foreign acquisition. Later, the Outer Continental Shelf Lands Act of 1953 would give a congressional authorization to this taking. In essence, the United States had claimed possession of the bottom lands of the continental shelf out to a depth of 200 meters and in some areas this extended the United States out as far as 60 nautical miles adding immensely to the size of the nation. In the same year, Congress also passed the Submerged Lands Act, which provided the states with management over their coastal bottoms but limited such management to 3 miles. Revenue from any leases passed to the federal government, and management of the outer continental shelf bottom was assigned to the Bureau of Land Management. The Minerals Management Service of BLM created in 1982 now carries out the leasing and development responsibilities for all federal lands. Revenues from such leases represent one of the largest nontax sources of federal funds (18). Not until 1978 was the Shelf Act amended to require that marine resources be protected during this process (17, 19).

Less understood, but more germane to fisheries issues, was the second Truman Proclamation of 1945, which dealt with the U.S. ability to make fisheries treaties and other international agreements and to create "conservation zones." The intent of the second 1945 proclamation was subsequently and incorrectly mixed with the first and interpreted by some as a unilateral decision to bring under complete U.S. control those previously nonregulated offshore fisheries resources. Resolution of the territorial sea issue in face of a plethora of then emerging nations required action and brought a new urgency to the issues of the management and allocation of marine resources (15, 16, 17). Regardless of international misperceptions, the two proclamations did not place control of free-swimming fisheries resources under sole U.S. custody and ownership.

The growing effort by developing and often newly independent nations to control fishing in the international waters adjacent to their own territory and other territorial issues caused the United Nations to convene the first Geneva Conference on the Law of the Sea (LOS) in 1958. Four conventions or multilateral treaties had been drafted by 1958 and were opened for signature, but only a nation signing any or all of the treaties could be considered bound by the contents. Not all did, but the conventions did set a world behavior norm and provided an impetus for the second LOS conference, convened in 1960.

These first four multilateral treaties were the Convention on the Territorial Sea and the Contiguous Zone, the Convention on the High Seas, the Convention on the Continental Shelf, and the Convention on Fishing and on the Conservation of the Living Resources of the High Seas. The first three codified generally accepted maritime practices and did not open new theoretical grounds. However, the fourth, the Convention on Fishing and on the Conservation of the Living Resources of the High Seas, was quite visionary and laid down a foundation in international law for subsequent events of the 1970s (10, 15–17).

Almost predictably, during the international discussions on the Convention on the Territorial Sea, nations were again unable to agree upon a width of the territorial sea, but they did formalize practices such as "innocent passage," "contiguous zones," the right of passage through straits, and the processes for determining waters internal to the territorial sea among other practices. In effect, the first three conventions codified a number of practices already being followed by many nations and provided a stability and a lever with which to prod recalcitrants. The fourth convention, however, went well beyond then current practice and gave to "coastal states" (nations) the responsibility for the wise management of resources in the international waters beyond their own territorial waters. Christy and Scott (17) considered it ". . . an example of deliberate international lawmaking by an international body to deal with problems foreseen rather than experienced. . . ." This convention would later provide an international justification to the unilateral actions of the United States just before as well as after the signing of the Magnuson Fisheries and Conservation Act of 1976.

Cooperation by American and Canadian scientists on fisheries research had predated World War II, and much of it centered on halibut and salmon on the Pacific coast and on the groundfish resources on the Atlantic coast. Although neither country then exercised federal management of the resources on the east coast, the binational halibut fishery on the Pacific coast had been under treaty-regulated management since 1923 (12). Moreover, both nations were also accumulating landing statistics on the harvest and were gathering considerable information on age structure and other materials needed for progressively more complex population dynamics studies on a number of species. The data sets resulting would become extremely valuable in later assessments. These studies, harvest data (20, 21, and previous years), and the research and landings data gathered by ICES in the northeastern Atlantic and North Sea on many of the same species would form much of marine fisheries information then available on a worldwide basis. Not until after World War II when the U.N. began to gather statistics worldwide was much known about the extent of commercial fishing throughout the rest of the world (see Ref. 22 for a current summary).

As mentioned, soon after the end of World War II, foreign fishing vessels from Western as well as Eastern Europe began to return to Georges Bank and the Gulf of Maine off New England and to the Grand Banks off Nova Scotia. Almost immediately, American and Canadian scientists, as well as the fishing industry, became increasingly restive as they saw developing signs of overfishing offshore on selected species. Politicians from both nations came under increasing constituent pressure asking for a process that would extend management beyond territorial limits.

In response in 1949, a multilateral fisheries treaty, the International Convention for the Northwest Atlantic Fisheries, was formalized and accepted by nine nations (United States, Canada, Norway, Portugal, Iceland, Denmark, Italy, Spain, and the United Kingdom). The signatory nations agreed to control the landings by the fishing vessels of their nationals according to a sustained yield principle. The regulatory body resulting, the International Commission for the Northwest Atlantic Fisheries (ICNAF), was housed in Halifax. It was responsible for gathering the best scientific evidence and then reaching consensus on appropriate regulations for a large area from southern New England to Georges Bank and northward and eastward to the seas between Greenland and Labrador. Scientists from signatory nations were to meet yearly in panels on a species-by-species and region-by-region agenda in which they considered management measures such as quotas, gear limitation, and other means of controlling harvest by vessels of their respective nations. Recommendations from the panels were then forwarded to the commissioners, who voted upon regulations that then went to each signatory government to implement. The scientific decisions were required by treaty to be based on a biologic concept described as "maximum sustained catch," in essence, maximum sustained yield (MSY). Although the commission determined fishing limitations for the entire ICNAF regulatory region, each signatory nation alone was able to regulate landings by its own vessels (23). With the possibility for bias in self-regulation and the flawed concept of MSY lay seeds for future trouble. Nonetheless, other nations signed on to the convention, and by 1975 some 22 nations, including the Eastern Bloc nations, were participating fully.

Maximum sustained catch or yield is an intuitively attractive term and would seem to be definable as the largest yearly harvest that would not harm a stabilized population. Conceptually, MSY in ICNAF made initial sense. To find MSY, one first determines from the stable population what is the yearly production surplus to that required to the maintaining of the population at its present size. You may then harvest that surplus, although you should bear in mind that harvesting more than that surplus will reduce the population and thus reduce the size of next year's surplus. On the other hand, you can invest the surplus and thus begin to rebuild a depleted population toward the carrying capacity of the biologic system. All very clean conceptually, but it was far less clean in a multispecies fishery immersed in a process of international decision making. As an end-all position, MSY soon came into trouble (24, 25).

Furthermore, by 1965 the smaller foreign trawlers were being replaced by large modern stern trawlers largely from the Soviet Union and the Eastern Bloc countries, and they entered the fisheries in the northwestern Atlantic with a vengeance. As a result, in 1967 the largest landing of haddock ever caught occurred in the western Atlantic, and the phrase "stable population" could probably not be applied to any fish stock. Because of community instability resulting from the excessive harvest, from then on, neither MSY nor any landing quotas derived therefrom were able to bring the ecosystem and its more than 30 stocks back into a rebuilding mode.

Although each of the Eastern European nations had become signatory members of ICNAF and most appeared to be following the rules, U.S. fishermen soon began contending that no matter what method was used to reduce harvest—national quotas, seasons, two-tiered quotas, mesh size regulation, even closed areas—nothing appeared to be halting decline in fish stocks. Privately, most federal fisheries workers agreed with the typical fisherman. As ICNAF's flawed reliance on single-species MSYs became more obvious, U.S. scientists looking at the entirety of the offshore ecosystem began recommending "ecosystem management" to describe its more holistic management approach to a multispecies resource (25). Ecosystem management would only become part of the environmentalist's lexicon when used by the U.S. Forest Service in the late 1980s to describe its holistic approach to national forest management (26, 27).

During the early 1970s, because of increasing U.S. pressure, ICNAF imposed more and more stringent rules. The scientific meetings between nations became more acrimonious as estimates of standing crop continued to trend downward, and each nation tried for a larger share of a smaller pot. Finally the U.S. industry convinced Congress that ICNAF, no matter how well intended, was a practical failure. Legislation was then proposed that embodied certain principles from the U.N.'s 1960 Convention on Fishing. Particularly germane was the concept that the "coastal state" (nation) had not only the right but also the responsibility to ascertain the total allowable catch not only off its coast but also at distances well beyond its territorial limits. Potential harvest beyond its needs would become available to other nations; in effect, the coastal state became the resource manager. This resulted in the Magnuson Fisheries Conservation and Management Act of 1976 (MFCMA), sometimes better known as the Two Hundred Mile Limit Bill. Although the United States neither extended its sovereignty nor assumed further territory, it did assume the exclusive right to manage most of the marine fisheries resources out to 200 miles (exceptions included the highly migratory fishes such as tuna). Not until 1987, at the end of his second term, did President Reagan sign a presidential proclamation raising the U.S. territorial limit to 12 miles, a proclamation not yet backed by congressional action. The Magnuson Act replaced all ICNAF regulations and took effect on March 1, 1977. In effect, ICNAF was abrogated by the United States. Shortly afterward, ICNAF was replaced by North Atlantic Fisheries Organization (NAFO) for management of resources in international waters beyond 200 miles and as a forum for international discussion.

The Magnuson Act created eight Regional Fisheries Management Councils composed of individuals with fisheries experience, nominated by governors, and selected by the Secretary of Commerce. The councils were charged with the responsibility of drafting management plans (MPs) appropriate to resources in their region and to do so on a stock-by-stock basis. Although initially, MPs were "single-species plans," it would soon become apparent that a single overarching MP was necessary if one were to manage a multispecies fishery such as that of the Gulf of Maine, Georges Bank, and the waters off southern New England. Nonetheless, on March 1, 1977, the fisheries within the 200-mile zone came under the management control authorized by various preliminary MPs first drafted by National Marine Fisheries Service (NMFS) scientists and then approved by the council and the Secretary of Commerce. Later the councils would acquire their own scientific staffs to process recommendations from NMFS into additional plans. By 1980, the New England fishery had two MPs, the Atlantic Groundfish Plan and the Atlantic Herring Plan, with others in preparation (28).

During the 1980s, each council separately, or in cooperation with an adjacent council prepared plans, had them accepted, and with the passage of time submitted amendments as necessary. By January 1994, 34 Fishery MPs (FMPs) and 5 preliminary MPs (PMPs) were in effect with others under development (21). The Code of Federal Regulations (CFR) (29) details federal regulations needed to implement plans.

According to the 1976 Act, MPs are to be based on the "best available scientific evidence" and from such evidence, a mathematical determination of MSY is to be created. Unfortunately, by 1977 the biologic flaws behind MSY were well known (24, 25). Regardless, MSY was determined. Next, its harvest level was lowered to a level based on ecologic, economic, and social considerations. The political act of blending these three disparate considerations together was at the heart of plan construction and the result was optimum yield (OY) (30). Next the council was to allocate OY among the various components of the U.S. fisheries—recreational, commercial, gill netters, large trawlers, small trawlers, and so on. Finally, if the U.S. fleet were unable to fully utilize OY, the Department of State would develop Governing International Fishing Arrangements (GIFAs) to permit foreign fleets to harvest the balance of OY. Thus, the United States acting as a coastal state would meet the intent of the 1960 Convention on Fishing and serve not as an owner of private property but as a prudent resource manager of a common property resource.

From its earliest meetings in 1976, the New England Council wrestled with a series of individual species MPs, each of which would appear to require an elaborate interaction with U.S. Coast Guard vessels and aircraft responsible for both monitoring and enforcement. The process became more and more involved, until finally, the single-species process reached a state of near collapse under the weight of its own complexities, and by 1979 management direction began to change toward a form of multispecies management. That was not easy.

By 1985, the FMP for the Northeast Multi-Species Fishery had been prepared and accepted by the New England Council. This had occurred in consultation with the Mid-Atlantic Council because certain species had distributions that overlapped the jurisdictions of the two councils. Although this FMP has subsequently been amended five times, with Amendment 5 taking effect 1 January 1994, the process of resolving uncertainty has still not occurred. By November 1994, Dr. Vaughn Anthony, chief scientist, NMFS, Woods Hole, contended that "it could take up to 15 years to rebuild cod, yellowtail and haddock stocks on Georges Banks—and only if urgent new measures are taken" (31). The battle over who gets to catch the last fish had still not ended for New England.

Success of the 200-Mile-Limit Theory: 1976–1982

Despite the difficulty the New England Council has had in deciding how to manage a multispecies resource, from the perspective of the U.S. fisherman, the first years of the Magnuson Act in New England from 1976 to 1980 in retrospect were a considerable success. The foreign fleets were now either under effective management control or had disappeared from sight. Unfortunately in 1976–1977, even before the Act's implementation, premature enthusiasm began spreading throughout the fishing industry. In response, the banking community eagerly lent money to build new vessels and to modernize, enlarge, and upgrade the New England fleet. For a time, the industry profited from the decrease in fishing pressure because the foreign fleet was gone. Gradually, however, the reduction in foreign pressure would be replaced by U.S. pressure.

Furthermore, considerable disagreement existed between the industry and scientists over the amount of fish that could safely be harvested while allowing stocks to rebuild. Coming into conflict were short-term and long-term objectives. In addition, there were several fishing seasons in which the opinions of the fishermen seemed more accurate than mathematical assessments created by biologists. Although assessments of currently available stock might not be seriously challenged, any forward projections of the rate at which resources would repair themselves were seriously challenged. Most of the groundfish species involved lay vast numbers of eggs, and any year class resulting therefrom might either be immense or vanishingly small depending on the vagaries of nature. Whether huge numbers of cod eggs will be swept off the banks to die in deep water or remain on the shelf to grow and ultimately enter the fishery as a dominant year class 3 years hence remains a chaotic mystery well beyond biologic predictability (32). The best predictor of an oncoming year class next spring still remains the number of juveniles collected during the previous fall oceanic survey (33). The success of spawning next spring can only be conjectured upon and thus becomes an uncertainty regarding the rate at which a population might rebuild. Anyone making a predictive stock assessment, particularly one extending years into the future and relying on spawning yet to occur must deal creatively with nonexistent facts. As a result, most predictive assessments are usually a contentious point between fishermen and managers (34, 35).

We Have Met the Enemy—Us: 1982 to Present

As the 1980s progressed, it became evident to more participants that the fishing industry in New England had grown to a size in which it had simply replaced the foreign fleet fishing power of the mid-1970s with that of its own. Furthermore, with every new device added to a wheelhouse, a Loran C or satellite navigation, or improved net construction and enhanced ship design, the ability of a single vessel to harvest groundfish faster and more efficiently gradually more than offset the absent foreign fleets. Concurrent with this resource pressure, the American diet also shifted somewhat from red meat to fish and markets and restaurants throughout the country featured Boston scrod or Boston bluefish at premium prices. By 1994 in central Ohio, three major grocery chains (Big Bear, Cub, and Kroger) sold haddock and cod at $10.00 a pound when they could get it. Furthermore, when they are available, these species share a display case with mahi mahi and yellowfin tuna as well as Dover sole daily flown in fresh. Yellowfin tuna reached $19 a pound in early 1994. Farm-raised catfish or ranched salmon from the growing aquaculture industry provide inexpensive alternatives at $4.00 a pound and are in direct competition with chicken and pork. Market shares for farm-reared catfish and similar products are expected to rise to 10% of the fish consumed in the United States by the year 2000.

One hundred million metric tons of commercial fishes is a generally quoted but conservative estimate of worldwide MSY (22). However, latest U.N. estimates of total world catch have been declining from this figure for several years, despite increasing fishing pressure. This is clearly a sign of widening overfishing. Although overfishing may be the prime reason, habitat loss in estuaries, salt marshes, and mangrove swamps as well as coastal pollution have also added to the decline (36). Meanwhile the world population increases, and the rising demand for fish protein is increasingly not being met unless a nation has the financial resources to import it and often from those most in need of it for their own population (22).

The Unresolvable Future

The future of commercial fishing and the industry's interaction with the American consumer contain a series of microproblems to be solved, and failure to solve them will lead inevitably to a much larger problem. Most now understand how an incremental loss of elements in coastal marshes and estuaries reduces the ability of a natural system to rebound. One could therefore easily infer that a single development project along an estuary followed by the next leads stepwise to serious habitat loss. Likewise, in fishing, the impact of a single vessel, followed by the effects of the second and so on, follows a similar logic. Although some of the fish species off New England may someday be cultivated, it seems likely that much of this natural resource will never be fenced in, owned, and thus prudently managed; the dilemma of common property resource management persists.

The U.N. 1960 Convention, however, addressed the belief that a coastal state has both the power and the responsibility to protect its coastal and offshore fisheries resources (10, 15, 16). Whether state and federal legislators and regulators can face fully this charge seems problematic. Closing the cod fishery off Newfoundland for 2 years or more by the Canadian government to permit the resource to recover is a strong response with serious regional financial, political, and social implications. A reduction by 50% proposed by the NMFS in the New England cod harvest has similar implications (37). Furthermore, closing the New England fishery without buying back boats probably would only lead to boats moving to another region. In essence, if the number of boats in a region remains the same or if the number declines but those remaining subsequently improve their fishing capacity, the decline will continue and nothing has changed.

Marine Fisheries Resources Within the 3-Mile Limit

Hitherto, this discussion has been directed to the management of waters beyond the territorial limit, but we need to examine state management agencies and their own power to regulate coastal fish stocks within the 3-mile territorial limit.

The concept of the freedom to fish in international waters has traditionally extended to marine territorial waters of the United States as well, and the states have been slow to exercise control over these resources. However, coastal and/or estuarine species residing within the 3-mile limit are under state agency jurisdiction according to the Magnuson Act.

Earlier arguments over whether state or federal jurisdiction controlled these resources have generally been resolved in favor of the state (38). However, the Magnuson Act does permit preemption of state management of a resource. That resource must extend into the 197-mile zone beyond territoriality and be covered by a MP. If the state takes action that will "substantially and adversely affect the carrying out of a fishery management plan" then federal preemption is possible. The federal government has been exceedingly reluctant to attempt such a takeover.

Accordingly, under usual conditions in coastal waters, each state agency decides whether to apply standard fishing controls and if it chooses, it may prohibit harvest, set catch limits, charge license fees, and so on. However, those species that are anadromous, wander offshore, or move laterally into the waters of adjacent states present additional jurisdictional problems. Anadromous species such as striped bass, Pacific and Atlantic salmon, American shad, and others that spawn in the streams of one state, leave parent streams as juveniles and then migrate onto the shelf present other concerns when they become vulnerable to either recreational or commercial fishing in the territorial waters of another. The problem of interstate research and management coordination for the Atlantic coast species has been under investigation since the 1940s in cooperative programs run by the U.S. Fish and Wildlife Service (USFWLS) and NMFS. These programs were often begun following political encouragement from the Atlantic States Marine Fisheries Commission (ASMFC).

In 1942, 17 jurisdictions along the Atlantic coast of the United States became linked through the Atlantic States Marine Fisheries Commission (ASMFC) when Congress in 1942 authorized its existence under the Interstate Commerce Clause of the U.S. Constitution. Until recently, the commission served primarily as a forum for interstate dialogue and, when necessary, lobbied on behalf of various federal anadromous and coastal fisheries programs being run either by USFWS or NMFS. Similar commissions were also created for states along the Gulf of Mexico and the Pacific coasts (39).

In 1979, ASMFC changed its purpose and became the forum through which states began developing cooperative management programs for fisheries resources in waters that are primarily territorial, and there are now 19 such plans in place. One, the Bluefish Management Plan of 1988, was accepted by the three Atlantic Coast Regional Fisheries Management Councils. Primary legislation and regulations concerning harvest remain with the states except in the Striped Bass Plan. The Atlantic Striped Bass Conservation Act of 1984 provided federal primacy over state control and allows for federal regulations to supersede state regulations not issued in compliance with the Act (39).

An effort to bring state-controlled coastal fishing for weakfish under similar federal control was proposed in 1990. Later, federal legislation was proposed for all species and in 1994 it passed (31). The Atlantic Coastal Fisheries Cooperative Management Act of 1994 makes the ASMFC responsible for the management of coastal resources, and state regulations will be required to comply. This process is not unlike federal standards requirements imposed on state environmental protection agencies and on the surface mine reclamation activities monitored by the states. State regulation of coastal fisheries resources along the Atlantic coast has long been a complex maze of rules that changed once across a state line, and many of these regulations had no biologic intent. This direction seems inevitable.

Methods of Assessment

The quality of the stock assessment lies at the heart of the controversy over how much pressure the resource can stand and how soon it will recover given a particular regime of management. A number of books review assessment techniques (14, 35, 40–50, among others). A number of shorter papers have also reviewed the process or reported on research results (36, 37, 51–68). One can begin a review of assessment techniques by reading Ricker (44) and then Hilborn and Walters (35). The papers by Anthony and by Sissenwine look in detail at the changes in the New England fishery under MFCMA. Schnute and Richards (62) summarize needed progress in the assessment process.

Whatever assessment techniques are chosen, effective resource decisions inevitably depend on knowledge of the size and dynamics of the resource. A trawler captain cannot merely look out of his pilothouse and assess the resource, nor, obviously, does a rational assessment emerge solely from a computer model. In this regard, the process differs from assessing one's personal finances. One can make sound decisions based on the size of one's checkbook balance. A checkbook balance is simple, and assessing the resources of a forest is only slightly more difficult. Trees can be counted, measured, scaled, and an estimate can be made relatively easily of the value of a harvest this year versus delaying harvest until next year. Unfortunately, fisheries resources, because they are hidden under water are far more difficult to assess. Further, since marine fisheries resources are spread over vast distances, and because most data needed for assessments must be gathered from ships, the cost of these assessments is very high. Furthermore the trawler captain does not collect the needed data; that is a responsibility of NMFS, and it is done completely at taxpayer expense. Because of such obstacles, assessments are not only expensive but become highly complex, requiring elaborate, statistically valid, sampling procedures that are often not easily justifiable to a layman. Unlike the other examples, stock assessment of marine fisheries resources is uncertainty personified. Finally, even the assessment of standing stock of fishes is relatively simple when compared with predictive assessments that must describe the direction of change in stock size over the coming 5 to 10 years under a given management regime.

An early assessment method called yield per recruit (YPR) can be related to an algebraic method first used by Thompson and Bell (69). Ricker (44) uses this method to demonstrate changes in the harvest one might expect from several populations of 1000 recruited fish exposed to different levels of pressure (fishing mortality) during their lives in a fishery. This example uses average weight at age for a population that declines according to a known (and fixed) rate of instantaneous total mortality (Z), a term composed of the sum of instantaneous natural mortality (M) and instantaneous fishing mortality (F). The larger the fishing mortality, the fewer fish reach older ages when each would weigh more. As the population declines numerically with age, fewer, though heavier, fish are alive to be harvested. A simple totaling of the weight of fish harvested when plotted against increasing values of F traces a convex curve, and the coordinates of the apogee become MSY on the Y-axis and optimal fishing mortality on the X-axis. This method, though essentially algebraic, shows the impact of increasing levels of instantaneous fishing mortality upon a stable population. Although this process easily creates a convex curve with MSY at the apex, it also assumes constant levels of growth between year classes and a constancy in year-class size (1000 in Ricker's example). These two conditions are scarcely ever met in reality. Although perhaps useful in a classroom to demonstrate the interactions between growth, natural and fishing mortality, this method fails significantly in practice.

A more elaborate YPR method, developed by Beverton and Holt (14), is more mathematically elegant in comparison, but it, too, suffers, because any YPR method assumes a near constancy in recruited year-class size year after year. However, like the Thompson and Bell method, it is useful in enabling one to foresee the effect of overfishing, though it is not particularly useful in setting criteria for preventing it. This is particularly so in assessing a species with great variation in recruitment, the rate at which individuals of a particular year-class enter the fishery at a size to be captured. Further, both methods depend on a population stability not evident in the current rapid decline in stock size. Both also require one to be able to effectively divide Z (Beverton and Holt (14) and Ricker (44)) into its components F and M. Any systemic failure resulting in an over- or underassessment of F will provide incorrect evaluations.

Parenthetically, F is related more to the capacity of the fleet to remove a fractional portion of the remaining population than it is related to the number of vessels present in the fleet. Accordingly, the agility of the industry to improve the catchability of an individual boat by improved navigational devices, improved nets, and, ultimately, through improvement of new vessels replacing older vessels changes F without necessarily changing the number of vessels in the fishery. Thus, F is related more directly to the ability to catch effectively than it is to the number of units at work at any instant, and, with the passage of time, the catchability factor of each unit inevitably increases, making it difficult to control fishing by limiting the number of licenses issued. This argument is fully developed in Ricker (44) and in Hilborn and Walters (35). Nonetheless, YPR was at the forefront of assessment technology until the late 1960s when the virtual population analysis (VPA) technique came to the fore (Ricker [44], Pope [70]).

Gulland (41) credited Jones (71), Gulland (72), and Murphy (73) with independently developing a quantitative method that can compute actual stock size, although by indirect means, through the use of Baranov's catch equation (74). Gulland (41) calls this technique "cohort analysis," but others use VPA. It has also been applied to freshwater situations (lake trout by Fry (75) and largemouth bass by Cole (76)). Various modifications of this system now exist including catch-age analysis (CAGEAN), first developed for assessment of Pacific halibut but now even being used to assess walleye population size in Lake Erie (77–79).

To be effective, cohort analysis requires detailed information on the total harvest of a species and good sampling and aging techniques that will break down the harvest into size of the component year classes or cohorts. Any systematic inaccuracies in aging or in deliberate or accidental misreporting of the harvest weakens this process. Further, any failure to account effectively for the accidental harvest of juveniles subsequently to be discarded from another-directed fishery also weakens this assessment. The process itself begins with an assessment of Z derived from the harvest and an estimate of F as it was applied to the population during the harvest. From a forward calculation, it is then possible to estimate actual fishable stock size and to look for oncoming changes in year classes. As a result, recommendations can be made for the harvest potential on a species-by-species basis. This method is currently used for most major stock assessments that form the basis for MPs under FMCA. Unfortunately, one cannot manage the 150 major species groups off the United States by this process because of the limitation of the cost of data acquisition and time constraints. However, FCMA does require "the best available scientific advice" as the basis for plans, and, in general, cohort analyses constitute that "best available information."

The requirement for ongoing data collection on the more than 40 species and species groups in the northeastern fishery is a real cost of management that is buried within the NMFS budget. One might wonder why the harvesters of the resource ought not to be required to pay a significant part of that cost instead of this cost being passed on to the general taxpayer. Further, because funds sufficient to cover the true cost of these management requirements have not been appropriated, NMFS apparently has shifted personnel and other resources from other legislatively mandated duties. More of this will be considered later.

If one is to provide yearly stock assessments, NMFS must gather catch data from the commercial and recreational harvest as well as any significant by-catch. One must reduce the number in the total harvest into estimations of year-class strengths and do so in a timely fashion such that assessments can be developed for council consideration. These yearly assessment and status reports need also to be made available for public review and use during council hearings and any subsequent challenges (80–82). After the council conducts public hearings and reaches its conclusions, it forwards those decisions via the NMFS regional office to the Secretary of Commerce, where, if accepted, they are then translated into regulations via the Federal Register for implementation.

Thus, the quantitative process begins with a valid estimate of the condition of fish stocks and their ability to sustain harvest of a given level. Inevitably, some fishermen will believe that NMFS biologists have been prevented politically from speaking the truth. Unfortunately, given the nature of the methods now available for assessment, no one will ever know the truth in time to affect optimal regulations, a process that builds uncertainty on either side of the issue.

Another part of the assessment process has been the larval fish and juvenile fish surveys and fall standing crop surveys conducted across the Atlantic shelf as well as through Georges Bank, the Gulf of Maine, and elsewhere nationally. Even omitting the Gulf of Maine, Georges Bank alone is larger than the state of Massachusetts and has a number of different bottom habitats, not all of which are uniformly selected by the over 100 species of fish on the Bank. Of this number, nearly 40 are commercially harvested. To estimate the current standing crop of juveniles soon to recruit into the fishery as well as the current adult population, NMFS has conducted bottom-trawling surveys each autumn since 1963. These surveys usually occupy some 300 trawling stations on Georges Bank and in the Gulf of Maine (33). These surveys are still being conducted from the R/V *Albatross* and R/V *Delaware,* two National Oceanic and Atmospheric Administration (NOAA) research vessels now nearly 35 years old and due for replacement.

Year after year, using a process of stratified random sampling, trawling stations within different bottom types are conducted using half-hour hauls with calibrated gear. The task is to estimate species rank and abundance and to assess the size of the population of juvenile cod, haddock, yellowtail flounder, and other species that resulted from previous spring spawnings. If each half-hour trawling were to sweep an area 30 m wide by 6400 m long, each sample would cover 0.192 km^2 of sea bottom. If all 300 stations each autumn were made only on Georges Bank—some are also made in the Gulf of Maine—they would sample no more than 57.6 km^2 of bottom. Georges Bank is larger than Massachusetts (27,340 km^2), but had this level of sampling been applied to Massachusetts, only an area 0.21% of the total state would have been sampled. Although this level of effort has been shown to be statistically adequate, fishermen have frequently challenged its outcome on commonsense grounds. Despite concern about sampling adequacy, autumn surveys of juveniles have nonetheless provided the best evidence for estimating size of an oncoming year class. Further, these surveys appear to provide valid estimations of the fish community and have developed evidence of long-term changes in species composition (83). Spiny dogfish, little skate, and winter skate constituted more than 71% by weight of the total catch of the spring 1994 bottom-trawl survey of the northeast U.S. fishery ecosystem. Atlantic cod, haddock, and yellowtail flounder accounted for only 3% of the survey catch. This biomass ratio (24:1) continues a recent trend to more sharks and rays and fewer groundfish (83).

In effect, the data set from these surveys has become one of the longest-running ecologic data sets in existence. Begun in 1963, it is a rigorous and statistically correct survey and has validated many of the changes in the fish community subsequently agreed upon by the commercial industry (20, 33, 59, 80, 81, 82, 84, 85). The groundfish survey process developed by the Population Dynamics Branch of the NMFS Laboratory at Woods Hole has thus provided a long-term view of a complex ecosystem from which certain elements have been overharvested. This analysis has led inexorably to a more holistic view of the management of an ecosystem; unfortunately, how to assess and manage such a multiple-species fishery is both mathematically and biologically very complex (35, 57). The need for multiple species assessment was emerging during creation of the 1976 Magnuson Act, but the Act was initially single-species in its conceptualization (56).

Initially, if single-species assessments were needed, NMFS efforts could be directed at accumulating sufficient growth and harvest data on the 40 species of commercial importance on Georges Bank. In 1975–1976, based on work done to satisfy ICNAF negotiation requirements, NMFS prepared several PMPs for the New England fishery, which then went into place with the implementation of the Magnuson Act. During the next 6 years, NMFS, as well as NEFMC and its management staff, and several consultants created additional drafted single-species plans while amending existing PMPs. Drafts for MPs for ocean perch and pollock were shelved in 1980 with the realization that the complexities of management and implementation would grow exponentially as the details of each single-species plan were layered on top of the one previous. By 1980, it was apparent that a single multispecies plan would be the only possible solution. This marked a major change in direction from the single species outlook that began during ICNAF when, from the U.S. perspective, the predominant worries were cod, haddock, and yellowtail flounder. This need for multispecies management also foresaw a further need for an integrated management approach that considered the ecologic ramifications of predator-prey relationships and the probability that irreversible community changes would occur as the heretofore dominant species (cod, haddock, and yellowtail flounder) were pushed to near commercial extinction. The list went on. These questions opened new assessment requirements and moved beyond the YPR and cohort analysis determinations that were single-species in nature. This also changed the nature of the evidence needed by councils and foresaw a need to look at the entire ecosystem of Georges Bank. Fortunately, one of the continuing missions of the NMFS Laboratory at Woods Hole had been the continuing accumulation of such basic ecology.

In 1985 NEFMC, in consultation with Mid-Atlantic Fishery Management Council, approved a document entitled "Fishery Management Plan. Environmental Impact Statement. Initial Regulatory Flexibility Analysis for the Northeast Multi-Species Fishery," which was the multispecies management plan (86). Unfortunately this document and the four amendments that followed only succeeded in deferring a solution to the problem of stocks that were declining faster than the regulations could catch up with them. In October 1993, Amendment 5 was approved by the council and on January 1994 was signed into existence by the Secretary of Commerce. Among other restrictions, landings of cod were to be cut in half (37). The economic and social ramifications of this to the New England fishing industry were expected to be severe. By late 1994, further efforts to close down the fishery were in progress. Meanwhile the Canadian government had moved even more vigorously and placed a 2-year moratorium on cod, an effort that could have a nearly catastrophic impact on small Canadian fishing villages.

As prices for Atlantic cod and haddock in midwestern U.S. fish markets approach $10.00 per pound, increasing prices only heighten motivation to go to sea for yet one more landing. Meanwhile the biologist remains confident that the assessments are essentially correct, while the boat owner sees family expenses and goes back for one more chance to make a living fishing. On January 1, 1994, when the Secretary of Commerce approved Amendment 5 to the Groundfish Plan, he also approved an infusion of $2.5 million to help rebuild and reshape the industry. Whether current efforts to put up to $30 million federal funds into efforts to buy back boats, provide offset funds permitting owners to tie up and stay home will bring about restoration of the New England fishing industry is not known. The cost to reshape the New England fishery alone will greatly exceed this figure and do little to address the economic and social issues in the towns and cities affected. It will indeed be painful for a community to decide that "we have truly met the enemy and it is us," to modify Pogo's words.

Throughout the life of the Magnuson Act, it has been under close scrutiny, and many studies have suggested changes some of which have been implemented in numerous Amendments (43). In 1993, the American Fisheries Society released a position paper (87) listing 11 points requiring attention in any reauthorization of the Act. The National Research Council (43) issued a report with seven recommendations. The following brings together their contentions with those of the author.

Controlled Access

The current Magnuson Act assumes "open access" but permits "limited entry" as a management strategy, although it requires that any effort to limit new vessels from entering be justified, something which is politically difficult (88). In 1976, the Act was primarily focused on repairing the resource by excluding, or at least controlling, foreign fishing. To fill the gap, new U.S. vessels were expected to enter the fishery. They did it so effectively that, by 1990, of the 156 stocks of U.S. living marine resources studied, 43% were now being overfished and yet, once NMFS assessments reached councils, the councils came under political pressure to raise the allowable catch. Even were a council to propose a form of limited entry to exclude new fishermen in order to control total effort, or in some other fashion bring about "limited access," it must anticipate that the current fishermen will improve the fishing capability of their licensed vessels. As the industry improved technically and the wheelhouse of a modern fishing vessel acquired more sophisticated electronics, the effort of the same boat inevitably increases. Making everyone fish the old-fashioned way only increases inefficiency, a cost inevitably passed on to the consumer—a modern version of the Luddite approach to an industry. Buying out licensed vessels by using funds raised by charging substantial license fees has not been tried. The current solution has been to recommend that each vessel be provided with individual transferrable quotas (ITQ) that could be captured or the right sold to another vessel.

Conservation First, Allocation Second

The initial Act described a process for developing final management plans (FMPs) that, seen from a distance, should have worked. Explicit in the Act was the concept of MSY, and controlling effort is implicit to MSY. From MSY flowed total allowable catch (TAC), a reduced yield, and from TAC came OY, a "sensible yield" based on biologic, social, and economic constraints. If true, then conservation preceded allocation and next came the allocation of the allowed balance amongst the various users. Early during allocation negotiations, sport fisheries were initially underrepresented on councils and did not do well in the allocation process, but this has been corrected in part. Far worse, several councils tended to override "best scientific evidence" and recommended harvests larger than appropriate. Once these recommendations reached the Secretary of Commerce, additional political pressure was then exerted to obtain secretarial concurrence, and the result thwarted the entire process. While stocks continued to decline, arguments became more strident politically: "shut us down and our coastal communities will become deeply depressed economically." Political pressure and self-interest are cited as forces causing the problem to worsen rapidly and whether science would ever prevail became a serious issue and particularly so when the science itself was questioned (43). Nonetheless, one-third of the 33 FMPs in force in 1993 had controls on access (67).

User Fees

Except for the nominal cost of licensing of boats to meet safety regulations, not only is access to the resource unlimited but there has been no cost of that access. Even private use of the public western rangelands is more expensive and recovers some share of the cost of management. By contrast, not only is the owner of the resource allowing it to be harvested for free but the owner is also paying all costs of the management of the resource. In response, one could expect to hear that under the spirit of the 1960 LOS agreements, the United States is *not* the true resource owner but only its manager. On the other hand, our willingness to pay the cost of management must confer some measure of ownership and justify recovery of some of the cost from the boat owner. In essence, the entire costs of data gathering, ship-time costs of NOAA research vessels and of Coast Guard surveillance vessels and aircraft, analysis, reporting, the cost of the councils, and other expenditures have been transferred to the U.S. taxpayer, and there is no cost to the user of the resource. As the budget for NMFS declines ever more severely, its inability to provide effective information becomes only more obvious and thus less credible. The Sport Fishing Institute (89) describes a requested increase of $6 million over the $224 million appropriated for FY 93.

> Even with the requested increases NMFS will lack the necessary resources to *adequately* address critically important issues. . . . Nearly 43 percent of the marine fish stocks for which assessments have been conducted are presently overexploited. The state of over half the remaining stocks is unknown.

Further, these requirements from the Magnuson Act are not the only NMFS responsibility—marine habitat, whales, endangered species, only add to an agency's overfilled plate.

Risk-Averse Decision Making

Decisions concerning harvest levels of fisheries resources can best be described as uncertainty personified. Assumptions about the size of a multispecies resource and its ecologic interactions, even the biologic dynamics of a single species and its interactions with a vast changing habitat, make for expensive shipboard research. Out of the process of acquiring biologic data comes a shadowy view of what is likely to be going on but one well hidden under water, and the chaotic ecologic process involved perhaps exceeds the capacity of current science to make accurate projections beyond several years into the future (8, 32, 35, 56, 90, among others). Nonetheless, assessment techniques have improved markedly and have overcome some of this mystery. Yet, of the 13 stocks now being regulated in the multispecies MP, each is overfished. Advice from NMFS to the New England Council is to cut effort in half if these are to be allowed to rebuild to a level permitting MSY (37). MSY is defined as "the largest average annual catch or yield that can be taken over a significant period of time from each stock under prevailing ecological and environmental conditions" (50 CFR 602.11[a]). Whether MSY is an adequate foundation for optimum yield was the question raised by Sissenwine (25), and the answer remains "no" as it was then, but the alternative foundation is only imperfectly being laid.

Overfishing

Overfishing is defined in the Act as a "level or rate of fishing mortality that jeopardizes the long-term capacity of a stock or stock complex to produce MSY on a continuing basis . . ." (91).

> The only direct control available under the Act is to adjust fishing mortality, which may be accomplished in several ways (e.g., by establishing or adjusting time/area closures or limits of catch, mesh size, vessel days, or the number of vessels entering the fishery. . . .

(50 CFR 602) This standard of the seven against which councils must measure their proposed MPs is reasonably clear and constitutes a way of reducing long-term impact by imposing short-term pain. It thus has been politically difficult to meet.

National Oversight

Absent from the original act and its subsequent amendments was any national review of ongoing practices and proposed changes. The American Fisheries Society proposed "A national board . . . to review specific issues of regional or national importance. . . ." Although a change in the process is clearly needed, such a board might only accumulate spokesmen for the same pressure groups and fail at the onset to affect badly needed change. The National Research Council report (43) also called for a number of revised responsibilities.

Leadership by the Secretary of Commerce

An appointment to a seat on a council occurs following gubernatorial nomination and are thus political from the start. In the past, each appointee has been considered likely to be beholden to the industry segment from which he or she came, and thus all management decisions made by a council are the political amalgam of these biases. Somehow the charge to councils must be changed to make them "trustees of the resource" rather than negotiators on behalf of their own biases. These biases are not "conflict of interest" in normal legal terms, but they limit the perception a member has of the whole problem before a council. Finally, given the many other responsibilities assigned to the Secretary of Commerce, approval of MPs must be rather low on the priorities of that office.

NMFS Budget

Chandler (92) uncharitably begins a review of NMFS with a quote from an unidentified congressional staff member: ". . . the National Marine Fisheries Service is one of those agencies whose whole is less than the combined total of its composite parts." He then incorrectly states that 2 million square miles were assigned to NMFS by the Magnuson Act in 1977; the Act assigned MP development to the councils and not to NMFS. A council proposes the management strategy and when the MP is found to be in accord with the seven standards (93) and with congressional intent, the plan *becomes* the management. A MP must conform to the seven federal standards in the law and will be scrutinized by a NMFS regional director, but ultimately each FMP reaches the Secretary of Commerce for final approval. The NMFS role is thus advisory, not regulatory. Few other federal agencies, except the Bureau of Land Management, have been as severely criticized for failure to do what they neither have the legal mandate nor resources to do. Unlike both the U.S. Forest Service and the Bureau of Land Management, which Congress does charge with regulatory authority, NMFS, at best, can only recommend management proposals to the councils.

Conflict of Interest

The potential for conflict of interest by council members is more a bias in favor of one's segment of the industry than a true conflict of interest. However, gubernatorial appointments could almost be ranked by the amount of political pressure one's constituents might raise on their behalf. Although this might pit each harvesting group against every other, it did not make for a broad, unbiased view of the need for reducing fishing pressure. Early on, the recreational element was vastly underrepresented and the conservation/environmental group had no representation. How the Act will be restructured to allow industrial participants without repeating this error is not yet clear.

Fisheries Expertise

The term "fisheries expertise" is a requirement for those nominated to councils and inevitably brings to council those representing a segment of the industry with a clientele to speak for. If the resource is in fact being managed in perpetuity for the public at large, a significant number of citizens without such clientele should also serve on councils. A number of conservation organizations and coastal universities have competent individuals able to speak for the resource.

Reappointment of Voting Members

The American Fisheries Society position is that voting members should be automatically renominated to insulate them from political pressure from the governor's office and others in the state. This may well only transfer the political sensitivity from the governor's office to the secretary's office, but it would allow a highly qualified and outspoken individual to be selected for a second term.

Habitat Conservation

In an era of "ecosystem management" and other efforts to look at the entire habitat rather than a species-by-species management process, the Magnuson Act has become dated. Toward this, The American Fisheries Society recommended that an eighth standard be added to the law against which MPs are evaluated: it would address the issue of conservation and management of entire habitats. While this added authority is doubtless needed, it will cast an entirely different perspective on the intent of the Act. The National Research Council (43) also speaks of the failure of the Act and of the FMPs emanating from it to look at the broader aspects of commercial fishing on the ecosystems affected. Such ecosystem management represents a new philosophy that is not reflected in congressional intent behind the original Act.

Secretarial Review

Public hearings and subsequent filings on elements within MPs and amendments have often been acrimonious, but they provide a fair way for citizen input. Behind-the-scenes lobbying even at the secretary's level has often offset this fairness.

Coordination with Other Acts

As the Magnuson Act is implemented, it converges, conflicts with, and overlaps a number of other well-intentioned congressional mandates from the Endangered Species Act to the Marine Mammal Protection Act and beyond. The CFR (94) describes in anguishing detail the process for implementing the Magnuson Act and its subsequent MP. Interleaving congressional intent behind this act with other acts impacting resources (e.g., Marine Mammal Protection Act), their conservation and management is clearly part of the cost of operating this unusual business in which the resource is claimed by the United States, is owned in perpetuity for the national good, and is harvested rent free, henceforth to be sold to the U.S. consumer.

Highly Migratory Species

In 1976, highly migratory species, defined in the Act as some tunas and several billfish species, were excluded from the control of councils. The tuna industry's contention then was that tunas by moving through the jurisdiction of many nations should only be subject to international treaty control. The industrial interest in Pacific tunas left somewhat underprotected other tunalike species in the western Atlantic, particularly the king mackerel and bluefin tuna. In a 1990 amendment to the Act, highly migratory species were included if the species entered the waters of two or more of the councils, something likely to happen only in the Atlantic. The king mackerel declined sharply before it and the Spanish mackerel were covered by the Coastal Migratory Pelagic Resources of the Gulf of Mexico and South Atlantic Management Plan (95). At greater risk, the Atlantic bluefin tuna remains under international management by International Commission for the Conservation of Atlantic Tunas (ICCAT).

Anadromous Species

Pacific salmon, American shad, and Atlantic salmon, among others, spawn in the freshwater streams of North America and as juveniles move offshore often beyond 200 miles. The U.S. position has always been that the country of origin owns the resource throughout its life on the high seas. Several treaties are now in place to bring this problem to resolution.

Interaction with Other Legislation

Various conservation groups have asked that the bluefin tuna be treated as an endangered species. Although NMFS has been focused on the biologic research and data collection required for implementation of the Magnuson Act, it has other duties, not the least of which is the development of recovery plans for endangered marine species as required by the Endangered Species Act (ESA). They share this responsibility with the USFWLS, but by 1989 NMFS had only completed six approved recovery plans of the 21 species then placed under its jurisdiction (92). The sea turtle controversy in the South Atlantic and Gulf States put NMFS and the councils in a quandary—how to regulate shrimp harvest while reducing to zero the killing of endangered sea turtles as required by ESA and regulations thereto. NMFS and the two affected councils were accused of failure to discharge this duty because of political pressures from commercial fishermen. Countering arguments are lack of funds and a need to address the consequences of the Magnuson Act.

The bluefin tuna problem has not been solved by the Magnuson Act or the ESA. Michael J. Bean, Environmental Defense Fund (96): "... the failure of the law (Endangered Species Act) is funding. Congress never appropriated enough money to let the Fish and Wildlife Service make more than a small dent. . . ." The USFWLS budget for implementation of the Act is $30 million per year. For similar ESA responsibilities, NMFS is budgeted only $10 million. An inspector general's report estimated that $4.6 billion would be needed to restore to health currently listed species. If we continue to insist that no decision to develop or harvest can occur until adequate knowledge of the impact exists, then the costs of acquiring such data on which to make decisions can only be assigned to and borne by a public agency or by the potential user. Whoever does the task, the costs will be similar and will ultimately be passed to the consumer.

In 1994, MFCMA was due for reauthorization by Congress but was not acted upon and remains viable only under a continuing resolution. Although its flaws have been emphasized in this Chapter, one of the initial intents to control foreign fishing has been met. However, "the system of regional management councils established by the MFCMA has not been successful in preventing over-exploitation of fish stocks" (97). Clearly the new direction will be successful only if it contains fishing effort at some scientifically acceptable level and if it controls total exploitation of the ecosystems under its jurisdiction. What began in 1976 as a species-by-species management process has become one of great interactive complexity and has been called by some "ecosystem management." Capturing this elusive concept and putting it into federal law will be an uncertainty of a differing sort.

Conclusions

Probably nothing is conceptually as simple to understand and yet so difficult and complex to manage as a marine fisheries resource. Many of the species now under the umbrella of U.S. management are poorly known biologically, additional data are very costly to acquire, and thus the ability of a busy manager to assess highly diverse coral reef fish communities or even the complex changes underway on Georges Bank or in the Gulf of Alaska nearly beggars the imagination. Other species are quite well known, and yet the dynamics of their changing population structure also requires elaborate and costly monitoring. The term "best available scientific information" used in the Magnuson Act describes the standard for making harvesting decisions and yet, in the final analysis, that information, more often than not, is extremely thin. Nonetheless it becomes the basis for all future determinations leading to an allowed harvest. Lacking better information, aggressive harvesting practices are only irresponsible. The failure to practice risk-adverse decision making in light of the lack of full knowledge will only continue stock declines to the point perhaps beyond the capacity to return and we will continue to live with uncertainty in the marine fisheries decision-making process.

References

1. Graham M. Modern theory of exploiting fishery, and application to North Sea trawling. Conseil Int. Expl. Mer 1935;10:264–274.
2. Graham M. The fish gate. London: Metheun, 1941.
3. Gordon HS. The economic theory of a common property resource: the fishery. J Pol Econ 1954;62:124–142.
4. Hardin G. The tragedy of the commons. Science 1968;162:1243–1248.
5. McGoodwin JR. Crisis in the world's fisheries. People, problems and policies. Stanford, CA: Stanford University Press, 1990.
6. Matthews DR. Controlling common property: regulating Canada's east coast fishery. Toronto: University of Toronto Press, 1993.
7. Hanna SS, Smith CL. Attitudes of trawl vessel captains about work, resource use, and fishery management. N Am J Fish Mgt 1993;13:367–375.
8. Wilson JA, Acheson JM, Metcalfe M, Kleban P. Chaos, complexity and community management of fisheries. Mar Pol 1994;18(4):291–305.
9. Brittin BH. International law for seagoing officers. 5th edition. Annapolis, MD: Naval Institute Press, 1986.
10. Burke WT. The new international law of fisheries: UNCLOS 1982 and beyond. Oxford: Oxford University Press, 1994.
11. Bean MJ. The evolution of national wildlife law. Revised Edition. New York: Praeger, 1983: 255–260.
12. Bell FH. The Pacific halibut. The resource and the fishery. Anchorage: Alaska Northwest, 1981.
13. Anderson ED. International council for the exploration of the sea. Fisheries (Bethesda) 19(9):33.
14. Beverton RJH, Holt SJ. On the dynamics of exploited fish populations. UK Min Fish Invest (Ser 2) 19. London: HHMSO, 1957
15. Johnston DM. The international law of fisheries. A framework for policy-oriented inquiries. New Haven: New Haven Press, 1987:157–287.
16. Johnston DM. The international law of fisheries. A framework for policy-oriented inquiries. New Haven: New Haven Press, 1987:332–334.
17. Christy FT, Scott A. The common wealth in ocean fisheries. Baltimore, MD: Johns Hopkins University Press, 1965:153–191.
18. U.S. Government. The United States Government manual. Washington, DC: Superintendent of Documents, 1986:329–330.
19. Bean MJ. The evolution of national wildlife law. Revised Edition. New York: Praeger, 1983: 177–178.
20. National Marine Fisheries Service. Fisheries of the United States, 1992. Current Fishery Statistics No. 9200. U.S. Department of Commerce, NOAA, 1993.
21. National Marine Fisheries Service. Fisheries of the United States, 1993. Current Fishery Statistics No. 9300. U.S. Department of Commerce, NOAA, 1994.
22. World Resources Institute. World resources 1992–93. New York: Oxford University Press, 1992:178–179.
23. ICNAF. Report of the First Annual Meeting with appendices. Report No. 1, 1951.
24. Larkin PA. An epitaph for the concept of maximum sustained yield. Trans Am Fish Soc 1977;106(1):1–11.
25. Sissenwine MP. Is MSY an adequate foundation for optimum yield? Fisheries (Bethesda) 1978;3(6):22–24, 37–42.
26. Forest Service. An ecological basis for ecosystem management. USDA-FS Gen. Technical Report RM-246, 1994.
27. LeMaster DC, Parker GR, eds. Ecosystem management in a dynamic society. Proc Conf West Lafayette, IN, 1991.
28. Penrose NL. Fishing in the 80s: a New England industry in transition. NOAA/SeaGrant, University of Rhode Island. Marine Memorandum 67, 1981.

29. Code of Federal Regulations. Title 50 of CFR, Chapter VI—Fishery conservation and management, National Oceanic and Atmospheric Administration, U.S. Department of Commerce, part 600 to end, 1993.

30. Roedel PM, ed. Optimum sustainable yield as a concept in fisheries management. Special Publication No. 9, Am Fish Soc, 1975.

31. *National Fisherman* November 1994:18.

32. Gleick J. Chaos: making a new science. New York: Viking Press, 1987.

33. Clark S. Application of bottom-trawl survey data to fish stock assessment. Fisheries (Bethesda) 1979;4(3):9–15.

34. Pierce DE, Hughes PE. Insight into the methodology behind National Marine Fisheries Service stock assessments or how did you guys come up with those numbers, anyway? Boston: Massachusetts Division of Marine Fisheries, 1979.

35. Hilborn R, Walters CJ. Quantitative fisheries stock assessment: choice, dynamics and uncertainty. New York: Chapman and Hall, 1992.

36. Schnute JT. A general framework for developing sequential fisheries models. Can J Fish Aquat Sci 1994;51:1676–1688.

37. Anthony VC. The state of groundfish resources off the northeastern United States. Fisheries (Bethesda) 1993;18(3):12–17.

38. Bean MJ. The evolution of national wildlife law. Revised Edition. New York: Praeger, 1983: 9–40.

39. Young-Dubovsky C. Atlantic Coastal Fisheries Cooperative Management Act. Fisheries (Bethesda) 1993;18(10):27–30.

40. Edwards EF, Megrey BA, eds. Mathematical analysis of fish stock dynamics. Paper. American Fish. Soc., 1989.

41. Gulland JA. Fish stock assessment. A manual of basic methods. Vol. 1. New York: Wiley, 1983.

42. Nagasaki F, Chikuni S. Management of multispecies resources and multi-gear fisheries. Experience in coastal waters around Japan. FAO Fisheries Technical Paper 305, Rome, 1989.

43. National Research Council. Improving the management of U.S. marine fisheries. Committee on Fisheries, Ocean Studies Board, Commission on Geosciences, Environment, and Resources, National Academy of Sciences. Washington, DC: National Academy Press, 1994.

44. Ricker WE. Computation and interpretation of biological statistics of fish populations. Fish Res Board Can Bull 1975;191:236–238.

45. Rothchild BJ, ed. Global fisheries. Perspectives for the 1980s. New York: Springer, 1983.

46. Rothchild BJ. Dynamics of marine fish populations. Cambridge, MA: Harvard University Press, 1986

47. Royce WF. Introduction to the practice of fishery science. San Diego, CA: Academic Press, 1982.

48. Royce WF. Fishery development. Orlando, FL: Academic Press, 1987.

49. Saila SB, Recksiek CW, Prager MH. Basic fishery science programs. A compendium of microcomputer programs and manual of operations. Amsterdam: Elsevier, 1988.

50. Sparre P, Venema SC. Introduction to tropical fish stock assessment. Part 1. Manual. FAO Fisheries Technical Paper 306/1. Rome, 1992.

51. Collie CJS, Sissenwine MP. Estimating population size from relative abundance data measured with error. Can J Fish Aquat Sci 1983;40:1871–1879.

52. Anthony VC. The New England groundfish fishery after 10 years under the Magnuson Fishery Conservation and Management Act. N Am J Fish Mgt 1990;10(2):175–184.

53. Gabriel WL, Sissenwine MP, Overholtz WJ. Analysis of spawning stock biomass per recruit: an example for Georges Bank haddock. N Am J Fish Mgt 1989;9:383–391.

54. Ludwig D, Hilborn R, Walters C. Uncertainty, resource exploitations, and conservation: lessons from history. Science 1993;260:17, 36.

55. Massachusetts Offshore Task Force. New England groundfish in crisis—again. Publ. 16551-42-2090-1-91-CR, 1990.

56. McGlade JM. Integrated fisheries management models: understanding the limits to marine resource exploitation. In: Edwards EF, Megrey BA, eds. Mathematical analysis of fish stock dynamics. Paper. American Fish. Soc., 1989.

57. Murawski SA, Finn JT. Biological bases for mixed-species fisheries: species co-distribution in relation to environmental and biotic variables. Can J Fish Aquat Sci 1988;45:1720–1735.

58. Nelson RS. Overfishing and the 602 guidelines: can the Magnuson Act work? Fisheries (Bethesda) 1993;18(10):36–37.

59. National Marine Fisheries Service. Report of the Twelfth Northeast Regional Stock Assessment Workshop (12th SAW), Spring 1991. NOAA/National Marine Fisheries Service, Northeast Fisheries Center, Woods Hole, MA, 1991.

60. Pinhorn AT, Halliday RG. Canadian versus international regulation of northwest Atlantic fisheries; management practices, fishery yields, and resource trends, 1960–1986. N Am J Fish Mgt 1990;10(2):154–174.

61. Rosenberg AA, Fogarty MJ, Sissenwine MP, Reddington JR, Shepherd JG. Achieving sustainable use of renewable resources. Science 1993;262:828–829.

62. Schnute JT, Richards LJ. Stock assessment for the 21st century. Fisheries 1994;19(11):10–16.

63. Schoning RW, Jacobsen RW, Alverson DL, Gentile TH, Auyong J. Proceedings of the National Industry Bycatch Workshop, February 4–6, 1992. Newport, OR: Publ. Natural Resources Consultants, 1992.

64. Sissenwine MP. Perturbations of a predator-controlled continental shelf eco-system. In: Sherman K, Alexander LM, eds. AAAS Selected Symposium 99. Variability and Management of Large Marine Eco-systems. Boulder CO: Westview, 1984:55–85.

65. Sissenwine MP. Why do fish populations vary?. In: May RM, ed. Exploitation of Marine Communities. Berlin: Springer-Verlag, 1984:59–94.

66. Sissenwine MP, Marchessault GD. New England groundfish management: a scientific perspective of theory and reality. In: Frady T, ed. Fisheries management: issues and opinions. Fairbanks: University of Alaska, Sea Grant Report 85-2, 1985:255–278.

67. Sissenwine MP, Rosenberg AA. Marine fisheries at a critical juncture. Fisheries (Bethesda) 1993;18(10):6–14.

68. Swartz SL, Sissenwine MP. Achieving long term potential from U.S. fisheries. Sea Technol 1993; Aug 41–45.

69. Thompson WF, Bell FH. Biological statistics of the Pacific halibut fishery. 2. Effect of changes in intensity upon total yield and yield per unit of gear. Rep Int Fish (Pacific Halibut) Comm 1934;8.

70. Pope JG. An investigation of the accuracy of virtual population analysis using cohort analysis. Int Comm NW Atlantic Fish Res Bull 1972;9:65–74.

71. Jones R. Estimating population size from commercial statistics when fishing mortality varies with age. Rapp. P.-V. Reun. CIEM, 155:210–214.

72. Gulland JA. Estimation of mortality rates. Annex to Arctic Fisheries Working Group Report. Paper presented to ICES Annual Meeting, 1965.

73. Murphy GI. A solution of the catch equation. J. Fish Res Bd Can 1965;22(1):191–202.

74. Baranov FI. On the question of the biological basis of fisheries. Nauchn Issled Ikhtiologicheskii Inst Izv 1918;1:81–128. (In Russian.)

75. Fry FEJ. Statistics of a lake trout fishery. Biometrics 1949;5:27–67.

76. Cole CF. Virtual population estimations of largemouth bass in Lake Fort Smith, Arkansas, 1957–1960. Trans Am Fish Soc 1966;95(1):52–55.

77. Deriso RB, Quinn TJ III, Neal PR. Catch-age analysis with auxiliary information. Can J Fish Aquat Sci 1985;42:815–824.

78. Knight R (chair). Report of the Lake Erie Walleye Task Group. March 1993. In: Lake Erie Committee 1993 Annual Meeting Minutes. Ann Arbor, MI: Great Lake Fishery Commission, 1994:3–26.

79. Ohio Department of Natural Resources. Strategic Plan for Lake Erie Walleye (Stizostedion

vitreum vitreum). In: Strategic plan 1990–1995. Columbus: Ohio Department of Natural Resources, Division of Wildlife, 1990:92–102.

80. National Marine Fisheries Service. Our living oceans: the first annual report on the status of U.S. living marine resources. NOAA Tech. Memo NMFS-F/SPO-1. Silver Springs, MD, 1991.

81. National Marine Fisheries Service. Status of fishery resources off the northeastern United States for 1992. NOAA Technical Memorandum NMFS-F/NEC-95, 1992.

82. National Marine Fisheries Service. Our living oceans: report on the status of U.S. living marine resources: 1992. NOAA (National Ocean. Atmos. Ad.) Tech. Mem. NMFS-F/SPO-15, Silver Springs, MD, 1993.

83. Fisheries. Cartilagenous fishes still dominate finfish biomass. Fisheries (Bethesda) 19(11): 42.

84. Grosslein MD. Groundfish survey program of BCF Woods Hole. Comm Fish Rev 1969; 31(8):22–35.

85. Grosslein MD, Laurec A. Bottom trawl surveys design, operation and analysis. CECAF/ECAF Ser 1982;81/22.

86. New England Fishery Management Council. Summary of proposed Amendment #5 to the Northeast Multispecies Fishery Management Plan, 1993.

87. American Fisheries Society. Reauthorization of the Magnuson Act. An American Fisheries Society legislative briefing statement. Fisheries (Bethesda) 1993;18(10):20–26.

88. Rettig RE, Ginter JJC, eds. Limited entry as a fishery management tool. Proc. National Conference, Denver. University of Washington Press, 1978.

89. Sport Fishing Institute. Funding support for fisheries budget sought by aquatic resource community. SFI Bull 1993;444:1–3.

90. Wilson JA, French J, Kleban P, McKay S, Townsend R. Chaotic dynamics in a multiple species fishery: a model of community predation. Ecol Model 1991;58:303–322.

91. Code of Federal Regulations. Title 50 of CFR, Chapter VI—Fishery conservation and management, National Oceanic and Atmospheric Administration, US Department of Commerce, 1992:602(c).

92. Chandler AD. The National Marine Fisheries Service. In: Audubon wildlife report 1988/1989. New York: Academic Press, 1988:3–98.

93. Code of Federal Regulations. Title 50 of CFR, Chapter VI—Fishery conservation and management, National Oceanic and Atmospheric Administration, US Department of Commerce, 1992:Section 602.

94. Code of Federal Regulations. Title 50 of CFR, Chapter VI—Fishery conservation and management, National Oceanic and Atmospheric Administration, US Department of Commerce, 1992.

95. Code of Federal Regulations. Title 50 of CFR, Chapter VI—Fishery conservation and management, National Oceanic and Atmospheric Administration, US Department of Commerce, 1992: Section 642.

96. Horton T. The Endangered Species Act. Too tough, too weak, or too late? Audubon 1992; 94(2):68–74.

97. Sport Fishing Institute. SFI alerts Clinton administration of the key issues facing sport fishing. SFI Bull 1993;441:1–4.

Scientific Uncertainty and Water Resources Management

9

Larry W. Canter

Water resources management is a broad term encompassing one to all of the following: water usage and wastewater disposal; planning and implementation of development projects involving dams and reservoirs, flood control, or irrigation; design and conduction of remediation projects for contaminated surface water and/or groundwater resources; identification and adoption of pollution prevention and water-quality protection strategies; and operation of institutional programs providing overall leadership and direction in management efforts. Scientific tools and measures involved in water resources management include hydrological and water-quality field studies; laboratory studies of chemical transport and fate in the surface water and subsurface environments; acute and chronic toxicity studies; mathematical modeling of hydrodynamic, biotic, and abiotic processes; and multi-criteria decision methods (also called decision analysis methods).

Uncertainties due to lack of sufficient information are associated with all aspects of water resources management, including the use of scientific tools and measures. The basic cause of uncertainty is the underlying variability of hydrologic and geophysical processes (1). Variability also exists in biotic processes involving nutrient cycling and energy transfers, and abiotic processes such as adsorption and precipitation. Sufficient data are needed to describe the natural stochasticity of all of these processes; however, deficiencies in historic information typically limit such descriptions.

Another cause of uncertainty in water resources management is governmental program separation into several basic groups (1). For example, water quantity concerns could be addressed by separate agencies dealing with surface water and groundwater resources, and water-quality issues could be addressed by a single agency or several agencies organized around pollution source categories. Local, regional, state, and federal agencies may be involved in different management roles relative to water resources. There are no simple solutions to the problem of multiple agency involvements; therefore, efforts must be made to achieve coordination and integrated decision making. This will be further addressed in a subsequent section.

Four approaches (or perspectives) can be taken toward dealing with uncertainty in water resources management (1):

a. It could be ignored; however, this is unwise
b. It could be avoided by incremental implementation of development plans; however, while this may reduce the potential impacts of uncertain events, it cannot eliminate all problems
c. Attempts could be made to reduce uncertainty by information gathering at the research and data collection preplanning stages; however, it should be noted that uncertainties may be reduced, but they cannot be completely eliminated
d. Uncertainty could be viewed as a chance occurrence and then addressed by incorporating risk in the planning process.

This chapter is written based on perspectives c and d; included are examples of uncertainties related to water quantity (flooding) in a typical water resources planning study, and water-quality concerns in surface water and groundwater systems. Because risk and uncertainty are often addressed together in water resources planning, a section on relevant definitions and concepts is included along with a section on applications to planning efforts. Brief information is then included on mathematical models and associated uncertainties. Features of institutional programs are then described along with examples of uncertainties related to key program elements. Finally, systematic approaches that could be used to reduce uncertainties related to various aspects of water resources management are described.

Example of Uncertainty Related to Flooding

The first example addresses uncertainty regarding flooding and associated damages. Uncertainty can exist due to the stochastic nature of flooding, including information on magnitudes and probabilities of occurrence. The major sources of hydrologic uncertainty include (2) data availability; data error; accuracy and imprecision of measurements and observations; sampling uncertainty, including the choice of samples and appropriate sample size; selection of an appropriate probability distribution to describe the stochastic events; estimation of the hydrologic and statistical parameters for use in models; low-probability flood extrapolation, such as tail problems of frequency curves; modeling assumptions; and the characterization of river basin parameters.

Flood frequency analysis may involve either (a) statistical data fitting or (b) derived distributions for regions that lack extensive hydrologic records. Uncertainties exist regarding both approaches. For example, future watershed development, climatic trends, randomness of events, and other similar factors are difficult to forecast with any degree of accuracy (2). Additional sources of uncertainty can include rainfall analysis, development of synthetic storms, standard project and probable maximum storms, antecedent moisture conditions, soil type, land cover, hydrograph analysis (particularly difficult for ungaged basins), rainfall-runoff relationships, current watershed development, flood hydrograph routing, use of steady flow and rigid boundary assumptions, number and quality of cross sections, energy losses and surface roughness coefficients, and flow around obstacles.

Damages associated with flooding are generally estimated as a function of the depth of floodwater. Additional factors that can contribute to damage uncertainty include the duration of the flood, sediment load, energy (waves, velocity, etc.), presence of ice, debris, and water quality (2).

Land use analysis is required to estimate the damages that will occur as a result of different flood events, both now and in the future. An inventory of the number, type, value, and susceptibility to flooding of structures in the floodplain is essential; however, the inventory may be based on only a sample of the environment (2). The sample surveyed may be limited by time and money constraints rather than the statistical properties of the sample; thus uncertainty will be introduced into the results. Future land use estimations can also cause additional uncertainty.

Information on property values is also required in the development of flood damage estimates. In addition to uncertainty related to the value of property, there can also be uncertainty about the value of contents in homes and buildings, the flood stage at which damage begins, first floor elevations of structures, responses to flood forecasts and warnings, flood fighting efforts, cleanup costs, and business losses (2).

In summary, while flood control projects have an extensive history in water resources planning, and while methods or guidelines have been developed to address hydrologic and socioeconomic information needs, there are many uncertainties that remain. These uncertainties are a result of the stochastic nature of flooding and the variability of the natural and man-made environments that may be affected by the flooding.

Examples of Uncertainties Related to Water Quality

Water-quality concerns encompass pollution or contamination issues associated with different media such as surface water, groundwater, and/or coastal water; and the associated consequences in terms of subsequent water usage and deterioration of resource features, recreational usage, and productivity. Examples of consequences include fish kills, reductions in opportunities for water usage for domestic or industrial purposes, and excessive aquatic plant growth. Due to necessary quality requirements for specific uses, and the wide variation of such quality requirements depending on the usage, water-quality deterioration can also lead to increased water treatment costs prior to its usage.

Traditional surface-water-quality problems focused upon the discharge of untreated domestic sewage into streams, rivers, and lakes; attention for the last several decades has also been given to the discharge of toxic organic contaminants and metals from industrial plants. Thermal discharges from power and/or industrial plants have caused concomitant reductions in the dissolved oxygen of receiving waters. Of recent emphases have been nonpoint sources of water pollution as represented by agricultural areas, industrial areas, and urban runoff waters. Agricultural nonpoint sources of pollution are of concern due to their potential introduction of nutrients from fertilizers or pesticides into surface water or groundwater resources. While it may be easy to list the types of pollution sources, their technical evaluation may be fraught with uncertainties. For example, statistically based information is needed on pollution source flow rates, stream or groundwater flow rates, and the types and quantities of pollutant releases. Information on toxic pollutants may range from minimal to completely absent.

Additional groundwater pollution sources include (3) septic tank systems, land disposal of municipal or industrial sludge, hazardous and nonhazardous industrial waste disposal via landfilling, municipal solid waste disposal via landfilling, leaking underground storage tanks, open dumps, surface impoundments, accidental spills, mining operations, injection wells, highway salt and other deicing chemicals, and various activities associated with resource extraction related to the oil and gas industry. As noted, uncertainty may occur in technical evaluations requiring quantitative information on (a) pollutant types and quantities and associated release rates from the sources and (b) transport and fate processes in the subsurface environment.

Water-quality concerns and pollution issues also have been identified along the coastlines of the United States, with many of the sources of pollution representing those that are common for both surface water and groundwater resources. Of particular importance in coastal areas is that these locations, which include estuaries, often represent highly productive zones in terms of fish, shellfish, and other organisms consumed by humans for food purposes. These zones are influenced by variabilities in hydrodynamic and biophysical processes; there are many data deficiencies and uncertainties related to the natural fluctuations of these processes.

Emphases on water-quality problems and associated risk concerns have changed over time. For example, one study denoted the changes in risk considerations in addressing water-quality/pollution problems in the twentieth century by comparing three historic issues; one involved protecting drinking-water supplies from pollution in the 1900s to 1920s where the primary concern was bacteriologic contamination; one addressed industrial wastewater and public health concerns in the 1940s to 1960s, wherein the focus was potential releases of organic chemicals and metals; and the third related to land disposal of hazardous waste and associated groundwater contamination (e.g., Love Canal) that characterized the 1970s and 1980s (4).

As a further illustration of changing perspectives, the concerns associated with waterborne disease agents, and which are appropriate risk indicators, can be noted. For example, microbiologists have recognized the growing risk of pathogens such as *Campylobacter, Aeromonas,* and *Cryptosporidium,* and enteroviruses. However, the public may be concerned about exposures to any disease-causing organisms. According to Rose, the common waterborne diseases such as cholera and typhoid have been controlled in the United States since 1970; however, between 1970 and 1985, 502 waterborne disease outbreaks occurred (5). Over half (52%) were attributed to *Giardia,* while 38% were of unknown etiology. *Giardia,* a protozoan which causes diarrhea, is now the most frequently identified cause of waterborne disease in the United States (6).

Enteric viruses have also been recognized as an increasing risk source for waterborne disease, with over 100 types of these viruses identified. Within the last two decades there has been increasing attention to the reclamation and reuse of water that has been previously used and polluted. While the practice of water reuse has long existed in the United States, the key difference represents the fact that the reused water may now be utilized near the point where treatment is provided. For example, both Arizona and Florida have mandated wastewater reuse in order to protect groundwater and provide for long-term water preservation. Such reuse presents the risk of over 100 possible types of enteric viruses found in sewage, as well as bacteria and the protozoa *Giardia* and *Cryptosporidium,* all of which are important causes of waterborne disease (7).

Risk and Uncertainty: Definitions and Concepts Related to Water Resources Planning

Risk and uncertainty are often considered together in water resources planning. For example, Kindler has noted that appropriate treatment of uncertain, unknown, and risky factors is imperative in water resources planning (1). Uncertainty basically refers to lack of certainty, and this can occur as a result of inadequate or imprecise information or the complete absence of information (2). The information that may be needed may be unknown (i.e., unavailable) or unknowable.

Different players may be involved in water resources planning; examples include hydrologists, hydrogeologists, environmental engineers, chemists, biologists, planners, geographers, and economists. Each group can have different perspectives about risk and uncertainty as related to the specific planning effort.

Accordingly, the following scientifically based definitions for risk and uncertainty in water resources planning efforts are presented (2):

Risk: The potential for realization of unwanted, adverse consequences; estimation of risk is usually based on the expected result of the conditional probability of the occurrence of the event multiplied by the consequence of the event, given that it has occurred.

Uncertainty: Uncertain situations are those in which the probability of potential outcomes and their results cannot be described by objectively known probability distributions, or the outcomes themselves, or the results of those outcomes are indeterminate.

To illustrate the relationship between risk and uncertainty, it is helpful to consider them as locations along a continuum of knowledge. Figure 9.1 displays this continuum with its extremes of complete certainty and complete ignorance (2, p. 14).

Risk and uncertainty analysis in water resources planning refers to the process of establishing information regarding acceptable levels of risk and uncertainty for an individual, group, society, or the environment. The most important outcome of risk and uncertainty analysis in the water resources planning process is improved decision making. To achieve this involves identifying and managing the risk and uncertainty in data and assumptions, and in the alternative plans and projects formulated using these data and assumptions (2). Examples of approaches for minimizing risk and uncertainty in water resources planning include (a) increasing design safety factors or (b) increasing the quantity and quality of information via techniques such as expanding the data base, eliminating or minimizing measurement errors, and using traditional statistical analyses (2). Additional information on minimizing risk and uncertainty will be presented in a subsequent section.

Water Resources Planning

The planning process as applied to water resources is comparable to other types of resource or development planning. It involves a logical series of steps, beginning with identification of needs, proceeding to recommendations for action, and culminating in implementation and monitoring. For example, Grigg has

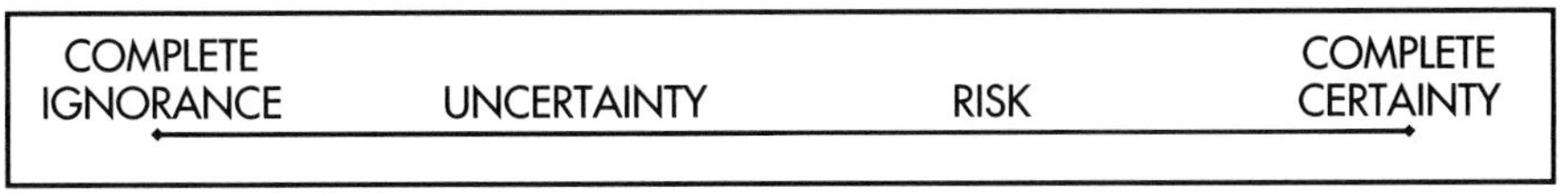

COMPLETE IGNORANCE	UNCERTAINTY	RISK	COMPLETE CERTAINTY

Figure 9.1 Continuum of knowledge.

developed a model of the problem-solving process as shown in Figure 9.2 (8). The model begins with recognizing the problem and making a commitment to its solution. The next step involves formulating goals and delineating goal achievement measures. The creative step involves developing (formulating) the alternative solutions for the problem. The step where impacts or consequences are assessed includes several possible activities: environmental impact studies, assessment of risks, and the formal location of an economically optimum solution. The decision-making phase can involve public hearings, environmental mediation sessions, or other consensus-building approaches. Implementation represents the response phase directed toward the problem. The final phase involves monitoring the operation of the program and making necessary adjustments. The process depicted in Figure 9.2 can be reduced to four critical phases: (a) a problem identification, commitment, and goal-setting phase; (b) the process of studying and analyzing; (c) the decision-making and implementation phase; and (d) the long-term follow-through phase.

The planning process is conceptually compatible with the planning process described for water resources projects promulgated by various federal agencies. For example, the federal process consists of the six major steps and associated iterations that can help sharpen the planning focus or change its emphasis as new data are obtained or as the specification of problems or opportunities changes or becomes more clearly defined. The steps include (9):

1. Specification of the water and related land resources problems and opportunities (relevant to the planning setting) associated with the federal objective and specific state and local concerns.
2. Inventory, forecast, and analysis of water and related land resource conditions within the planning area relevant to the identified problems and opportunities.
3. Formulation of alternative plans.
4. Evaluation of the effects of the alternative plans.
5. Comparison of alternative plans.
6. Selection of a recommended plan based on the comparison of alternative plans.

Table 9.1 summarizes examples of activities related to risk and uncertainty analysis in association with the six steps in the planning process (after 2). In conjunction with such analyses, it may be helpful to consider a taxonomy of sources of risk and uncertainty in water resources planning; an example taxonomy is shown in Table 9.2 (2, p. 50).

An early activity in conjunction with steps 1 and 2 involves the delineation of planning goals and objectives; such goals and objectives can be subject to uncertainty due to conflicting purposes and needs of various water users (1). The following substeps related to the identification of planning objectives illus-

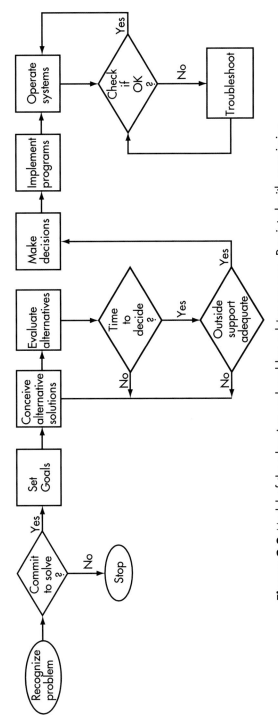

Figure 9.2 Model of the planning and problem-solving process. Reprinted with permission from Grigg, NS. Water resources planning. New York: McGraw-Hill, 1985.

Table 9.1 Risk and Uncertainty Analysis Applied to Water Resources Planning

PLANNING STEP	EXAMPLES OF ACTIVITIES RELATED TO RISK AND UNCERTAINTY ANALYSIS
1 Specification of problems and opportunities	Critical elements for a good risk and uncertainty assessment are (1) problem identification, (2) understanding public views, (3) understanding public attitudes about risk and uncertainty, and (4) establishing specific risk and uncertainty study objectives. Planners should clearly state the problem, without reference to a solution, in an objective manner that enjoys public support, if possible. Gather as much information as possible regarding the problem, but expect uncertainties to arise in the initial assessments of water problems. Make efforts to understand local concerns and viewpoints and to ascertain public attitudes about risk and uncertainty. Surveys, questionnaires, focus groups, etc., are tools for minimizing uncertainty regarding public views. Identify the range of alternatives proposed by the public. Attempt to define three vital concepts regarding risk: (1) acceptable levels of risk, (2) residual risk, and (3) risk transfers. An example of risk transfer is induced flooding that results from construction of a flood control project; the risk to one community is reduced at the cost of an increased risk to another community.
2 Inventory and forecast (without plan condition)	The three fundamental elements of the risk and uncertainty analysis are identification of (1) key risk and uncertainty issues, (2) methods used to address risk and uncertainty, and (3) multiple without project condition scenarios. Identify natural sources of risks and uncertainties related to floods, droughts, hurricanes, earthquakes, etc.; also identify social or human-related sources such as population growth, land use, zoning regulations, economic institutions, willingness to pay, risk preferences, technological hazards, and pollution. The most critical task is identifying "key" variables that can affect forecasts and plan formulation and are not known with certainty. Methods (or models or approaches) used in forecasting should be identified, and their limitations or uncertainties should be noted. Examples of approaches include (a) adoption of forecasts made by other agencies or groups; (b) use of scenarios based on differing assumptions regarding resources and plans; (c) use of expert group judgment via the conduction of formalized Delphi studies or the usage of the nominal group process; (d) extrapolation approaches based on the use of trends analysis or simple models of environmental components; and (e) analogy and comparative analyses that involve the use of "look-

alike" resources or projects and the application of information from such look-alike conditions to the planning effort. Finally, planners should identify and consider multiple without project scenarios; such scenarios could be based on different assumptions related to key variables.

3 Formulation of alternative plans (evaluation)

The critical elements in this step are (1) formulating a set of alternatives, (2) screening the alternatives, and (3) risk management. A broad perspective should be used in formulating alternative plans. An explicit description of how risk and uncertainty issues affected the formulation of alternatives should be developed. The formulated alternatives (plans) should be screened relative to their completeness, effectiveness, efficiency, and acceptability. The screening process should also address each plan's contributions to the risk and uncertainty objectives. In this context, the extent of risk management potentially afforded by each plan could be articulated.

4 Comparison of alternative plans—detailed evaluations

The critical elements of the risk and uncertainty analysis are (1) evaluation of each alternative's contribution to the planning objectives, (2) avoiding the appearance of certainty, and (3) transition in focus to implementation issues. The assessment of risk and uncertainty can be accomplished through methods that include, but are not limited to, sensitivity analyses, more and better data, classical statistical analysis, risk to cost trade-off analysis, professional judgment, subjective probabilities, qualitative forecasts, and a summary of assumptions. The evaluation should, if possible, include quantitative estimates of each alternative's contribution to the risk and uncertainty planning objectives. It would be desirable if these objectives could address the following: minimizing risk to life, health, and safety; minimizing risk to the environment; minimizing residual risks; and, minimizing the uncertainty of project impacts. Planners should avoid the appearance of certainty in written or verbal presentations by using careful language. Qualifying wording such as "dependent on," "subject to," and "assuming" are appropriate. Also, if benefit-cost ratio (BCR) data are presented, a range of values should be noted for each alternative. The evaluation should clearly describe the conditions for the maximum and minimum BCRs. This step should also give attention to implementation issues relative to the plans; such issues could include potential adverse impacts, residual risk, increases in existing risks, creation of new risks, and transferred risks.

Table 9.1 continued

PLANNING STEP	EXAMPLES OF ACTIVITIES RELATED TO RISK AND UNCERTAINTY ANALYSIS
5 Comparison of alternatives— detailed analysis	The principal elements of the risk and uncertainty analysis are (1) quantifying the cumulative effects of risk and uncertainty, (2) comparing the risk and uncertainty aspects of the alternatives, and (3) displaying the results. During this step the planner must bring all the risk and uncertainty analysis together in a meaningful and understandable way. Cumulative levels of risk and uncertainty associated with the alternatives should be identified and quantified, if possible. This step should focus on a comparison of risk and uncertainty evaluated in previous steps. In comparing alternatives, planners should clearly document the use of subjective judgment in determining the impacts associated with implementation of that alternative. Where objective estimates of risk have been used the associated technique should be identified. Displaying the results of the various comparisons is vital in the planning process. Summary tables and other formats such as decision trees could be useful. Decision makers should be presented with a clear understanding of what is uncertain in each alternative and what the implications of that uncertainty are. They should also be informed of the risks related to each alternative and the trade-offs among them.
6 Plan selection	Risk and uncertainty management is the focus of the plan selection step. Some general questions that could be considered to help guide the management of an identified risk might include the following: (1) Is the risk significant? (2) What are the mitigation alternatives? (3) What are the costs and benefits? (4) What are the legal, social, and political ramifications? and (5) What are the implementation and performance issues? Methodologies to integrate judgmental aspects with empirical approaches to evaluate trade-offs among alternatives could be utilized; these are referred to as decision analysis methodologies. Once a plan is selected, the description should address the risk and uncertainty played in the selection process should be described; the description should address the risk and uncertainty objectives, major risk and uncertainty issues, and the cumulative effects of risk and uncertainty.

Source: Reference 2.

Table 9.2 Potential Taxonomy of Sources of Risk and Uncertainty in Water Resources Planning by U.S. Army Corps of Engineers

	TIME	SPATIAL	SOCIAL/CULTURAL
Assessment			
Data			
not available			
insufficient			
sampling problems			
Theory			
not available			
insufficient			
incorrect			
Methods			
not available			
not well developed			
measurement problems	Past Present Future	Study area Region State Nation	Public Non-Federal partner State Corps of Engineers
Management			
Perceptions			
uninformed			
unknown			
Values			
conflicting			
unknown			
Objectives			
not clear			
conflicting			
unknown			
Technology			
alternative actions			
alternative decision			
process			

Source: Reference 2.

trate places where uncertainties related to public viewpoints, baseline data, and future conditions may arise (10):

a. Identify public concerns via determining the views of various publics on problems and needs to be addressed.
b. Analyze resource management problems and determine whether they are related to water resources and can be achieved by water resources development or management.
c. Define the study area based on steps a and b.
d. Describe the baseline conditions in the study area; the information needs for such a description are summarized in Table 9.3 (10).
e. Identify future conditions in the study area without any specific project taking place to address the identified needs; this identification should include consideration of the view of various segments of the public as well as views of professional planners, and the projections currently used by federal, state, and local planning agencies.
f. Establish planning objectives based on considering steps a–e; these objectives should be refined as the planning proceeds.

The formulation of alternatives (step 3 in the basic planning process) involves developing alternative plans that address the identified planning objectives. One aspect of formulation involves considering a potentially wide variety of technical and institutional measures for managing water and related land resources (10). A broad range of measures should be examined to identify those that can address one or more of the planning objectives; representative measures for water resources planning are listed in Table 9.4 (10). Alternative plans are formulated by combining different measures into resource management systems.

Several alternative plans are typically identified early in the planning process, but some may be dropped as the study progresses and plans are refined. Alternatives should be screened based on their completeness, effectiveness, efficiency, and acceptability (9). Completeness refers to the extent to which a given alternative plan provides and accounts for all necessary investments or other actions to ensure the realization of the planned effects. Effectiveness is the extent to which an alternative plan alleviates the specified problems and achieves the specified objectives or goals. Efficiency is the extent to which an alternative plan is the most cost-effective means of alleviating the specified problems and realizing the specified opportunities. Finally, acceptability is the workability and viability of the alternative plan with respect to acceptance by state and local entities and the public, and compatibility with existing laws, regulations, and public policies (9).

Steps 2, 4, and 5 in the basic planning process involve forecasts of future conditions for the without-plan and with-plan conditions. Such forecasts can

Table 9.3 Information Needs for Describing the Baseline Conditions for a Study Area

---

a. Description of the resource base includes a brief summary of climate, geology, and topography; human and natural resources, both physical and biologic; demographic, cultural, and aesthetic characteristics; land use, particularly in the floodplain and possible reservoir areas; transportation networks; financial resources; and economic activity including manufacturing, trade, and agriculture.

b. Description of significant environmental elements in the study area involves locating and identifying those characteristics deemed to be aesthetically, ecologically, or culturally important. To identify these characteristics involves analyzing such factors as soils, water, air, plants, animals (including people), and cities; forces such as wind and tides; human activities; conditions such as light, temperature, pollution, and humidity; and processes such as mining. Elements to be considered are those that are important to society at the time of the study and those projected to be important in the future, including elements that are critical in terms of scarcity, fragility, and lack of resiliency, or that would otherwise be sensitive to change.

c. Description of existing public and private programs for planning and managing resources in the study area should include identification of existing management systems and facilities and any under construction, funded for construction, or approved for construction.

d. Identification of institutions dealing both directly and indirectly with resource management in the study area should include information on existing jurisdictional, functional, and financial arrangements in the study area.

e. Analysis of adequacy of information collected about base conditions must be made early in the study so that any additional data needed can be developed in a timely manner at reasonable cost.

Source: Reference 10.

involve the use of analogies, mathematical models, and expert judgment. Uncertainties related to mathematical modeling are addressed in the next section.

Finally, step 6 involves the comparison and evaluation of viable alternative plans. This requires the amalgamation of engineering, economic, and environmental information on each plan along with value judgments reflective of professional and societal perspectives. The use of systematic decision-making techniques can facilitate the amalgamation process and reduce the associated risk and uncertainty. Information on such techniques is included in a subsequent section.

Table 9.4 Examples of Conceptual Management Measures for Water Resources Planning Objectives

PLANNING OBJECTIVE	MEASURE
Flood-damage reduction	Reservoirs
	Levees
	Channel modification
	Flood warning and floodplain evacuation
	Flood proofing
	Floodplain zoning
	Watershed management
Water supply	Surface water reservoirs
	Groundwater reservoirs
	Interbasin transfers
	Precipitation augmentation
	Wastewater reclamation
	Water conservation
	Reallocation
	Watershed management
	Water harvesting
	Desalinization
	New technologies
Water-oriented recreation	Reservoirs
	Regulated streamflows
	Parks
Water-quality improvement	Streamflow augmentation
	Land-use zoning
	Improved agricultural practices
	Pollutant diversion
	Selective withdrawal from reservoirs
Fish and wildlife preservation	Streamflow augmentation
	Gravel management
	Fish hatchery
	Wildlife areas
	Wildlife habitat improvement
Hydroelectric power	Reservoirs
	Stream diversion
	Run-of-river projects
Groundwater overdraft correction	Recharge
	Restriction of pumping
	Modified cropping pattern

Source: Reference 10.

Mathematical Modeling

Mathematical modeling is a fundamental tool used in various facets of water resources management. For example, models can be used for (a) forecasting surface runoff and possible flooding as a function of rainfall distribution and intensity; (b) forecasting flood elevations in geographic areas prone to flooding; (c) development of statistical presentations of streamflow during base-flow (low flow), average, and high-flow conditions; (d) describing groundwater flow and the influence of pumping rates on the drawdown of groundwater in the vicinity of wells; (e) quantifying the influence of saltwater intrusion in coastal areas with excessive groundwater usage; (f) forecasting water-quality changes in rivers or estuaries downstream of municipal or industrial wastewater discharges, or nonpoint runoff from urban, industrial, or agricultural areas; (g) development of nutrient balances and information on nutrient cycling rates in lakes and coastal zones; (h) forecasting pollutant transport and fate in the subsurface environment, including the unsaturated and saturated zones; (i) delineating water-quality changes that will occur in man-made reservoirs; and (j) evaluating the potential effectiveness of remediation measures for contaminated groundwater. Modeling is also used in environmental impact studies conducted prior to project decision making. Examples of specific types of models used in impact studies for predicting aquatic effects are listed in Table 9.5 (11). These examples illustrate that models may be needed for both quantitative and qualitative concerns, including ecologic consequences in water resources management.

All mathematical models are based on assumptions; thus uncertainty can be introduced. For example, Anderson and Burt noted the following about hydrologic modeling, and the statement is also relevant to the modeling of water quality and aquatic ecosystems (12):

> All models seek to simplify the complexity of the real world by selectively exaggerating the fundamental aspects of a system at the expense of incidental detail. In presenting an approximate view of reality, a model must remain simple enough to understand and use, yet complex enough to be representative of the system being studied.

Hydrologic and biophysical models can be divided into three types (12):

1. *Black-box models* contain no biophysically based transfer function to relate input to output; they depend on establishing a statistical correspondence between input and output.
2. *Conceptual models* occupy an intermediate position between the deterministic approach and empirical black-box analysis; they are formulated on the basis of a simple arrangement of a relatively small number of components, each of which is a simplified representation of one process element in the system being modeled.

Table 9.5 Summary of Techniques for Predicting Aquatic Effects

Prediction of Discharges into the Aquatic Environment

■ Mathematical models for predicting runoff
  rural runoff models
  urban runoff models
  combined runoff models

Prediction of Hydraulic Effects in the Aquatic Environment

■ Experimental methods
  physical hydraulic models
■ Mathematical models
  dynamic models for
    rivers
    estuaries and harbors
    coastal waters

Prediction of Water Quality Effects in the Aquatic Environment

■ Experimental methods
  physical hydraulic models
  in situ tracer experiments
■ Mathematical models
  river
    simple mixing models
    dissolved oxygen models
    complex river water quality models
  estuaries
    1-d dispersion models
    2-d dispersion models
  coastal waters
    2-d dispersion models
    oil spill trajectory models
  lakes
    simple mixing models
    nutrient budget models
    dynamic lake cycling models

Prediction of Higher-Order Effects on Plants and Animals

■ Techniques for predicting effects on recreation
  empirical models
■ Mathematical models
  pollutant distribution
  population models
  productivity models
  habitat availability models

Source: Reference 11.

3. *Deterministic models* are based on complex physical, biologic, or chemical theory; however, despite the simplifying assumptions necessary to solve flow or quality equations, such models may have large demands in terms of computational time and data requirements; in addition, they may be costly to develop and operate.

It is beyond the scope of this chapter to include a detailed review of mathematical models. Rather, summary comments will be made on some categories of models. For example, simple mixing models could be used to approximate water-quality conditions. More sophisticated surface water quality and quantity models range from one-dimensional steady-state models to three-dimensional dynamic models which can be utilized for rivers, lakes, and estuarine systems (12–15). Groundwater flow models have been recently modified to include subsurface processes such as adsorption and biologic decomposition (16, 17).

Simple mixing models are based on the assumption of uniform mixing of a contaminant in a defined area of the receiving environment. Examples of simple mixing models include (18) (a) river mixing and dilution models, which assume uniform mixing of conservative contaminants across the river cross section and, therefore, straightforward dilution of the effluent flow in the river flow; and (b) lake mixing models, which assume uniform mixing of a conservative contaminant over the total depth of unstratified lakes, or over the depth of the upper or lower layer (depending on the discharge location) in stratified lakes. Simple mixing models can be used for estimating "worst-case," "average," and "best-case" conditions.

Steady-state dispersion models provide steady-state predictions by taking into account aquatic or subsurface processes that influence the behavior of contaminants. Examples of processes that may be addressed include advection (i.e., transport) in one, two, or three dimensions; diffusion in one, two, or three dimensions; transfer or partitioning between components of the environment (including living subjects); and physical, chemical, and biologic transformation or removal ("sinks"). Examples of steady-state dispersion models include (a) dissolved oxygen models for rivers based on Streeter-Phelps, (b) simple segment models for rivers and estuaries, (c) lake circulation models, and (d) groundwater flow and time of travel models based on Darcy's law. Because these models provide steady-state predictions, they can also be used for the same purposes as simple mixing models.

Complex dynamic models have fairly extensive input data requirements and depend on computer solution. Examples of complex models include river basin flow and quality models such as QUAL-IIE, marine and estuarine dispersion models, watershed runoff models, biologic population or productivity models, nutrient cycling and eutrophication models, and groundwater flow and solute transport models. The basic equation solved by QUAL-IIE is the one-dimensional advection-dispersion mass transport equation, which is numerically integrated over space and time for 15 water-quality constituents. The constituents modeled by QUAL-IIE include (19) dissolved oxygen, carbonaceous biochemical oxygen demand, temperature, algae as chlorophyll *a*, nitrogen species (organic, ammonia, nitrate, and nitrite), phosphorus species (organic and dissolved), coliforms, one nonconservative constituent, and three conservative constituents. The key advantage of dynamic models is that their computer usage means that many more factors can be taken into account in modeling environmental behavior; their main disadvantage is the extensive requirements for input data and calibration.

Aquatic ecosystem models range from index approaches based on habitat characteristics to energy system diagrams. Examples of index models include the Habitat Evaluation Procedures developed by the U.S. Fish and Wildlife Service (20), and the Habitat Evaluation System developed by the U.S. Army Corps of Engineers (21). Both models involve the calculation of an index for baseline conditions incorporating both quality and quantity information. Prediction of changes involves determination of the index under future with and without conditions. Numerous aquatic species population models, as well as productivity models, have been developed based on empirical approaches involving statistical correlations (22). For example, the instream flow incremental methodology (IFIM) can be used to assess how much water could be extracted from a river at various times of the year without adversely affecting the fishery resource (23). The approach is based on the concept that a particular species can be correlated with specific habitat requirements such as water quality, velocity, depth, substrate, temperature, and cover. If these requirements are known then an assessment of the habitat suitability for a particular species can be made by measuring the quality of the available habitat and projecting what it will become under future conditions. Finally, the most sophisticated aquatic ecosystem models are those involving energy system diagrams; these have been utilized in some resource management studies (24).

Numerous uncertainties can arise regarding mathematical models and their usage. Examples of such uncertainties can be considered in relation to model limitations as follows (12):

1. Limitations due to inadequacies of current theory or to failure of the model to incorporate certain elements of current theory
2. Limitations caused by the scarcity of appropriate field data for model calibration and operation
3. Limitations caused by the inadequacy of computer capacity for modeling of complex hydrodynamic or biophysical problems
4. Limitations of calibration procedures

5. Limitations in selected management applications where operational constraints have not been fully incorporated into utilized models

Institutional Aspects of Water Resources Management

An effective and integrated ongoing water resources management program at the local, regional, or national scale should include multiple elements encompassing both technical and policy issues. Such a program must address both surface and groundwater resources, water demand and usage limitations, pollution sources and control, and pre–decision analysis associated with land-use planning and development activities, including water resources planning. Table 9.6 includes 12 elements that should be incorporated in water resources management programs. It should be noted that while the nonprioritized listed elements may not be all inclusive, they do represent the majority of relevant concerns. Uncertainties can be associated with each element, and examples are listed in Table 9.6. Brief descriptions of each element are in the following paragraphs (25).

Surface water and groundwater resources may be interrelated in given geographic areas. During periods of high surface water flows alluvial groundwater systems can be recharged, while during low surface flow periods groundwater may be the dominant contributor to the stream base flow. Water usage from either of the resources may have implications for both the abstracted and interrelated resource. Pollutant discharges can also influence the quality characteristics of both surface and groundwater systems.

Statutory authority refers to the legal authority vested in the entity with the responsibility for water resources planning and management. In addition to the federal government, many states and even local governmental agencies may have laws, regulations, or policies that are also directly or indirectly pertinent. It is highly unlikely that one agency at any governmental level would have the responsibility for all relevant laws, regulations, or policies, thus it becomes vitally important that institutional coordination occur on a routine and planned basis.

Water usage standards refer to specific limitations on the quantities of withdrawal from surface water or groundwater systems or both. Usage limitations or restrictions may be based on the surface water flow conditions, or on the amount of land ownership in relation to groundwater abstraction. The more arid the region (and when alternate surface water or groundwater supplies are not readily available), the more important it becomes to have carefully developed regulatory approaches for water usage.

Surface water and groundwater quality standards are also vital to an effective water resources management program. Standards are necessary for development and use of surface water or aquifer classification systems, pollution source evaluations, interpretation of water-quality monitoring data, and the

Table 9.6 Elements and Examples of Uncertainties in an Integrated Water
Resources Management Program

ELEMENT	EXAMPLES OF UNCERTAINTIES
1. Recognition of relationships between surface water and groundwater resources	Lack of definitive quantitative information on base flows and water table fluctuations during low- and high-flow conditions in surface streams
2. Statutory authority and water usage and quality standards	Overlapping responsibilities for water management among agencies; uncertainties related to protection factors included in quality standards
3. Resource characterization studies (quantity and quality)	Lack of data for statistical delineation of natural flows and quality variations
4. Resource usage studies	Variations in seasonal and annual water usage as a function of climatic conditions, population growth, and development activities
5. Water usage conservation program	Number of users participating in conservation program
6. Resource pollution studies	Relative importance of point and nonpoint pollution sources for surface water and groundwater resources
7. Pollution source control and resource protection program (point and nonpoint sources)	Lack of information for evaluating the effectiveness of control and/or protection program
8. Technically based decision making regarding resource development, usage allocations, wastewater (point and nonpoint) discharges, and permitting program	Incomplete information regarding environmental impacts of choices; economic analyses with limited emphasis on environmental costs; understanding and reflecting value judgments of different publics
9. Monitoring and enforcement program	Absence of quality assurance/quality control program for water sampling and analytical procedures
10 Emergency response program	Lack of information on specific pollutants and their transport and effects characteristics, and appropriate remediation measures

Table 9.6 continued

ELEMENT	EXAMPLES OF UNCERTAINTIES
11. Institutional coordination program (related to number 2)	Inability to keep current on policy changes in various program operations of relevant agencies
12. Public awareness, public participation, and education program	Effectiveness of techniques used to communicate risk and uncertainty information to various publics

issuance of permits (26). Surface water quality standards can be related to beneficial use designations of stream segments in a river system. In recent years the number of constituents addressed in surface water quality standards has been increasing; and the numerical standards (in terms of maximum allowable concentrations) have been decreasing. Groundwater quality standards have tended to be fewer in numbers in terms of the constituents addressed; however, with the current emphasis on synthetic organic and metal constituents, numerical standards for these components are becoming necessary.

Discharge standards are also pertinent for point and nonpoint sources releasing effluents into surface bodies of water, and possibly groundwater systems. Publicly owned treatment works (POTWs or municipalities) are subjected to secondary treatment standards (or even tertiary treatment standards depending on the location). Industrial discharges are subject to pretreatment standards if connected to POTWs, or industry-specific effluent standards if discharging directly into a receiving body of water. Pollution prevention measures relative to stormwater runoff (nonpoint source) from industrial and urban areas can also be utilized in water-quality programs.

Resource characterization studies must be planned and conducted in order to have a technical (and quantitative) baseline for subsequent decision making. Such studies should address flow variations and quality characteristics of surface bodies of water, and developable groundwater supplies and their associated quality characteristics. Resource usage studies are also vital in water resources management. Historic patterns of the usage of both surface and groundwater resources should be established. Comparisons of current usage patterns with applicable usage limitations (based on resource availability and/or applicable governmental laws, regulations, or policies) may be necessary in developing an optimum and equitable management plan. Depending on water usage patterns, it may be necessary to develop a water conservation and/or reuse program which can be implemented on an as-needed basis.

Resource pollution studies should be directed toward identifying both point and nonpoint sources of pollution in relevant surface and groundwater systems. With expanding documentation of pollution resulting from man-made sources, it is important to prioritize the source categories and individual contributors. Specific source categories may be the administrative responsibility of different agencies; accordingly, institutional coordination would be needed. In addition, source control measures, or preventive measures, directed toward minimizing future surface water or groundwater pollution from existing and/or new sources is important in water-quality protection. Ongoing programs of source control and permitting will be necessary.

Examples of surface water pollution sources in the United States subjected to controls and treatment measures via permitting programs include municipal and industrial wastewater point source discharges, and nonpoint discharges (stormwater) from urban and industrial areas. Wastewater treatment is typically required for point sources, and the utilized technologies can be considered in terms of primary, secondary, and tertiary (advanced) schemes.

An example of resource protection is the wellhead protection program initiated by the 1986 Safe Drinking Water Act. The "wellhead protection area" (WHPA) means the surface and subsurface area surrounding a water well or wellfield, supplying a public water system, through which contaminants are reasonably likely to move toward and reach such water well or wellfield. The emphasis in the program is on limiting those activities within the WHPA which have greater risks of causing groundwater contamination (27). Factors basic to risk delineation include but are not limited to the radius of influence around a well or wellfield, the depth of drawdown of the water table by such well or wellfield at any given point, the time or rate of travel of various contaminants in various hydrologic conditions, and the distance from the well or wellfield.

An important element in an integrated water resources management program is technically based decision making. Such decision making is required relative to water resources development planning as mentioned earlier, and for surface water or groundwater usage allocations, waste load allocations and permits relative to point/nonpoint pollutant discharges, or technologies for pollution source control, or environmental remediation projects. Risk assessment studies or environmental impact assessment (EIA) studies can be used to provide basic information for decision making. Tools that can provide a structure to technically based decision making include multiattribute or multicriteria decision models, multiattribute utility measurement, and/or decision analysis. Examples of such tools will be presented in the next section.

Continuing environmental monitoring will also be a requirement in an integrated water resources management program. The monitoring should be targeted on critical water resources (in terms of quantity or quality or both) and on key pollutant sources. Data organization and presentation can include the use of numerical indices that represent a composite of measured quality and quantity parameters. Monitoring results should be integrated with an enforcement program that can include periodic inspections of pollutant discharges, annual or less frequent environmental regulatory auditing, renewable permits after 3 to 5 years, and the preparation of annual monitoring and enforcement reports.

Surface water and groundwater protection programs should also address responses to both acute and chronic pollution problems. Acute problems are reflected by accidental chemical spills where immediate and appropriate remedial action measures are necessary to minimize water pollution. A comprehensive water pollution protection strategy should include a contamination response program based on a pollution incident tracking program.

Institutional coordination is needed between various governmental entities involved in water resources management. Overlapping responsibilities typically result from single-purpose laws and administrative responsibilities. Since multiple agencies may be involved in collecting relevant information, there is a major need for the provision of information access between agencies.

A final element for a water resources management program is associated with education and training of staffs involved in program planning and implementation, as well as the general public. Technical staffs, as well as policy and legal staffs, are necessary in the planning and implementation of management programs. In addition to technical and policy-related education needs, there is a major need for increasing public awareness and participation relative to the importance of surface water and groundwater as a resource and appropriate measures for protection and remediation.

Approaches for Addressing Uncertainty

Several approaches are available for reducing uncertainty related to water resources management. Examples include the use of interdisciplinary efforts, the solicitation of public involvement in decision making, the use of effective risk communication techniques, the recognition of limitations in modeling, and the incorporation of systematic decision tools.

Utilize Interdisciplinary Efforts

An interdisciplinary effort is required in water resources planning; it can also facilitate the various facets of water resources management. The use of many scientific disciplines in an ongoing, interactive approach is desirable as a way of overcoming lack of knowledge by any single professional. Examples of pertinent disciplines include hydrologists, chemists, biologists, hydrogeologists, environmental engineers or scientists, geographers, planners, policy specialists, economists, and attorneys. An interdisciplinary effort should not be limited to the expertise immediately available in the planning or management agency. As necessary for a particular study or issue, agency expertise may be supplemented by knowledge and skills from cooperating agencies, universities, consultants, and other sources (28).

Solicit Public Involvement

Water resources planning and/or management agencies should invite the early and continuing involvement of governmental entities at the federal, regional, state, and local levels; national, regional, and local public and private organizations and groups, including conservation groups and Indian tribes; and individuals. Public involvement in water resources management is appropriate for the following reasons related to reducing uncertainty (28):

1. First, the public is the basic source, and in many cases the only source, of knowledge and opinions that are needed to make the planning or management process work.
2. Second, as a reviewer of the results of planning or management efforts, the public will have opportunities to ensure that their views have been properly incorporated; understand the implications of their views on decisions; and react to possible decisions in a way that will facilitate their appropriate modification.

Use Effective Risk and Uncertainty Communication Techniques

Perceptions of risk and uncertainty related to water resources management can vary between planners, decision makers, and various groups or individuals within the general public. Effective risk (and uncertainty) communication could be used to reduce incorrect perceptions based on limited information. Risk communication can be defined as a process of interaction over time between senders and receivers of information about a risk (29). For example, communication planning related to toxic chemicals in the water environment (and inferred actual or potential water quality problems) can be enhanced by realizing that risk communication should be looked at not only in terms of how accurate, detailed, or intelligible the information is, but also in terms of how the information will be interpreted. Specific points to be considered in planning a risk and uncertainty communication program (as well as factors that influence the public perception of risks) are (29):

1. Reception of information about risk will vary from community to community, among various publics within any community, and through time. People's acceptance of the risk information they are given, while clearly affected by their attitudes about the risk itself, is also affected by the local context in which the risk situation is embedded.
2. Receivers bring cultural assumptions and inputs of individual knowledge and experience to the communication interaction. The receiver inputs will act as filters, making it unlikely that there will be a one-to-

one correspondence between the message transmitted and the message actually received. What is said is not necessarily what is heard, and what is "correct" is not necessarily what is believed.

3. Many messengers, both official and unofficial, are involved in presenting information to the public about a given risk.

As a result of these and other influences, risk communication in water resources planning needs to be carefully considered. Two broad objectives that might be considered are (a) to provide the audience with a better understanding of the risk and uncertainties surrounding planning alternatives and thus to stimulate cogent and informed discussion—and ultimately a defensible resolution—and (b) to communicate risk in order to encourage appropriate behavior by individuals and communities (30). Ten practical guidelines related to the content of risk communication efforts are summarized in Table 9.7 (30).

Recognition of Limitations in Modeling

As described in an earlier section, there are many limitations and associated uncertainties in the use of mathematical modeling in water resources management. A desirable approach for reducing uncertainty would include delineating the assumptions and limitations of utilized models and expressing model results within a range (or including error bars). Sensitivity analyses of modeling efforts can facilitate the expression of the results.

Utilization of Decision Analysis Methods

Risk or uncertainty should be incorporated in the decision-making process for new development projects and ongoing management strategies. For example, when choosing between different projects or different features of one project, it is not sufficient to consider only the expected outcome of each project, or of each feature. The risk associated with each alternative must be taken into account. This implies that a trade-off must be established between the utility of the expected outcome of each alternative and their respective degrees of risk (31).

To serve as a more specific example, in comparing alternatives in water resources planning, the following four principles of evaluation should be utilized (32):

1. Compare the future impacts with the alternative to the future impacts without the alternative (the "with and without" principle).

2. Consider all the significant externalities—that is, all the ramifications of the alternatives outside the project. To do this, the area (geographic, political, etc.) that is directly included in the planning study should be clearly defined. The effects of project implementation on the outside areas are called externalities. In some studies external economies,

Table 9.7 Guidelines for Risk Communication

OBJECTIVE	GUIDELINES
Increase understanding	■ Uncertainty should be expressed in a variety of ways using such physical analogies as wheels of fortune and jars containing colored balls. Care should be taken to avoid encouraging thinking of independent events as cyclic, with fixed return intervals.
	■ Disagreements among experts should be made explicit and not concealed. If possible, the range of opinions should have probability weights attached to the alternative possibilities.
	■ The decision problem outcomes should be framed in at least two ways, one stated as losses from best case and one as gains from worst case.
	■ Stress the analogy of the project to an insurance policy against catastrophic loss whenever this is appropriate.
	■ Provide information that allows the audience to assess the risk (at least in terms of threats to life and limb) of the contemplated project relative to other activities and programs, both individual and collective.
Encourage action	■ The campaign should effectively convey a message about the seriousness of the risk. This effort should not, however, be allowed to degenerate into a scare campaign, for the behavior triggered by fear is likely to be counterproductive.
	■ The program should provide social reinforcement of risk reduction behavior especially at the local level. This will cultivate strong group interest and moral commitments within the community.
	■ The campaign should make an attempt to convince the consumers that their actions aimed at reducing risk will help to mitigate risk impacts.
	■ Risk reduction efforts requested by the campaign should be equitable. All members of the community should be required to make sincere efforts to reduce the risk.
	■ The specific strategies of the campaign should rely, to the extent possible, on providing a feedback on risk reduction efforts and providing economic and social incentives for doing so.

Source: Reference 30.

which increase the benefits, have been added and external diseconomies, which increase costs, have been neglected. Be consistent and include all direct (or primary) externalities in the evaluation.

3. Evaluate the alternatives for future flexibility. This is one of the best ways for dealing with risk and uncertainty.

4. Conduct a sensitivity analysis. For example, a plan that has lower net benefits than another plan but is more insensitive to future population size might be preferable in a rapidly changing area.

Decision trees can be useful tools for addressing risk and uncertainty in water resources planning. By definition, a decision tree is a schematic tool for evaluation of sequential decision problems. An example is shown in Figure 9.3 (2, p. 58). Decision trees consist of (a) decision points: specific points of time when a decision must be made are shown as decision points; (b) event points: a number of states of nature that may occur are shown as event points; (c) probabilities: the known or subjective probabilities of events are presented above each of the event branches; and (d) conditional payoffs: the conditional payoff of each event branch is estimated and recorded at the end of each branch (2).

Several decision analysis methods have been developed for the comparative evaluation of alternatives in water resources planning. An early example, called the water resources assessment methodology (WRAM), was promulgated by the U.S. Army Corps of Engineers (33). Key elements included the selection of an interdisciplinary team; selection and inventory of assessment variables (environmental factors); impact prediction, assessment, and evaluation; and docu-

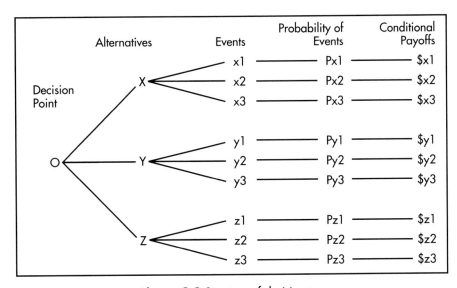

Figure 9.3 Structure of decision tree.

mentation of the results. The element including decision factor importance weighting and impact scaling was called impact prediction, assessment, and evaluation. The weighted rankings technique (unranked paired-comparison technique) was used to determine the relative factor importance coefficient (FIC) for each assessment variable. Importance weight assignments are required for each study and should reflect the importance of the variables in the given geographic location.

Impact scaling in the WRAM was accomplished through the use of functional graphs (relationships), linear proportioning, or the development of alternative choice coefficients (ACCs). Linear scaling is associated with considering the largest impact and calculating the relative proportions of each change based on assigning the largest impact a scale value of 1.0. It should be noted that changes could be either beneficial or detrimental, and linear scaling should delineate between them. Development of ACCs involved the comparative evaluation of the impacts of each alternative on each decision factor (assessment variable). The final decision matrix developed from the WRAM involved summing the products of the FICs times the ACCs. The best choice would be the alternative plan with the highest product summation.

The first example of the usage of WRAM was on the Tensas River Basin project (a flood control project) in eastern Louisiana (34). Existing data was used along with site visits by an interdisciplinary team. Assessment variables were selected and weighted relative to their importance in assessment and evaluation within each of four decision accounts: national economic development (NED), environmental quality (EQ), social well-being (SWB), and regional development (RD). The projected effects on each variable were then scaled across the alternative plans and the without-project condition. The weights and scales were then multiplied for each variable within an account and then added for each alternative plan. Although no best alternative was selected, the strengths and weaknesses of each alternative were revealed, thus allowing for trade-offs.

The WRAM was developed with an orientation to water resources projects; however, this approach could be used for other decision making in water resources management. Examples include decision making regarding relevant standards, pollution source prioritization, control technology selection, and evaluation of groundwater remediation alternatives.

In the late 1970s and early 1980s the concept of "commensuration" was introduced into the water resources planning vocabulary. Commensuration refers to the process of measuring different things by a single standard or measure or "common units" (35). The common units serve as the basis for trade-off analysis. The four essential components of commensuration are (35) (a) to identify the factors that are to be commensurated; (b) to determine whose value judgments about those factors are to be considered; (c) to discover what those value judgments are; and (d) to combine the judgments of all of the selected individuals into a single collective set of judgments. Three examples of developed commensuration methodologies will be briefly described.

Mumpower and Bollacker developed the Evaluation and Sensitivity Analysis Program (ESAP), which is a computerized environmental planning technique for the evaluation of alternative water resource management plans (36). ESAP is based on a weighting-scaling approach to impact assessment and alternative evaluation. The evaluation of the desirability or acceptability of an alternative is based on a systematic combination of information about the impacts on the natural and cultural resources and information about the importance (weight) and preferred levels of such resources.

The analyses performed by ESAP provide the user with two basic types of information: (a) an evaluation of alternatives for each individual or group considered and (b) an analysis of how the individuals or groups differ in their evaluation of alternatives (36). The analysis requires the user to provide a hierarchical description of the evaluation problem, which includes the variables or resources affected by the alternatives, projected levels of the variables or resources for each of the alternatives, and information about the weights or importance values and the variables for each of the groups under consideration. If the user wishes to examine the effects of uncertainty on evaluation, data describing uncertainty must also be input into the program.

The ''cascaded-trade-offs method'' is a multiple-objective, multiple-publics method for evaluating alternatives in water resources planning (37). The method can be used to arrive at an overall ranking of planning alternatives on the basis of public values. The key feature of the method is that it provides a basis for trade-offs across both issue dimensions (decision factors) and publics. The method shows not only how to deal with trade-offs where trade-offs must be made, but also how to identify or create situations where trade-offs do not need to be made. In addition, it provides for an examination of the extent to which the overall ranking of the alternatives is sensitive to the addition of mitigation measures and to uncertainty in the data.

Finally in terms of commensuration approaches, Brown and Valenti (38) developed the multiattribute trade-off system (MATS). MATS is a computer program designed to help planners evaluate multiattribute alternatives to reach a judgment of each alternative's relative worth or desirability. MATS leads the user through a series of questions (a trade-off analysis) that focuses on the relative importance of various characteristics of the alternatives. The program documents the judgments that lead to the development of a policy for evaluating alternatives. The program can then apply this evaluation policy to up to 40 alternatives.

A recent trend in decision making in water resources management involves the usage of computer software. For example, Torno, et al. developed a training manual and computer program for the evaluation of the environmental impacts of large-scale water resources development projects (39). Sufficient information is included to enable the knowledgeable user to perform the evaluation on virtually any water body of interest. An interactive computer program has been developed to simplify application of the methodology.

In a more generic sense, a computer model has been developed that, with user input, can be used to help determine the relative weights of evaluation parameters used to evaluate projects or decision choices under consideration, and to determine the utility function for each of the attributes (40). A unique feature of this model, which can be applied in many water resources management decisions, is that it incorporates uncertainties of three types: (1) those dealing with the factor (parameter) weights; (2) those dealing with the worth of each project with regard to each factor; and (3) those dealing with the utilities of the attributes. This model called ("d-SSYS" for decision support system) may be applied to any problem of competing alternatives. It is a true decision support system; that is, it is an interactive system that provides the user with easy access to decision models and data in order to support semistructured decision-making tasks.

In addition to these examples, still other tools are available. For example, in a recent study conducted at the University of Oklahoma, 14 multiattribute decision-making techniques were reviewed in order to identify and develop a comprehensive set of applicable techniques to be used in a proposed generic decision-based methodology; the list of techniques is in Table 9.8 (41). Each identified technique was addressed by a brief description of the process by which the alternatives are evaluated, the resulting outcome, the type and amount of information and data required by the evaluation process, the conditions of applications and basic assumptions, and, when available, references to applications in environmental decision making.

Summary

Water resources management encompasses decisions related to quantity-related issues, such as river flows, water supply, and flooding, and quality-related issues, such as pollution sources, pollutant cycling, and quality degradation. Management activities can range from resource development planning to enforcement of water usage or water-quality standards. Natural variabilities in hydrologic and biogeochemical processes, and the absence of sufficient information, are fundamental causes of uncertainty in management efforts. The inclusion of risk and uncertainty analysis in water resources planning is critical for improved decision making. Uncertainties and limitations of mathematical models for surface and subsurface hydrologic systems, and for aquatic ecologic

Table 9.8 Examples of Multiattribute Decision-Making Techniques

OBJECTIVE OF TECHNIQUE	EXAMPLES OF TECHNIQUES
Elimination of the nonfeasible set of alternatives (also called sequential elimination techniques)	(1) Conjunctive screening (2) Conjunctive ranking (3) Disjunctive screening (4) Lexicographic screening (5) Compensatory screening
Elimination of the dominated set of alternatives	(6) Noninferior curve technique (7) Indifference map technique (8) Reasonable social welfare function (RSWF) (9) Outranking approaches
Evaluation of the nondominated set of alternatives	(10) Utility theory and decision analysis[a] (11) Compromise programming (12) Displaced ideal technique (13) Cooperative game theory (14) Analytic hierarchy process (AHP)

[a]Includes four models—the additive model, the multiplicative model, the quasi-additive model, and the hierarchical additive model.

systems, should be recognized. Institutional and technical uncertainties can arise in various facets of ongoing surface water and groundwater programs focused on effective water-quantity and water-quality management. Uncertainties in water resources management can be reduced by more extensive information coupled with the use of interdisciplinary efforts, the incorporation of public values in decision making, and the utilization of systematic decision analysis tools for evaluating multiple choices in development or management strategies.

References

1. Kindler J. Planning and decision-making framework. In: Thanh NC, Biswas AK, eds. Environmentally-sound water management. Delhi: Oxford University Press, 1990:59–86.
2. The Greeley-Polhemus Group Inc. Guidelines for risk and uncertainty analysis in water resources planning. Vol. I. IWR Report 92-R-1. Fort Belvoir, VA: U.S. Army Corps of Engineers Institute for Water Resources, 1992:5–15, 21–34, 38–41, 50, 56–61, 69–71.
3. Loehr RC. Groundwater contamination—the problem and potential solutions. Nat Forum 1989;69(1):26–28.
4. Tarr JA, Jacobson C. Environmental risk in historical perspective. In: Covello VT, Johnson BB, eds. The social and cultural construction of risk. Boston: D. Reidel, 1987:317–344.
5. Rose JB. Emerging issues for the microbiology of drinking water. Water/Eng Mgt 1990:23–29.

6. Rose JB, Haas CN, Regli S. Risk assessment and the control of waterborne giardiasis. Am J Public Health 1991;81(6):709–713.

7. Rose JB, Gerba CP. Assessing potential health risks from viruses and parasites in reclaimed water in Arizona and Florida, USA. Water Sci Technol. 1991;23:2091–2098.

8. Grigg NS. Water resources planning. New York: McGraw-Hill, 1985:27–34.

9. Principles and guidelines. Economic and environmental principles and guidelines for water and related land resources implementation studies. Washington: Executive Office of the President, 1983.

10. Petersen MS. Water resource planning and development. Englewood Cliffs, NJ: Prentice-Hall, 1984:54–66.

11. Environmental Resources Limited. Prediction in EIA. London: Environmental Resources Limited, 1984:290.

12. Anderson MG, Burt TP. Modelling strategies. In: Anderson MG, Burt TP, eds. Hydrological forecasting. New York: Wiley, 1985:1–13.

13. Henderson-Sellers B. Water quality modeling. Vol. IV. Decision support techniques for lakes and reservoirs. Boca Raton, FL: CRC Press, 1991.

14. James A, ed. An introduction to water quality modeling. West Sussex, England: Wiley, 1993.

15. U.S. Army Corps of Engineers. Water quality models used by the Corps of Engineers. Information exchange bulletin E-87-1. Vicksburg, MS, 1987.

16. Domenico PA, Schwartz FW. Physical and chemical hydrogeology. New York: Wiley, 1990.

17. Water Science and Technology Board. Ground water models—scientific and regulatory applications. Washington: National Academy Press, 1990.

18. Environmental Resources Limited. Environmental impact assessment—techniques for predicting effects in EIA. Vol. 2. London: Environmental Resources Limited, 1982.

19. Ray C. Use of qual-I and qual-II models in evaluating waste loads to streams and rivers. In: Biswas AK, Khoshoo TN, Khosla A, eds. Environmental modeling for developing countries. London: Tycooly, 1990:65–79.

20. U.S. Fish and Wildlife Service. Habitat evaluation procedures (HEP). ESM102. Washington, 1980.

21. U.S. Army Corps of Engineers. A habitat evaluation system for water resources planning. Vicksburg, MS: Lower Mississippi Valley Division, 1980.

22. Starfield AM, Bleloch AL. Building models for conservation and wildlife management. New York: Macmillan, 1986.

23. Brookes A. Channelized rivers—perspectives for environmental management. Chichester, England: Wiley, 1988:67–71.

24. Odum HT. Systems ecology: an introduction. New York: Wiley, 1983.

25. Canter LW. Water resources management: fundamental considerations and linkages with land planning. Workshop on environmental planning, Milano, Italy, 1991.

26. Canter LW. Functions and activities of ground water protection: implications for institutional coordination. Environ Prof. 1986,8(3):219–224.

27. U.S. Environmental Protection Agency. Guidelines for delineation of wellhead protection areas. Washington, 1987.

28. U.S. Army Corps of Engineers. Policy and planning guidance for conducting civil works planning studies. ER 1105-2-100. Washington, 1990.

29. Fessenden-Raden J, Fitchen JM, Heath JS. Providing risk information in communities: factors influencing what is heard and accepted. Sci Technol Hum Values 1987,12(3–4):94–101.

30. Russell C. Guidebook for risk perception and communication in water resources planning. Part I. Underpinnings and planning applications. IWR Report 93-R-13. Fort Belvoir, VA: U.S. Army Corps of Engineers, Institute for Water Resources, 1993:55, 73.

31. Organization for Economic Co-operation and Development. Management of water projects—decision-making and investment appraisal. Paris: OECD, 1985:52–59.

32. Helweg OJ. Water resources planning and management. New York: Wiley, 1985:246–247.

33. Solomon RC, et al. Water resources assessment methodology (WRAM): impact assessment and alternatives evaluation. Y-77-1. Vicksburg: U.S. Army Engineer Waterways Experiment Station, 1977.

34. Richardson SE, et al. Preliminary field test of the water resources assessment methodology (WRAM)—Tensas River, Louisiana. Y-78-1. Vicksburg: U.S. Army Engineer Waterways Experiment Station, 1978.

35. Lord WB, Deane DH, Waterstone M. Commensuration in federal water resources planning: problem analysis and research appraisal. Denver: U.S. Bureau of Reclamation, 1979.

36. Mumpower J, Bollacker L. User's manual for the evaluation and sensitivity analysis program. E-81-4. Vicksburg: U.S. Army Engineer Waterways Experiment Station, 1981.

37. Anderson BF. Cascaded tradeoffs: a multiple objective, multiple publics method for alternatives evaluation in water resources planning. Denver: U.S. Bureau of Reclamation, 1981.

38. Brown CA, Valenti T. Multi-attribute tradeoff system: user's and programmer's manual. Denver: U.S. Bureau of Reclamation. 1983.

39. Torno HC, et al. Training guidance for the integrated environmental evaluation of water resources development projects. Paris: UNESCO, 1988.

40. Klee KJ. D-SYSS: A computer model for the evaluation of competing alternatives. EPA 600/S2-88/038. Cincinnati: U.S. Environmental Protection Agency, 1988.

41. Lahlou M, Canter LW. Alternatives evaluation and selection in development and environmental remediation projects. Environ Impact Assess Rev 1993;13:37–61.

Scientific Uncertainty and the Environmental Impact Assessment Process in the United States

10

Larry W. Canter

The practice of environmental impact assessment is typically identified as beginning with the passage of the National Environmental Policy Act (NEPA) in the United States, with the effective date being January 1, 1970. Environmental impact assessment (EIA) can be defined as the systematic identification and evaluation of the potential impacts (effects) of proposed projects, plans, programs, or legislative actions relative to the physical-chemical, biologic, cultural, and socioeconomic (including health) components of the environment. The primary purpose of the EIA process, also called the NEPA process, is to encourage consideration of the environment in planning and decision making and to ultimately generate projects, plans, programs, or actions that are environmentally compatible.

Near the end of the first decade of preparing environmental assessments (EAs), which represent preliminary studies, and environmental impact statements (EISs), which represent comprehensive studies, increased attention was given to the use of scientifically based approaches. This attention was reflected in paragraph 1502.24 of the Council on Environmental Quality (CEQ) regulations issued pursuant to NEPA (1):

> Agencies shall insure the professional integrity, including scientific integrity, of the discussions and analyses in environmental impact statements. They shall identify any methodologies used and shall make explicit reference by footnote to the scientific and other sources relied upon for conclusions in the statement.

The emphasis on scientific approaches focused attention on the fact that there are many uncertainties associated with the EIA process (2). Uncertainty has many definitional considerations ranging from doubt, or lack of certainty, to a science-based definition referring to the estimated amount by which an observed or calculated value may depart from the true value. Uncertainty, as used herein, refers to insufficient information on the project and environmental setting, inaccuracy in impact prediction, or lack of knowledge in impact interpretation, the effectiveness of mitigation measures, and decision making.

Uncertainty is directly and indirectly addressed in the CEQ regulations (3). Paragraph 1508.27 suggests that the following concept be utilized in impact significance determination—consider the degree to which impacts are uncertain or involve unique or unknown risks. Uncertainty is indirectly referenced in relation to incomplete or unavailable information (§1502.22) as follows (3, p. 942):

> When an agency is evaluating reasonably foreseeable significant adverse effects on the human environment in an environmental impact statement and there is incomplete or unavailable information, the agency shall always make clear that such information is lacking.
>
> (a) If the incomplete information relevant to reasonably foreseeable significant adverse impacts is essential to a reasoned choice among alternatives and the overall costs of obtaining it are not exorbitant, the agency shall include the information in the environmental impact statement.
>
> (b) If the information relevant to reasonably foreseeable significant adverse impacts cannot be obtained because the overall costs of obtaining it are exorbitant or the means to obtain it are not known, the agency shall include within the environmental impact statement: (1) a statement that such information is incomplete or unavailable; (2) a statement of the relevance of the incomplete or unavailable information to evaluating reasonably foreseeable significant adverse impacts on the human environment; (3) a summary of existing credible scientific evidence which is relevant to evaluating the reasonably foreseeable significant adverse impacts on the human environment; and (4) the agency's evaluation of such impacts based upon theoretical approaches or research methods generally accepted in the scientific community. For the purposes of this section, "reasonably foreseeable" includes impacts which have catastrophic consequences, even if their probability of occurrence is low, provided that the analysis of the impacts is supported by credible scientific evidence, is not based on pure conjecture, and is within the rule of reason.

This chapter addresses uncertainty in the EIA process, with particular emphasis on practices in the United States. The information can also be applied to EIA practices in the more than 75 countries that have promulgated NEPA-type laws. The chapter includes sections on the increasing need to address uncertainty, a conceptual model for planning and conducting impact studies, techniques for impact prediction, scientific uncertainties in impact prediction, categories of uncertainty, and addressing uncertainty in the EIA process.

Increasing Need to Address Uncertainty

As the EIA field evolves and matures, several studies and changes in emphases are bringing increased attention to uncertainty in the EIA process. Examples of two comprehensive studies and four changing emphases will be utilized to illustrate the increased attention. The changing emphases are due to impact studies related to hazardous waste site cleanups, strategic environmental assessments, life cycle assessments, and the increased usage of EAs in lieu of EISs.

A review of 11 EIA case studies related to highways and dams in Canada, Finland, the Federal Republic of Germany, The Netherlands, Norway, and the United States was conducted in the mid-1980s to analyze practical experience with EIA and to draw lessons from that experience (4). One group of resultant recommendations related to the content of EISs in terms of minimum elements and other pertinent topics. The minimum elements included project setting (purpose and need), a description of the proposed project, a description of the existing environment, reasonable alternatives, including the do-nothing alternative, an assessment of the environmental impacts of the proposed project and the alternatives, and summary. Other pertinent topics that were recommended to be addressed, as appropriate, included the scoping process, a monitoring program, mitigation measures, gaps in knowledge, and uncertainties. The delineation of uncertainties was identified as an aid to more informed decision making.

The Senior Advisors to the Economic Commission for Europe recently conducted a study of impact prediction methods in EIA; and over 20 technically focused recommendations were promulgated. The following recommendation related to uncertainty considerations (5, p. 5): "in environmental predictions, the identification and quantification of the sources of uncertainty should be an important step in the application of methods. The results of environmental predictions should indicate the margin of uncertainty involved." This recommendation highlights the fact that scientific predictions can be subject to many questions and assumptions, even though such predictions are scientifically based and mathematically complicated.

During the decade of the 1980s, considerable attention was given, and continues to be given, to the cleanup of uncontrolled hazardous waste sites in the United States. These sites are subjected to risk assessment studies, called integrated exposure assessments. Figure 10.1 contains an overview of the integrated exposure assessment (IEA) process (6); this process is functionally similar to the EIA process. The concluding element in the IEA process is an uncertainty analysis; in this analysis each step in the assessment is reviewed to identify any uncertainties involved and to evaluate their separate and cumulative impact on assessment results. Uncertainties may result from the use of default values for various input parameters, from the use of simplified estimation procedures in contrast to rigorous computer analysis or monitoring-based analysis, from an

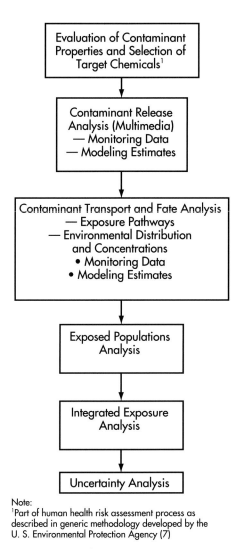

Note:
[1]Part of human health risk assessment process as described in generic methodology developed by the U. S. Environmental Protection Agency (7)

Figure 10.1 Overview of the integrated exposure assessment process (6).

inability to accurately define exposed populations, and other factors. The uncertainty analysis provides information for decision makers who must evaluate the results of the IEA with regard to implications for potential risks associated with the uncontrolled site and appropriate remedial technology selection (6).

Strategic environmental assessments (SEAs) refer to applying the EIA process to earlier, more strategic tiers of decision making for policies, plans, or programs (8). SEAs in the United States practice are typically referred to as programmatic EISs or tiered EISs. It is expected that increased usage of SEAs will occur, particularly in the EIA process in developed nations. The levels of uncertainty in SEAs will probably be greater than for project-specific impact studies; accordingly, uncertainty issues will need to be systematically addressed in SEAs.

Life-cycle impact assessments are being suggested as useful studies relative to projects or generated products (9, 10). Such assessments could include project decommissioning or "cradle-to-grave" considerations for generated products, including their disposal. The Society of Environmental Toxicology and Chemistry (10) has indicated that a critical need exists for methods and approaches for factoring uncertainty into the life-cycle impact assessment process; further, the extent of uncertainty in all steps of the analysis needs to be addressed.

In EIA practice in the United States, the purpose of an EA is to establish whether or not significant impacts would be expected as a result of a proposed project or action. If such impacts are anticipated, then an EIS should be prepared; if not, then a Finding of No Significant Impact (FONSI) is appropriate. Many uncertainties can be associated with identifying significant impacts in EAs; some examples include (11) (a) the analysis of the impacts of accidents, (b) the assessment approach with incomplete or unavailable information, (c) the uncertainty associated with assessment methods, and (d) the consideration of cumulative impacts. Such uncertainties could be dealt with by defining thresholds that trigger public review of an EA before the decision to prepare a FONSI or EIS is made. For example, accident probabilities could be used to trigger EA review, as could model predictions whose error bars exceed significance levels. If such uncertainties are not properly addressed, then inappropriate decision making could occur regarding resultant FONSIs or EISs. Between 30,000 and 50,000 EAs are prepared annually in the United States as compared to 400 to 500 EISs. Therefore, addressing uncertainties in the preparation of EAs is very important in the overall EIA process.

Model for Planning and Conducting Impact Studies

A 10-step or 10-activity model, as shown in Figure 10.2, can be used for planning and conducting impact studies leading to EAs or EISs (12). The model is flexible and can be adapted to target important concerns related to specific projects in unique locations. The focus in this model is on single projects, although it could also be applied to plans, programs, policies, or other actions such as granting permits. It is assumed that an appropriate interdisciplinary team has been assembled to work on the EIA study. The acronyms in the model are described in the following paragraphs. The connections between the four groups of steps or activities suggest interrelationships and the need for feedback. Although the model is straightforward, there are many uncertainties, particu-

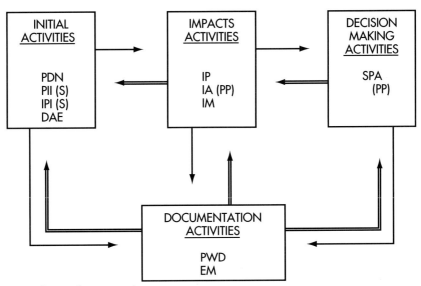

Note: Definitions of terms are in the text.

Figure 10.2 Conceptual framework for environmental impact studies.

larly as related to impact prediction. Table 10.1 summarizes examples of uncertainty associated with each of the 10 activities.

The first activity would be to determine the features of the proposed project, the need for the project, and the potential alternatives that either have been or could be considered for the subject project (called PDN for project description and need). Key information that would be needed relative to this activity includes

1. Type of project, project life, and how it functions or operates in a technical context.
2. Proposed location for the project.
3. The time period required for project construction.
4. The identified need for the proposed project in the particular location wherein it is proposed.
5. Alternatives that have been or should be considered; examples of generic alternatives for projects include site location, size, design features, and pollution control measures, and timing relative to construction and operational issues.

There may be uncertainty associated with the specific project regarding size and detailed design and operational features, with this type of information having implications relative to environmental impacts. Uncertainty may also exist regarding the alternatives to be evaluated, and the amalgamation of information

Table 10.1 Examples of Causes of Uncertainties Related to Various Activities in the EIA Process

ACTIVITY*	EXAMPLES OF CAUSES OF UNCERTAINTIES
PDN	Incomplete information on need for project, design features of project (size and pollution control measures), and viable alternatives.
PII	Required time schedules for compliance with environmental media laws.
IPI	Relationships between changes in the biophysical and socioeconomic environments, and methods to use for cumulative impact assessment.
DAE	Ranges in natural variability of extant environmental data; absence of site-specific data.
IP	Selection of appropriate models to predict changes; absence of baseline data for model calibration; inadequate models for quantifying or even descriptively addressing impacts.
IA	Absence of quantitative criteria for impact interpretation; conflicting societal viewpoints on significance of environmental changes.
IM	Limited information on effectiveness of planned mitigation measures; lack of project proponent commitment to implementation of mitigation measures.
SPA	Importance considerations for identified decision factors; lack of comprehensive information on all alternatives being evaluated.
PWD	Need to communicate relative risk of proposed project and alternatives.
EM	Lack of information for statistical design of monitoring systems to incorporate environmental changes from project implementation.

*Definitions are in text.

on selected features of the alternatives that could be used as a basis for selecting the proposed action.

The second activity should focus upon pertinent institutional information (PII) related to the construction and operation of the proposed project. PII refers to a multitude of environmental laws, regulations, policies, and executive orders related to the physical-chemical, biologic, cultural, and socioeconomic environment. For example, there are over 50 federal statutes in the United States that have direct or indirect relevance in the planning and conduction of impact studies (12). In addition to federal statutes, additional institutional requirements may exist at both state and local levels and they may have relevance for specific projects.

One of the approaches which can be used to aid the PII activity is scoping

(S). As part of the scoping process, which would include contacts with regulatory agencies and other interested publics, the identification of PII could occur. In summary, PII can aid in the interpretation of existing quality conditions relative to the environment, and it can serve as a basis for interpreting the anticipated impacts (effects) of the resultant project.

Depending on the impact issues related to the project being analyzed, uncertainty may exist relative to pertinent laws and regulations; of particular concern may be the time schedule required for regulatory compliance. One example of current interest in the United States is the multiple requirements of the Clean Air Act Amendments (CAAA) of 1990 and the associated time schedules for compliance. This problem has been exacerbated by slippages in the time schedules specified in the CAAA. As a further example, many environmental media laws are becoming more comprehensive in terms of the number of pollutants to be addressed and requirements for monitoring and evaluation.

The third step or activity is to identify potential impacts (IPI) associated with the subject project. This early qualitative identification of anticipated impacts can facilitate the focus of subsequent steps; for example, it can aid in describing the affected environment and impact prediction. The identification of potential impacts can be an outcome of the scoping process. An appropriate task is to identify generic impacts related to the project type being analyzed. In many cases pertinent impacts can be identified by applying a generic checklist. Other tools and techniques include the identification of potential impacts via the preparation of simple interaction matrices, with such matrices consisting of a list of project actions displayed against a list of environmental factors. Network diagrams and cause-effect (or consequence) diagrams can also be used, and overlay mapping techniques may be helpful. Uncertainties can develop during the use of any of these tools and techniques.

The fourth activity is focused on the preparation of a description of the affected environment (DAE). This is placed fourth in the process so as to enable the selective identification of pertinent environmental factors for the study in progress. A selective approach involves identifying key environmental factors anticipated to be changed by the proposed project, and then preparing appropriate descriptions of existing conditions relative to these factors. This selective approach can be facilitated by scoping and the use of one or a combination of checklists, interaction matrices, impact networks, and overlay mapping.

There will always be uncertainty related to the extent of information necessary for describing the affected environment. A fundamental problem in many studies is that extant environmental information is not as site specific as needed to conduct a truly systematic, site-specific scientific study. Uncertainties may also arise in conjunction with sample collection methods, analytical procedures, and quality assurance/quality control (QA/QC) measures.

The fifth and technically most difficult and challenging activity is termed impact prediction (IP). Impact prediction basically refers to the quantification, where possible, or at least the qualitative description, of the anticipated impacts of the proposed project on selected environmental factors. It should be noted that the terms "predicting impacts" and "forecasting impacts" are typically considered to be synonymous. In contrast, Culhane, Friesema, and Beecher (13) have suggested that "forecasting impacts" is more appropriate for impact studies since (a) predict means to foretell with the precision of calculation, knowledge, or shrewd inference from facts or experiences, whereas (b) forecast, in contrast, suggests that conjecture rather than real insight or knowledge is probably involved. Thus, they suggest that anticipated changes resulting from a project that are addressed in an EIS should be more appropriately termed "forecasts." However, this terminology is not universally accepted, and predicting and forecasting will be considered herein as synonymous terms.

In impact prediction it would be desirable to specify the range in magnitude of effects (impacts) and the possible errors in the prediction of effects. However, uncertainties in impact prediction have lead to qualitative descriptions of impacts and the use of phrases such as "might" or "possible" (14). Accordingly, assumptions related to impact predictions should be clearly delineated. Also, the probability of occurrence and prediction ranges should be noted, if possible. For example, a prediction could be that 95% of the time the noise level in an area would be between 65 and 70 dBA (15).

There are numerous uncertainty issues of concern in prediction; they range from the prediction approaches or models utilized, to basic assumptions for the predictions and the functioning of the biophysical and socioeconomic environments. Additional information on impact prediction is presented in the next section.

The sixth activity in the conceptual model is entitled impact assessment (IA). In the terminology used herein, assessment refers to the interpretation of the significance of anticipated changes resulting from the proposed project. Impact interpretation can be based on the systematic application of a definition of significance; in the United States such a definition is included in the CEQ regulations (3).

Uncertainty can also exist relative to the interpretation of the anticipated changes resulting from a proposed project. For impacts on resources wherein specific quantitative standards exist, interpretation can be based on comparing the resultant anticipated condition to the numerical standard. This approach can facilitate interpretation of impacts related to water, air, and noise quality; however, for many types of impacts, only qualitative criteria or policy directives exist. These latter concerns are related to impacts involving threatened or endangered species, historic and archeologic resources, visual quality, and many facets of the socioeconomic environment. Accordingly, there are many uncertainties related to applying professional and societal value judgments in determining the significance of anticipated impacts.

The seventh activity is associated with identifying and evaluating potential impact mitigation measures (IM). Mitigation has been defined in the CEQ regulations as follows (3): (a) avoiding the impact altogether by not taking a certain action or parts of an action; (b) minimizing impacts by limiting the degree or magnitude of the action and its implementation; (c) rectifying the impact by repairing, rehabilitating, or restoring the affected environment; (d) reducing or eliminating the impact over time by preservation and maintenance operations during the life of the action; and (e) compensating for the impact by replacing or providing substitute resources or environments. The definition suggests a sequential consideration of avoidance followed by sizing, followed by rectifying, and so on. While the IM step is identified as the seventh activity, there is no reason to wait until this point in the study to identify and evaluate impact mitigation measures. For example, these types of measures could be identified in a preliminary manner in association with the third activity on identifying potential impacts (16).

The costs and potential effectiveness of identified mitigation measures involve many uncertainties. Mitigation emphases are appropriate in the EIA process; however, uncertainty can be associated with the potential success of selected technical or scientific measures. An illustration is associated with the construction of wetlands and the use of mitigation banking to offset wetland or other types of habitat losses. The success rates of constructed wetlands are uncertain and controversial, thus, a decision to approve a project based on assuming mitigation success may be inappropriate.

The eighth activity involves selecting the proposed action (SPA) from alternatives that have been evaluated. For public projects there is considerable emphasis on the identification and evaluation of alternatives. In fact, the CEQ regulations in the United States indicate that the analysis of alternatives represents the heart of the EIA process (3). Environmental impact studies typically address a minimum of two alternatives, and they can include upwards of 50 alternatives. Typical studies address three to five alternatives. The minimum number usually represents a choice between construction and operation of a project versus project nonapproval.

For impact studies involving comprehensive comparisons of alternatives, such comparisons can be facilitated by the adoption of a decision framework, possibly via using multicriteria decision-making techniques. In this regard, there are necessary decisions relative to comparative factors and their relative importance, along with the ranking, rating, or scaling of the anticipated impacts of each alternative relative to the comparative factors. Finally, decisions are necessary regarding the aggregation of the information, whether such aggregation involves qualitative considerations or the development of composited quantitative indices. While there may be uncertainties associated with such decisions, they do represent a structured approach that can be useful in providing a systematic framework for decision making. Additional information on these tools is in a subsequent section.

The ninth activity involves preparing written documentation (PWD) related to the impact study. Written documentation could involve the preparation of an EA and an EIS. The most important point about PWD is that sound principles of technical writing should be utilized (17). These include the development of outlines, careful documentation of data and information, the liberal usage of visual display materials, and the careful review of written materials so as to ensure effective communication with both technical and nontechnical audiences. A standard format for EAs and EISs in the United States has been developed by the Council on Environmental Quality (3).

The final activity in the model is the planning and implementation of appropriate environmental monitoring (EM) programs, with this activity being particularly important for large-scale projects with potentially significant environmental consequences. Environmental monitoring may be necessary in establishing baseline conditions in the area of the project; however, of more potential relevance is longer-term monitoring in the environs of the project to carefully document impacts that are actually experienced (18). Such monitoring data can be useful in project operational decisions and the preparation of project audits. Environmental monitoring systems can be considered in two phases: (1) development of a monitoring system and (2) implementation and operation of the system. Planning is required for both phases, and detailed information on planning environmental monitoring programs is available elsewhere (18, 19).

While not specifically delineated in Table 10.1, several other types of uncertainty can be associated with the EIA process. One example is related to public participation (PP) through scoping or public information meetings. Uncertainty may exist regarding what publics should be or will be represented and the viewpoints that they may express on the project or impact issues.

Techniques for Impact Prediction

The key technical activity in the EIA process is the prediction of impacts (effects) for both the without-project and with-project conditions. Numerous technical approaches could be used. To serve as an example, the Principles and Guidelines of the Water Resources Council delineated several approaches which could be utilized for water resources projects (20). These approaches include (a) adoption of forecasts made by other agencies or groups, (b) use of scenarios based on different assumptions regarding resources and plans, (c) use of expert group judgment via the conduction of formalized Delphi studies or the usage of the nominal group process, (d) extrapolation approaches based on the use of trends analysis or simple models of environmental components, and (e) analogy and comparative analyses that involve the use of "look-alike" resources or projects and the application of information from such look-alike conditions to the planning effort.

The range of impact prediction techniques currently used in the EIA process is broad and encompasses the use of analogies through sophisticated mathematical models (21); the range is shown in Table 10.2. In a specific impact study, several prediction techniques may be required due to data availability or lack thereof and the applicability or nonapplicability of specific mathematical models. In addition, as greater attention is being given to the global environment and to potential global consequences of large-scale projects or activities, it is becoming increasingly necessary to consider mesoscale environmental consequences via the use of regional or global modeling.

Simple Techniques

Perhaps the simplest approach for impact prediction is to utilize analogies or comparisons to the experienced effects of existing projects or activities. This approach could be termed a "look-alike" approach in that information gathered from similar types of projects, in similar environmental settings, could be utilized to qualitatively (descriptively) address the anticipated impacts of the proposed project. Professional judgment would be necessary in using analogies for predicting specific impacts on the environment.

An inventory technique involves developing an inventory of environmental resources through either the assemblage of existing data or the conduction of baseline monitoring, with the presumption then being that the particular resources that are in the existing environment, or portions thereof, will be lost or degraded in quality as a result of the proposed project or activity (21). This could be perceived as a worst-case prediction and for certain types of resources in a geographic area, it would represent a reasonable approach.

Qualitative or semiquantitative impact predictions can be made via the use of checklists or interaction matrices. Checklists range from simple listings of anticipated impacts by project type to questionnaire checklists that incorporate detailed questions to provide a structure to impact prediction. Interaction matrices include simple x-y matrices to identify impacts and to provide a basis for further evaluation of such impacts in terms of their magnitude and importance. Stepped matrices can be used to delineate secondary and tertiary consequences of project actions. Networks (or impact trees or chains or consequence diagrams) represent a variation of matrices that can be systematically utilized to trace the consequences of a given project or activity.

Indices and Experimental Methods

Another category of impact prediction approaches involves the use of environmental indices (22). An environmental index refers to a mathematical or

Table 10.2 Impact Prediction Techniques Currently Used in the EIA Process

---

Simple Techniques

Analogy

Inventory

Checklists (simple or descriptive)

Matrices (simple, stepped, networks, impact trees, consequence diagrams)

Indices and Experimental Methods

Environmental media indices (air, surface water, groundwater vulnerability, noise)

Other indices (visual, quality of life)

Habitat indices (HEP, HES)

Experimental methods (laboratory, field, physical models)

Mathematical Models

Air quality dispersion

Hydrologic processes

Surface and groundwater quality and quantity

Noise propagation

Biologic impact (HEP, HES, population, nutrients, chemical cycling, energy system diagrams)

Archeological (predictive)

Visual impact

Socioeconomic (population, econometric, multiplier factors, health)

---

descriptive presentation of information on a series of factors that can be used for classification of baseline quality and sensitivity and for predicting the impacts of a proposed project or activity. The basic concept for impact prediction would be to anticipate and quantify (if possible) the change in the index as a result of the project or activity, and to then consider the difference in the index from the with and without project conditions as one measure of impact. Numerous environmental indices have been developed, including examples for air quality, water quality, noise, visual quality, and quality of life (12).

One type of index which has been widely used is based on habitat considerations; examples include the habitat evaluation procedures (HEP) developed by the U.S. Fish and Wildlife Service and the habitat evaluation system (HES) developed by the U.S. Army Corps of Engineers (23, 24). Both approaches involve the development of a numerical index to describe habitat quality and size.

Experimental methods range from the conduction of laboratory experiments to develop factors or coefficients for mathematical models, to the implementation of large-scale field experiments to measure changes in environmental features as a result of system perturbations. In addition, physical models or microcosms have been utilized to examine impacts related to hydrodynamic or ecologic changes within given environmental settings. Experimental methods are primarily useful for anticipating changes in physical/chemical components or biologic features of environmental settings.

Mathematical Models

Quantitative approaches for impact prediction involve the use of mathematical models. Numerous types of models have been developed to account for pollutant transport and fate. In addition, models are available for describing environmental features and the functioning of ecosystems. The purpose of this brief review is not to summarize the state of the art of mathematical modeling for impact studies but to provide some information as to the availability of types of models that can be used in such studies (21).

With regard to air-quality dispersion, numerous models exist for addressing point, line, and area sources of air pollution (25, 26). Within recent years models have been developed for long-range transport of pollution and for atmospheric reactions leading to photochemical smog (ozone) formation and acid rain. Many air-quality models are available in PC software through the U.S. Environmental Protection Agency, professional societies, and consulting firms; thus they represent a usable technology for many impact studies.

Surface water and groundwater quality and quantity models are also plentiful, with major research developments within the last decade accounting for solute transport in subsurface systems. Surface water quality and quantity models range from one-dimensional steady-state models to three-dimensional dynamic models for rivers, lakes, and estuarine systems (27, 28). Some groundwater flow models now incorporate subsurface processes such as adsorption and biologic decomposition (29). The International Ground Water Modeling Center, along with the Association of Ground Water Scientists and Engineers, has several hundred groundwater models in software form that can be used to address flow and pollutant consequences of given projects or activities. The U.S. Environmental Protection Agency and the U.S. Army Corps of Engineers are also sources of PC-based surface water and groundwater models.

Noise propagation models can be utilized for point, line, and area sources of noise generation (30, 31). These models range from simple calculations and the use of nomographs to sophisticated computer modeling for airport operations. Noise models exist for continuous or discontinuous noise sources, including instantaneous noise related to construction-related blasting. The U.S. Environmental Protection Agency, Federal Highway Administration, and Federal Aviation Administration can supply PC versions of various noise models.

Biologic impact prediction models used in the United States are typically characterized as involving habitat approaches; examples include HEP and HES (23, 24). As noted, these models involve the calculation of an index incorporating both quality and quantity information. Prediction of impacts involves determination of the index under baseline and future with and without conditions. Other biologic impact models include species population models developed based on empirical approaches involving statistical correlations (32). The most sophisticated models involve energy system diagrams; these types of models have been utilized in some impact studies.

Predictive modeling is also possible for ascertaining the potential for archaeologic resources being located in geographic study areas. Such modeling is based on evaluating a series of factors to indicate the likelihood of the presence of archaeologic resources; the factors are related to existing information on early occupations in the area, and other biophysical and sociologic factors (21). This type of modeling is often used to determine the necessity for planning and conducting archaeologic field surveys (33).

Visual quality has also been a subject of importance in selected impact studies. Visual impact modeling approaches have been developed by several federal agencies, including the U.S. Forest Service, U.S. Bureau of Land Management, U.S. Soil Conservation Service, and U.S. Army Corps of Engineers (34). These models typically involve evaluation of a series of factors, in some cases quantitatively and in others descriptively, and aggregation of the information into an overall visual quality index for the study area.

Impact prediction related to the socioeconomic environment is often associated with the use of human population and econometric models. Population forecasting can range in sophistication from projections of historic trends to the use of complicated cohort analysis models (21). Econometric models relate the population and economic characteristics of study areas so that interrelationships can be depicted between population changes and resultant changes in economic characteristics. In this regard, several input-output models can be used in impact studies (35). Impact predictions for other socioeconomic factors, such as housing and educational resources, can be accomplished through the use of multiplier factors applied to population changes.

Scientific Uncertainties in Impact Prediction

There are many potential uncertainties related to impact prediction, including the use of mathematical models. Some scientifically based uncertainties may be due to the following (36):

1. First, the environment is a complex, dynamic system involving interactions that are difficult to determine and often poorly understood.
2. Second, those changes that are of particular interest and relevance to a decision maker are often those that are impossible to quantify, such as the loss of an area of ecologic importance.
3. There is also the problem of where to stop since any major new development will give rise to associated developments over time and in larger geographic areas, and these secondary developments can also bring about consequential changes.

Regarding modeling, De Jongh (22) has indicated that all predictive methods involve some model of the environment, be it mathematical, physical, or conceptual. Uncertainties arise because these models cannot exactly reproduce what happens in the environment. Accordingly, three types of uncertainties may occur: (a) in any environmental system many different processes may affect one variable, and the model may not account for all influencing processes; (b) errors can arise from the nature of the model with respect to a particular process; for example, a decay relationship could be zero, first, or second order, inverse or exponential; and (c) resolution errors can arise from imprecise spatial and temporal resolutions of the selected model.

In some impact studies health risk assessments are used as a means of integrating human health impact concerns. There are many uncertainties related to health risk assessments; for example, the following can arise for waste disposal projects (37):

1. During the hazard identification stage: through incomplete characterization of the chemicals released from the site; failure to include key chemicals during indicator chemicals selection; and poor site investigations of landfills and contaminated land.
2. During the hazard analysis stage: failure to allow for variability in emission and discharge rates; incorrect characterization of the surrounding land use, or insufficient allowance for heterogeneity in land-use and activity patterns; failure to account for the most important exposure pathways; and mismatches between environmental fate and transport models and the biophysical environment they are meant to simulate.
3. During the risk estimation stage: in the applicability of animal toxicity data to humans; in the sensitivity between individuals within a population; in extrapolations from high to low doses to derive an RfD (reference dose below which adverse health effects are not observed) or a slope factor; and in the characterization of health effects from exposure to mixtures of chemicals.

Uncertainty factors have been used in health risk studies in relation to dose-response assessments. Specifically, the LOEL (lowest observed adverse effect level) is divided by a safety factor, also called an uncertainty factor, to determine the NOEL (no observed adverse effect level). Examples of uncertainty factors include (14)

1. Uncertainty factor = 10: use with valid experimental results from studies on prolonged ingestion by man, with no indication of carcinogenicity.
2. Uncertainty factor = 100: use when experimental results of studies of human ingestion are not available or scanty (e.g., acute exposure only); or with valid results of long-term feeding studies on experimental animals, or in the absence of human studies, valid animal studies on one or more species. No indication of carcinogenicity.
3. Uncertainty factor = 1000: use when there is no long-term or acute human data, and only scanty results exist on experimental animals with no indication of carcinogenicity.

A fundamental question relates to whether predicted (forecasted) impacts actually occur as a result of a project. Several studies have addressed this question; for example, Culhane et al (13) reviewed 29 EISs relative to the types of forecasted impacts, their characteristics, and their associated accuracy. The EISs included seven bridge and highway/road projects, three wastewater treatment plants or waste disposal projects, two airport projects, six water resources projects (small watershed, barge canal, two flood control, dredging, and dredged material disposal), three urban development projects, three park or forest management projects, three nuclear industry-related projects, one beach park expansion project, and one rural electric project. The most frequently occurring impact groupings involved social impacts and physiographic impacts. Based on this review, the authors described an ideal model for impact prediction by noting that the predictions should be quantified, their significance should be interpreted, and the certainty and uncertainty should be specified. Table 10.3 summarizes the characteristics of the 1105 forecasts in the 29 EISs in terms of the "ideal" EIS prediction (13). Unfortunately, the majority of the forecasts did not match the ideal conditions; for example, 959 out of 1105 forecasts used descriptor words to address the probability or possibility of the impacts.

Categories of Uncertainty in the EIA Process

There are no standardized categories of uncertainty related to the EIA process; however, several have been suggested. For example, Suter et al. described several types of uncertainty, including modeling errors, natural stochasticity, and

Table 10.3 Characteristics of Forecasts Closely Related to the Model of an Ideal EIS Prediction: Quantification, Significance, and Certainty

FORECAST CHARACTERISTICS	FORECASTS	
	N	PERCENT
Quantified	262	23.7
"No impact" forecast	123	11.1
Verbal unquantified forecast	720	65.2
	1,105	100
Significance of forecasted impact		
"High" (or synonym, explicitly stated)	32	2.9
"Moderate" (or synonym, explicitly stated)	8	0.7
"Insignificant" (or synonym, explicitly stated)	285	25.8
Quantified, without explicit significance	163	14.8
Vague/ambiguous significance statement	78	7.1
No explicit statement of significance	539	48.8
	1,105	100
Certainty about forecasted impact		
Quantified probability	1	0.1
Certainty guaranteed by situation	74	6.7
Impact conditional on intervening event	62	5.6
Probability implied by key words "will," "will not," "very likely," etc.	641	58.0
Possibility implied by key words "may," "could," "may not," etc.	318	28.8
	1,105	100

Source: Reference 13.

parameter errors (38). Petts and Eduljee suggested two categories of uncertainty in the EIA process (37):

1. Data uncertainty

- Project definition and characteristics
- Incomplete and/or irrelevant baseline information
- Model error
- Problems in defining dose-response relationships
- Inaccurate collection of data, for example, during measurement and sampling

Table 10.4 Examples of Impact Categories and Relative Uncertainties

IMPACT CATEGORY	EXAMPLES OF UNCERTAINTIES
Beneficial or detrimental	Societal viewpoints on beneficial vs. detrimental impacts can vary widely.
Reversible or irreversible	Lack of information on rates of terrestrial or aquatic ecosystem change as a result of perturbations.
Short term or long term	Uncertainty increases for predictability of long-term impacts.
Local or regional	Uncertainty increases for regional impact predictions.
Direct (primary) or indirect (secondary)	Uncertainty is greater for indirect impacts.
Single project or cumulative	Uncertainty is greater for cumulative impacts; examples include appropriate spatial boundaries and reasonably foreseeable future actions in the study area.

2. Decision uncertainty

- Failure to undertake an adequate scoping exercise
- Use of formalized scoring and weighting systems
- Data manipulation to meet different interests
- Pressure to use the "worst-case" scenario
- Other decisions; that is, while the EIA process for a particular policy, plan, or project is proceeding, many other decisions that may have an influence could be taken totally external to the process and by different parties
- The lack of strategic plans and policies, thus leaving the project impact study in a decision-making vacuum

A topic of increasing importance in the EIA process is cumulative impact assessment. In this regard, it is necessary to consider the impacts of the proposed project in the context of the impacts of other past, present, and reasonably foreseeable future actions that will occur within the study area for the proposed project. Uncertainty can exist regarding historic land usage and past and present projects and their impact-related characteristics. Uncertainty can be profound when attempting to ascertain future courses of action to be taken by various governmental agencies and the private sector.

To serve as a final illustration of categories, uncertainty could be considered relative to pertinent impact categories. Table 10.4 summarizes relative uncertainty depending on selected categorizations of anticipated impacts. For example, uncertainty related to anticipated impacts will increase as consideration is given to longer-term concerns versus short-term changes. Uncertainty will also increase as considerations shift from local to regional to national impact issues. Finally, while uncertainty exists regarding impacts associated with a specific project, it will be magnified by addressing cumulative impacts.

In summary, while there are no standardized categories of uncertainty related to the EIA process, it is helpful to consider various types of uncertainties and to be sure that these are addressed as appropriate in the EIA process.

Addressing Uncertainty in the EIA Process

Several approaches and tools and techniques can be utilized to address uncertainty within the EIA process. Recognition of uncertainty will be of benefit to decision makers, and where the level of uncertainty can be reduced the decision-making process will be facilitated. At a minimum, uncertainty should be acknowledged within the EIA process, with the following 11 suggestions representing a pragmatic approach for dealing with uncertainty, whether such uncertainty arises in conjunction with the EIA process or other studies (39):

1. Do not wait to be confronted. Acknowledge uncertainties up front.
2. Put bounds on uncertainty: say the level is between two known quantities; give error bars and confidence limits.
3. Make it clear that not all data are equally uncertain. Saying you are uncertain is not the same as claiming ignorance.
4. Say what is certain. Tell what you do know.
5. Say what has been done to reduce uncertainty. Talk about how you have resolved related uncertainties and how that has helped the process. Discuss research you have already done and the ways it has reduced uncertainty.
6. Say what you will do to reduce uncertainty further, when the results of studies, new information, etc., will be available.
7. If the remaining uncertainty is very small or very difficult to reduce further, say so.
8. Explain your cautiousness—and do not call your estimates or actions "conservativeness," a term people find confusing. Be overcautious until you are surer, and talk about the margins of safety you are folding into your risk estimates to make sure that uncertainty does not lead to a safety problem.
9. Do not hide behind uncertainty. If it is more likely than not that the problem is real, say so.
10. Acknowledge and apologize if the project sponsor has not been responsive to uncertainty concerns.

11. Never say, "There's no evidence of *X*" if you have not done the study that tests the possibility.

This list of pragmatic suggestions is not intended to be applied in every impact study; rather the list represents certain thoughts and concepts that could be applied as appropriate.

More explicit guidelines for planning and conducting impact studies would aid in decreasing uncertainty. Guidelines would facilitate decision making at the point of the deciding between preparing a FONSI or an EIS. In other words, if more explicit guidelines were developed for determining when significant impacts are expected to occur, including definitions of such impacts, this should reduce uncertainty in the process. However, while additional guidelines would be helpful, it should not be assumed that specific definitions could be developed that would be applicable for every impact study.

One area of uncertainty is associated with baseline information used to describe the affected environment; the fundamental need is to clearly establish the quality of the baseline data. One approach would be to focus on methods of data collection and analysis, and to ensure that appropriate QA/QC procedures have been used in the assemblage of such data. Further, the issue of natural variability of environmental conditions should be addressed within the context of available baseline data.

As noted, the study activity with the greatest level of uncertainty is impact prediction. Several approaches or tools can be utilized to reduce uncertainty in impact prediction. One approach would be to conduct pilot experiments and testing to establish waste streams from the proposed project (8). This approach would be valuable for projects with atmospheric, liquid, or solid waste streams. In addition, care should be taken to carefully identify anticipated impacts by reviewing analogs (case studies) of projects of similar type. This can be facilitated through the conduction of computer-based literature reviews.

Regarding the use of modeling in impact prediction, several positive things can be done to address uncertainty related to input data and actual predictions. Tables 10.5 and 10.6 address the handling of uncertainty for these two topics, respectively (2). Further, if quantitative modeling is utilized for impact prediction, sensitivity analysis could be used to assess the consistency in relationships between key input and output variables (15).

One issue often of concern in impact studies is related to the probability of accidents and their associated consequences. One approach for addressing this concern is through the conduction of accident and uncertainty analyses (8). In addition, contingency planning should be a part of the impact study process, with the emphasis on reducing risks related to an accident as opposed to only using a quantitative probabilistic approach for addressing the accident likelihood. Contingency planning could range from the development of a simple

Table 10.5 Techniques for Addressing Uncertainty in Input Data

EXAMPLES OF TECHNIQUES	OBJECTIVES IN USING TECHNIQUE
Handling uncertainty in measurement and analysis	To select measurement and analytical methods to achieve a required level of accuracy in terms of the bias and imprecision (i.e., total error) of data. These data may subsequently be used for prediction or for other purposes.
Sampling program design	To obtain information describing a system, with a required level of detail and accuracy by design of an appropriate sampling program in terms of the size, frequency, location and randomness (or otherwise), of sampling.
Sensitivity analysis	To identify those inputs that contribute most to uncertainty in prediction and so allocate priorities for further effort to improve the accuracy of predictions. Inputs can be ranked according to their priority for further research to improve the accuracy of predictions.

Source: Reference 2.

emergency response plan to a comprehensive plan that delineates duties and responsibilities for various individuals and organizations during and following several accident scenarios.

Reducing risks related to proposed projects can also aid in minimizing concerns relative to uncertainty. For example, De Jongh indicated that an emphasis on risk reduction may be more important than trying to address and resolve all problems related to uncertainty in the EIA process (2). Therivel et al. also noted the desirability of delineating risk minimization measures to avoid the worst possible outcome in the event of accidents related to proposed projects (8).

Interpretation of predicted impacts also has associated uncertainties due to the absence of quantitative standards for some factors or resources, and the need to apply both professional and societal value judgments in such interpretations. Accordingly, several approaches could be used to alleviate interpretation uncertainties. One approach would be to assess the effects of the most likely option; a second would be to assess the option likely to cause the most significant impact (the worst case); and a third would be to assess the effects of all realistic options. The second and third approaches could be used to acknowledge

Table 10.6 Techniques for Addressing Uncertainty in Prediction

EXAMPLES OF TECHNIQUES	OBJECTIVES IN USING TECHNIQUE
Scenario approach	To predict a range of possible outcomes (say maximum, minimum, most typical) taking into account input (variable and parameter) uncertainty.
Monte Carlo simulation	To predict the probability distribution of possible outcomes taking into account input uncertainty.
Constrained parameter approach	To predict the probability distribution of possible outcomes under different scenarios, taking into account input uncertainty.
First-order error analysis— propagation of error	To predict the uncertainty, i.e., the mean and variance of the outcome taking into account the mean and variance of uncertain inputs.
Generation of moments technique	To predict the uncertainty, i.e., mean, variance, skewness and kurtosis of the outcome, taking into account these characteristics of uncertain inputs.
Speculative simulation modeling	To predict the probability density function of expected outcomes taking into account uncertainty in inputs in conditions of sparse data.
Expert systems	To make predictions from a basis of lack of knowledge and data.
Probability encoding	To elicit expert opinion on the uncertainty associated with data in the form of a probability density function or cumulative probability distribution.

Source: Reference 2.

uncertainty in the EIA process (15). Further, precautionary approaches should be utilized in evaluating predicted impacts with higher levels of uncertainty (8). Such precautionary approaches should be conservative, thus aiding in minimizing concerns over levels of uncertainty. Such precautionary approaches could include continuing environmental monitoring and contingency planning.

Environmental monitoring can be used for several purposes during the construction, operational, and decommissioning phases for a project. The emphasis in EIA practice during the first two decades was primarily related to EIS preparation to achieve compliance with NEPA, thus allowing the proponent to proceed with the proposed project. Increasing attention is being given to the use of follow-on monitoring to document experienced impacts in relation to predicted impacts and to serve as a basis for project management and environmental auditing programs. Such monitoring could be utilized to refine prediction methods and models and thus improve impact predictability for future projects of similar type. In addition, if monitoring is associated with mitigation efforts or contingency planning, it can represent a positive approach for addressing uncertainty.

While the impacts of proposed projects need to be addressed by professionals having expertise in relevant substantive areas, it should be recognized that no individual has sufficient knowledge to address all impacts of concern for a proposed project. Accordingly, interdisciplinary teams can facilitate the application of scientific approaches in impact prediction and assessment. Use of specialized experts for particular issues may also be valuable; for example, projects involving uncertainties related to wetland losses can be aided by the expertise of one or more wetland specialists. In addition, expert systems are being developed for different facets of the EIA process. These systems incorporate the collective professional judgment of a number of individuals into a series of rules or heuristics; thus they can be used to facilitate the planning and conduction of impact studies.

As mentioned, uncertainty within the EIA process is also related to the evaluation of alternatives. Decision analysis has been suggested as a management tool for handling uncertainty associated with comparisons of alternatives or comparisons of mitigation measures for one alternative (2). Decision analysis can provide a structure to the decision process; in addition, expert opinion can be incorporated in the analysis.

To illustrate some approaches available through decision analysis, it is useful to perceive of trade-offs between a set of alternatives relative to a series of decision factors. Petersen has noted that in a trade-off analysis the contributions of various alternatives are compared to determine what is gained or forgone in choosing one alternative over another (40). Table 10.7 displays a hypothetical trade-off matrix for systematically comparing alternatives, or specific mitigation measures for one alternative, relative to a series of decision factors (12). The following approaches can be used to complete the trade-off matrix in Table 10.7 (12):

1. Qualitative approach, in which descriptive synthesized and integrated information on each alternative (or mitigation measure) relative to each decision factor is presented in the matrix.
2. Quantitative approach, in which quantitative synthesized and integrated information on each alternative (or mitigation measure) relative to each decision factor is displayed in the matrix.

Table 10.7 Example of Trade-off Analysis Structure for Decision Making

DECISION FACTORS	ALTERNATIVE PLANS			
	1	2	3	4
Degree of meeting defined needs and identified objectives				
Economic efficiency of plan				
Benefits/costs				
Excess benefits				
Internal rate of return				
Environmental cost-benefit analysis				
Environmental impacts				
Air quality				
Surface water quantity/quality				
Soil quality and groundwater quantity/quality				
Noise				
Ecosystems				
Habitat quantity/quality				
Threatened or endangered species				
Historic/archaeologic resources				
Socioeconomic characteristics				
Human health risks				
Public preference				

Source: Reference 12.

3. Qualitative/quantitative approach, in which a combination of (1) and (2) is used to complete the matrix.
4. Ranking, rating, or scaling approach, in which the qualitative or quantitative information on each alternative (or mitigation measure) is summarized via the assignment of a ranking, or rating, or scale value relative to each decision factor (the ranking or rating or scale value is presented in the matrix).
5. Weighting approach, in which the importance weight of each decision factor relative to each other decision factor is considered, with the resultant discussion of the information on each alternative or mitigation measure (qualitative; or quantitative; or qualitative/quantitative;

or ranking, rating, or scaling) being presented in view of the relative importance of the decision factors.

6. Weighting-ranking/rating/scaling approach, in which the importance weight for each decision factor is multiplied by the ranking/rating/scale of each alternative (or mitigation measure), then the resulting products for each alternative are summed to develop an overall composite index or score for each alternative; the index may take the form

$$Index\ j = \sum_{i}^{n} (IW)_i(R)_{ij}$$

where Index j = composited index for the jth alternative

n = number of decision factors

IW_i = importance weight of ith decision factor

R_{ij} = ranking/rating/scale of jth alternative for ith decision factor

If the qualitative or quantitative approach, or the combination of the two, is used for completion of Table 10.7, the included information should be based on impact predictions. Impact prediction information would also be needed as a basis for the approaches involving impact ranking/rating/scaling. If the importance weighting approach is used, the critical issue is the assignment of importance weights to the individual decision factors, or at least their arrangement in a rank ordering of importance. Examples of structured techniques that could be used in importance weighting include ranking, the nominal group process, rating, predefined importance scale, multiattribute utility measurement, unranked or ranked pairwise comparisons, and a Delphi study. These techniques have been used in importance weighting basic to numerous decisions on environmental issues.

Ranking/rating/scaling of each alternative for each decision factor is the second major aspect in the use of decision analysis (also called multiple-criteria decision making). Examples of ranking/rating/scaling techniques which have been used in impact studies include alternative profiling, use of a reference alternative, linear scaling based on maximum change, letter or number assignments into impact categories, evaluation guidelines, unranked paired comparisons, functional curves, and predefined impact rating criteria.

Finally, public participation can cause increased attention toward uncertainty within the EIA process. Opponents of a proposed project may emphasize the levels of uncertainty, while proponents may tend to dismiss or minimize uncertainty concerns. Accordingly, it may be necessary to utilize environmental mediation or negotiation techniques to facilitate communication regarding uncertainty and to reach a decision relative to the proposed project and mitigation measures. Several techniques for environmental mediation or negotiation have been developed, including the use of third-party negotiators to resolve specific issues.

Summary

Multiple issues related to uncertainty in the EIA process have received increased attention since the mid-1980s. Uncertainty in the EIA process can result from (a) minimal information on the proposed project and alternatives, (b) questions regarding relevant laws or extant baseline data, (c) inability to predict impacts using probabilities and accuracy qualifiers, (d) the need to apply professional and societal value judgments in impact interpretation, and (e) lack of information on the likelihood of success of proposed mitigation measures. Uncertainties related to impact prediction may be the most important; they can be caused by lack of understanding of environmental processes, inability to test or calibrate models, and unrealistic assumptions regarding the proposed project.

In order to accomplish improved decision making, it is desirable to recognize and minimize uncertainties within the EIA process. If uncertainties are overestimated in an impact study, then identification of excessive requirements for mitigation measures might occur. Overestimation could also increase perceived needs for an EIS in lieu of a FONSI. Conversely, if uncertainties are not appropriately identified, inadequate mitigation measures may be specified and a FONSI prepared. Accordingly, it is important for practitioners to recognize the causes of uncertainty and incorporate key approaches for reducing them. Examples of approaches for alleviation include the conduction of sensitivity analyses of impact predictions, the use of decision analysis techniques for alternatives evaluation, and the incorporation of environmental mediation in public participation programs.

References

1. Council on Environmental Quality. National environmental policy act—regulations. Federal Register 1978;43(230):55978–56007.
2. De Jongh P. Uncertainty in EIA. In: Wathern P, ed. Environmental impact assessment—theory and practice. London: Unwin Hyman, 1988:62–84, 306–308.
3. Council on Environmental Quality. 40 Code of Federal Regulations. Washington: U.S. Government Printing Office, 1987:929–971.
4. Economic Commission for Europe. Application of environmental impact assessment—highways and dams. ECE/ENV/50. Geneva, Switzerland: United Nations, 1987:vii, xvii–xviii.
5. Senior Advisors to ECE Governments on Environmental and Water Problems. Methods and techniques for prediction of environmental impact. ECE/ENVWA/21, New York: Economic Commission for Europe, United Nations, 1992.
6. U.S. Environmental Protection Agency. Superfund exposure assessment manual. EPA/540/1-88/001. Washington: Office of Remedial Response, 1988:2–3, 8–11, 36–42, 49–50.
7. U.S. Environmental Protection Agency. Superfund public health evaluation manual. Washington: Office of Emergency and Remedial Response, 1985.
8. Therivel R, Wilson E, Thompson S, Heaney D, Pritchard D. Strategic environmental assessment. London: Earthscan, 1992:13,27–29,152,157.

9. Society of Environmental Toxicology and Chemistry. A technical framework for life-cycle assessments. Pensacola: SETAC Foundation for Environmental Education, 1991.

10. Society of Environmental Toxicology and Chemistry. A conceptual framework for life-cycle impact assessment. Pensacola: SETAC Foundation for Environmental Education, 1993:110.

11. Hunsaker DB Jr. Environmental assessments: uncertainties in implementation. CONF-8708184—1. Oak Ridge National Laboratory, 1987:11–14.

12. Canter LW. Environmental impact assessment. New York: McGraw-Hill, 1995, in press.

13. Culhane PJ, Friesema HP, Beecher JA. Forecasts and environmental decision making. Boulder: Westview Press, 1987:81–97.

14. World Health Organization Regional Office for Europe and the Center for Environmental Management and Planning. Environmental and health impact assessment of development projects. London: Elsevier Science, 1992:27,71–72.

15. Glasson J, Therivel R, Chadwick A. Introduction to environmental impact assessment. London: UCL Press, 1994:122–124,252.

16. Canter LW, Robertson JM, Westcott, RM. Identification and evaluation of biological impact mitigation measures. Env Mgmt 1991;33:35–50.

17. Mills GH, Walter, JA. Technical writing. Dallas: Holt, Rinehart and Winston, 1978.

18. Canter LW. The role of environmental monitoring in responsible project management. Env Prof 1993;15(1):76–87.

19. Marcus LG. A methodology for post-EIS (environmental impact statement) monitoring. Circular 782. Washington: U.S. Geological Survey, 1979.

20. Water Resources Council. Economic and environmental guidelines for water and related land resources implementation studies. Washington, 1983:Chapter 3.

21. Canter LW. Environmental impact assessment. In: Liu D, ed. Environmental engineers handbook. Radnor, PA: Chilton, 1995, in press.

22. Ott WR. Environmental indices: theory and practice. Ann Arbor: Ann Arbor Science, 1978.

23. U.S. Fish and Wildlife Service. Habitat evaluation procedures (HEP). ESM 102, Washington, 1980.

24. U.S. Army Corps of Engineers. A habitat evaluation system for water resources planning. Vicksburg, MS: Lower Mississippi Valley Division, 1980.

25. Zanetti P. Air pollution modeling—theories, computational methods, and available software. New York: Van Nostrand Reinhold, 1990.

26. Turner DB, Workbook of atmospheric dispersion estimates. Second edition. Boca Raton: Lewis, 1994.

27. Anderson MG, Burt TP. Modelling strategies. In: Anderson MG, Burt TP, eds. Hydrological forecasting. New York: Wiley, 1985:1–13.

28. James A, ed. An introduction to water quality modeling. West Sussex, England: Wiley, 1993.

29. Water Science and Technology Board. Ground water models—scientific and regulatory applications. Washington: National Academy Press, 1990.

30. Magrab EB. Environmental noise control. New York: Wiley, 1975.

31. World Health Organization. Assessment of noise impact on the urban environment. Environmental Health Series No. 9, Copenhagen: Regional Office for Europe, 1986.

32. Starfield AM, Bleloch AL. Building models for conservation and wildlife management. New York: Macmillan, 1986.

33. King TF. The archaeological survey: methods and uses. Washington: Heritage Conservation and Recreation Service, U.S. Department of the Interior, 1978.

34. Smardon RC, Palmer JF, Felleman JP. Foundations for visual project analysis. New York: Wiley, 1986.

35. Canter LW, Atkinson SF, Leistritz, FL. Impact of growth. Chelsea, MI: Lewis, 1985.

36. United Nations Environment Program. Senior level expert workshop to evaluate benefits and constraints of environmental impact assessment process in SACEP countries. Bangkok, Thailand, 1987:35.

37. Petts J, Eduljee G. Environmental impact assessment for waste treatment and disposal facilities. Chichester: Wiley, 1994:123–124,238–240.

38. Suter GW, Barnthouse LW, O'Neill RV. Treatment of risk in environmental impact assessment. Env Mgt 1987;11(3):295–303.
39. Hance BJ, Chess C, Sandman PM. Industry risk communication manual: improving dialogue with communities. Chelsea, MI: Lewis, 1990:94–98.
40. Petersen MS. Water resource planning and development. Englewood Cliffs, NJ: Prentice-Hall, 1984:84–85.

Implications of Scientific Uncertainty for Environmental Impact Assessment: The International Environment

11

Mahmoud Kh. El-Sayed

Introduction

The environmental impact process is an important step in providing information on proposed projects to decision makers. Increasingly, many developing countries are placing greater importance on the requirements of the environmental impact process to provide information on activities that may have adverse consequences on human health or the environment at the national, regional, or global levels.

The scientific and technical aspects of the environmental impact assessment (EIA) process largely are based on perceived certainties regarding available data, information, and approaches. When uncertainties are inherent in proposed activities (as they always are), state-of-the-art EIA procedures need to develop methods and techniques to factor consideration of the uncertainties into the procedures to a greater extent. This probably will require methods to develop parameters or boundaries around our scientific information in order to provide levels of confidence in the information, or the development of precautionary approaches for assessment of the consequences of activities in order to take into account the uncertainties. It also might require revision of legislation and procedures of the EIA process in relation to scientific uncertainty. The need to include a greater consideration of scientific uncertainty into the process is especially great for activities that threaten global resources.

This chapter will focus on these issues by analyzing a specific case study for a global environmental problem, namely, assessing the consequences of climatic changes for the Nile Delta and Alexandria in Egypt.

Review of the Environmental Impact Assessment Process

Environmental impact assessment is a process analyzing the positive and negative effects of a proposed project, plan, or activity on the environment, or is the process of identifying, predicting, interpreting, and communicating the potential impacts that a proposed project or plan may have on the environment (1). The specific purpose of the assessment is to provide the decision makers with information allowing them to introduce environmental protection considerations in the decision-making process leading to the approval, rejection, or modification of the project, plan, or activity under examination.

An EIA consists not only of a written report (i.e., the EIA document) through which information is provided to the decision makers, but also procedural provisions ensuring that the decision makers take the information into consideration. Therefore, EIA should be viewed as an integral part of the project planning process, beginning as early as possible with identification of the potentially significant environmental impacts, continuing through the planning cycle and including, as much as possible, public participation (1). This simply means that the scientific and technical aspects of the EIA process *largely are based on perceived certainties regarding available data, information, and approaches.* Steps for applying the EIA process and the general guidelines for preparation of EIA are shown in Figure 11.1.

Problems of Scientific Uncertainty for Global Problems: Assessing the Consequences of Climatic Changes for the Nile Delta and Alexandria

The climate change effect is among man's potentially most pressing environmental problems that represents major scientific, environmental, economic, social, and political challenges. Changes in global climate between now and the middle of the twenty-first century are likely to be dominated by the influence of global warming due to the increasing concentrations of carbon dioxide and other greenhouse gases in the atmosphere (2).

Many important economic and social decisions being made today, such as water resources management, coastal engineering projects, urban and energy planning, nature conservation, and so on, are based on the assumption that past climate data provide a reliable guide to the future. This is no longer a safe assumption; climate change must be considered, particularly in view of the current population explosion, increasing of utilization of coastal areas for industry, agriculture, fisheries, and tourism (2).

Probably the most important aspects of climate change on the world ocean and coastal zones will be the impacts of sea-level rise. These impacts are categorized as follows:

- Socioeconomic impacts (threatened populations in low-lying areas and island nations), such as inundation, agricultural losses, erosion and tourism, increased potential of coastal flooding, water supply and quality, infrastructure.
- Ecologic impacts (alteration and degradation of the biophysical properties of beaches, estuaries, and wetlands).

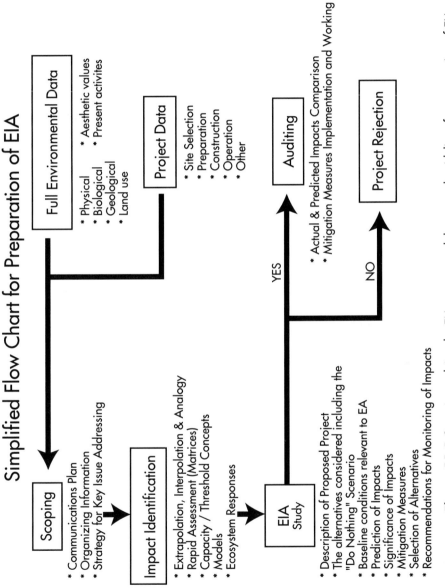

Figure 11.1 Steps in applying the EIA process and the general guidelines for preparation of EIA.

■ Physical aspects of shoreline retreat: inundation, erosion, and recession of barrier islands and sensitive areas of coral reefs and mangroves (3).

The Coastal Area in Egypt

The length of coastlines along the Mediterranean and the Red Sea in Egypt is 3000 km. The shelf area up to 200 m depth is 37.4×10^3 km²; and the area of the exclusive economic zone (EEZ) is 173.5×10^3 km². The population in coastal urban agglomerations is more than 6 million. The average annual marine catch is 49.5×10^3 metric tons. The annual production from offshore oil and gas is $26,927 \times 10^3$ metric tons and 130×10^9 m³, respectively.

The coastal area in Egypt has witnessed a major development and ambitious industrial, transport, and tourism plans during the last two decades. This in turn has exerted an enormous environmental pressure. The environmental problems facing this area are not only categorized under certainty, but uncertainty is highly involved. Industrialization, urbanization, oil and gas exploitation are among the examples that exerted enormous environmental pressure along the coastal area; however, the impact of climatic changes is expected to add pressures to the sensitive areas along the Mediterranean (Nile delta is the classical example) and the coral reef of the Red Sea.

The Nile Delta

Approximately 15% of Egypt's current GNP originates in the northern Nile delta coastal area under 2 m elevation (Fig. 11.2). Although this area has witnessed a rapid major development during the last two decades and will probably continue in the coming years; it is one of the world's most vulnerable areas to the impact of climate change and sea-level rise.

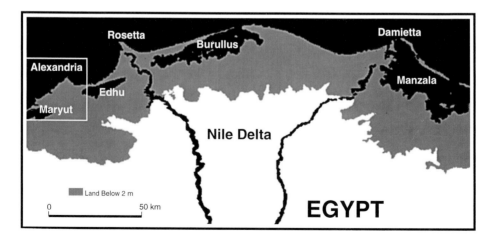

Figure 11.2 The northern Nile delta: setting and morphology. Inset Alexandria.

Detailed studies of this sensitive area concerning its setting, physical characteristics, and the possible impacts of climate change were presented in several publications (4, 5, 6). Therefore this chapter will focus on the most western part of the Nile delta (Alexandria), since this area is one of the most prosperous economic and social centers in Egypt, although it is vulnerable to the impacts of climate change.

Alexandria

Alexandria is located to the west of the Nile Delta, between lat. 31°08′, 31°26′N and long. 29°27′, 30°04′E. It extends for over 42 km from east to west (Fig. 11.3).

The city lies on a wide sandy bar, with calcareous sandstone and limestone elevations formed by Quaternary deposits between the Mediterranean and Lake Maryut to the south (Fig. 11.3). However, the coastal stretch, particularly to the east and west extremities of the city, consists generally of low-lying lands. Sea cliffs (about 5 m high) are common along this coast and form as waves erode carbonate coastal ridges. The shoreline is rocky in some places with narrow

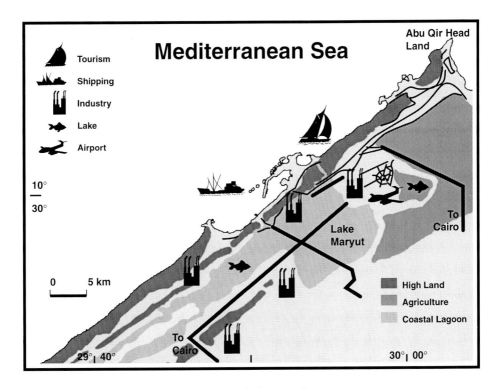

Figure 11.3 Alexandria: setting, morphology, and socioeconomic structure.

sand beaches in the embayments and wider ones in the remarkably low-lying lands. Ancient subsided shorelines are represented by a series of submerged hard grounds and emerged islets that extend parallel to the present shoreline (7). This region is probably dominated by the sedimentary and tectonic influence that governed the development of the Nile delta.

The Alexandria region can be subdivided into a series of distinct geomorphic areas, including beach, carbonate ridge, inland depression, drain and canal, low hill, reclaimed lagoon, and irrigated farmland (8).

Alexandria is exposed to a number of storm surges (NW and SW) which generally last from October to May. Wind speeds are generally between 2 and 50 km/h. Alexandria is rainy between December and February, receiving an average of 200 mm of rainfall per year. The general current pattern along the Mediterranean off Egypt is from west to east. Along the coast, the tidal current is weak relative to both littoral and coastal currents. The Alexandria coastline is microtidal, with waves approaching from the north and northwest. The average daily variation of the mean sea level along the Mediterranean off Egypt may approach 50 cm.

Relative Sea-Level Changes (Eustacy and Subsidence)

Along the Mediterranean coast of Egypt, it has been long recognized that nature-induced subsidence and eustatic sea-level changes are very influential in the development of this region.

Data revealed that 2000 years ago, sea level in the Mediterranean was 50 cm lower than at present (9); however, in the last 100 years it has risen by 1.3 mm/year. Tide-gauge data for the past 50 years from the Alexandria region indicate that relative sea level has continued to rise (10). Tide-gauge data for Alexandria were interpreted as recording both eustatic sea-level rise (to 0.1 cm/year) and tectonic land subsidence (to about 0.15 cm/year) (8).

Archaeologic and geologic evidences indicate late Quaternary tectonic activity in the Alexandria region. Observations on submerged Greco-Roman ruins in Alexandria and Abu Qir show that ruins as old as 2500 years along the coast had submerged 2 to 5.5 m (8, 11). Archaeologic evidence from Alexandria has revealed that a continuous subsidence of the city has occurred over the past 3000 years at an average rate of 1.2 mm/year, and that the tectonic factor is an important component of this submergence (11).

Social and Economic Structure

Alexandria, the second largest city in Egypt, comprises major economic activities. The national plans call for an expansion of the axes of the activities of Alexandria to the west and southwest by the year 2005 (Fig. 11.4). The present socioeconomic structure of Alexandria is briefly described as follows.

Population   The estimated population of Alexandria at present is 3,500,000, about 6.36% of the total population of Egypt. The expected population of the city in the year 2005 will be about 5,000,000.

Agriculture (Cultivated and Reclaimed Lands) and Irrigation   Low-lying arable lands around Alexandria make up the northwestern sector of the fertile Nile crescent (8). At present, about 42,000 ha are cultivated around Alexandria, about 1.42% of the total cultivated land in Egypt. An extensive network of irrigation canals covers large portions of the southeastern sector of Alexandria. The southwestern portion of the area is less extensively cultivated, but irrigation projects are expanding into that region (8).

Plans in Egypt call for massive expansion of reclaimed lands of about 850,000 to 1,250,000 ha in Egypt by the year 2000. The proposed reclaimed areas to the west and southwest of Alexandria are 58,000 ha.

Groundwater will be one of the highly exploited resources in the near future to meet the high demand for water for land reclamation in this area. The main delta aquifer is confined, increasing in thickness from about 250 m near Cairo to about 900 m in the north near Alexandria (12). The surface of the main water table is at or near sea level throughout the area of Alexandria. The depth of the water table varies from less than a meter to 50 m, depending on the season and topography (13).

Industry   Forty percent of Egypt's industry surrounds Alexandria, which are mostly chemical, petrochemical, refineries, textiles, food processing, steel and iron, cement, paper, and other industries. About 130,000 workers are engaged in the industrial activities in Alexandria; their activities in the different industrial

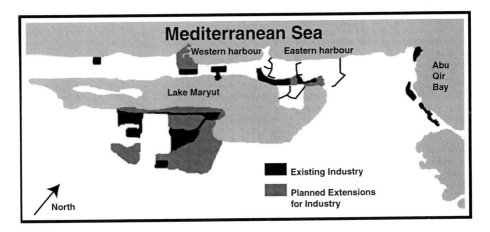

Figure 11.4 Comprehensive plan of Alexandria 2005.

sectors are as follows: textiles, 30%; food processing, 17%; machineries and metal products, 17%; chemicals, 15%; building materials and ceramics, 6%; papers, 8%; woods 5% and converting industries 0.7% (14).

Tourism   Alexandria is an important center for tourism. It has summer resorts, recreational areas and water sports facilities, historic and contemporary museums, and other touristic sites.

The beaches of Alexandria are the main summer and recreational resort in Egypt. They extend for some 50 km from east to west for Alexandria proper, and further westward to new flourishing tourist village areas. These beaches consist in most places of narrow sandy strips delimited landward by concrete seawalls supporting the sea promenade (Corniche). More than 3 million summer vacationers spend their holidays along these beaches.

Communications and Harbors   The main trade harbor along the Mediterranean off Egypt is the western harbor of Alexandria, which handles more than 80% of the country's trade. The eastern harbor of Alexandria is the largest yachting and fishing harbor. An efficient dense network of roads and railway connects the city with the whole country and with the delta region and adjacent agricultural lands. Alexandria has an international airport, which is the second largest airport in Egypt.

Impact of Climatic Changes (Sea-Level Rise) in the Northern Nile Delta and Alexandria

The lower Nile delta has large areas under 1 m elevation, with parts below sea level, including sizable coastal lagoons. Natural protection is afforded by high sand dunes on some stretches, but other areas are vulnerable to flooding by winter storm surges. Higher sea level will accelerate the present retreat of the actually eroded beaches (6).

A sea-level rise of 10–20 cm would certainly aggravate local problems such as wave attack on harbor installations, Nile mouth promontories, and shoreline stretches that are subject to flooding by storm surges. A rise of sea level to 30–50 cm would have more serious effects, imposing extensive measures of protection.

In a socioeconomic scenario on the likely impact of sea-level rise in the northern Nile delta, a 1-m rise in sea level would inundate 12 to 15% of Egypt's arable land and consequently will affect 15% of the nation's gross domestic product (GDP) (4). The projected rise in sea level together with subsidence and reduced sediment supply could cause a high relative rise of sea level that would submerge much of the delta region, within 50 km of the coast (15). The rise of sea level will manifest itself differently along the coastal stretch of the Nile delta.

Its influence will be mostly controlled by the morphologic setting and the existing and planned engineering countermeasures (5).

The above scenarios and many others (7,11,16) were mainly qualitative or at most semiquantitative in nature where their integral response is concerned, including socioeconomic aspects and feedback.

Alexandria

The most vulnerable areas to sea-level rise will be the low-lying lands in Alexandria, including beaches, harbors, and Lake Maryut and its surroundings.

The beaches of Alexandria are continuously inundated as a consequence of relative sea-level rise associated with the absence of littoral drift supply of sand. Inhabitants along the coastal area of Alexandria could hardly fail to notice the gentle erosion of beaches since the construction of the seawall and the seaside promenade in 1932. However, the existing hard and soft engineering defensive measures (e.g., seawalls, groins, breakwaters, and beach nourishment) will have a protective role along the coastal and low-lying lands of Alexandria.

The eastern and western harbors are in constant need of attention due to frequent overtopping by waves and the rise of sea level. Maintaining the efficiency of port structures and inland drainage system in the face of a sea-level rise of even 20–30 cm will require extensive adjustments (6).

As the industrial zone in western Alexandria is positioned on relatively high lands, it will be naturally protected from a sea-level rise. Newly reclaimed lands in southwest Alexandria will not be directly affected by sea-level rise, but arable land in the east side of the city will be vulnerable.

As a consequence of sea-level rise, saltwater would advance further inland in the northern delta aquifer. The urgent need of water in the future to meet the demands of land reclamation around Alexandria would, however, motivate the exploration and exploitation of groundwater up to a safe yield $(0.15 \times 10^9$ m^3/year (12)). This would accelerate the subsidence of the area. Even now a wedge of saltwater intrudes into the huge artesian delta aquifer at different depths.

The ecologic impact of global climatic changes and rising sea level on the lagoonal ecosystem as in Lake Maryut could be minimal and possibly self-adjusting. Fishing, if gradually adjusted to changes, should not be much affected (6).

The sea-level rise will cause flooding of the low-lying lands around Alexandria to the east and around Maryut Lake, and similarly on the beaches. Subsequently, parts of the city would become isolated from the mainland.

Fortunately, the expanding Abu Qir-Alexandria-El Dekheila-El Agami urban center is positioned on a carbonate ridge several meters above sea level. Expansion of irrigation projects into the desert (which is well above sea level) to the south and southwest of Alexandria is a reasonable course of action for future development of this region.

It is difficult, however, to quantify the negative impacts of a sea-level rise in Alexandria. The reason is that the existing socioeconomic structure will adjust itself following the gradual effect of the expected sea-level rise. Preparedness and mitigation of the gradual impact will be the concern of authorities.

The following are options to mitigate the negative impact of sea-level rise on the socioeconomic activities in Alexandria:

- Strengthening the existing coastal defense system of the beaches, harbors, and low-lying lands.
- Encourage the protection of Lake Maryut by hard engineering solutions, because this area is the weakest point of the defense system.
- Prevent highly developed capital investment projects from being established in low-lying lands around Alexandria, unless a predictive model of the likely impact of sea-level rise is developed.
- Rational water management, particularly for groundwater, should be founded on a thorough understanding of water availability and movement.
- Develop land-use strategies and encourage functional and inventory research in the area. In this respect, the formulation of a regional sea-level monitoring network and data base center is recommended.

Legislation and Institutional Frameworks Regarding the Environmental Impact Assessment Process in Egypt

The responsibility for carrying out EIAs depends on the national legislative requirements of individual countries and varies considerably from country to country in order to fit into the specific socioeconomic and political system of the country. The EIA requirements of projects or plans that may adversely affect human environment are embodied in the national legislation of many countries and in a number of international agreements. However, the procedures used to prepare such assessments vary considerably from country to country, and at present there is no agreement, on a global or regional level, on procedures that may lead to comparable results (1).

For example, the United States Congress was among the first to enact, in January 1970, a comprehensive environmental legislation, the National Environmental Policy Act (NEPA), using the concept of EIA and requiring a systematic interdisciplinary evaluation of the potential environmental effects of all major federally funded projects.

In Egypt, national environmental legislation, Law of the Environment (Law No 4/ 1994), has clearly required a comprehensive EIA study prior to the approval of any project with potential environmental impact. However, no explicit EIA procedure has been decided for the different activity sectors. Besides, the EIA has not included parameters or measures to develop precautionary approaches for assessment of activities to take into account any *uncertainties*.

The Egyptian Environmental Affairs Agency (EEAA) is the national body responsible for preparation of the general policy and plans for the protection of the environment and its development, beside the follow up of the policy and plans implementation in coordination with the agencies in charge of implementing these policies.

Availability and Reliability of Scientific Information and Data

Sea-level fluctuations are part of a natural cycle. However, in recent years scientists have noted trends that differ from the normal changes they have come to expect. In the last century the eustatic sea level has risen more than 30 cm, a rate faster than that observed at any time in past millennia. This phenomenon appears to be the result of a gradual increase in temperature due to the rapidly increasing levels of carbon dioxide in the atmosphere.

Assumptions for climate change state that a doubling of CO_2 will bring about a temperature increase of 1.5 to 4.5°C by 2030, 2050, or 2075 according to actions taken by governments to control and reduce the emission of gases into the atmosphere. Predictions for sea-level rise (cm) following these assumptions were as follows (17):

	YEAR		
	2025	2050	2075
Optimistic	+13	+24	+40
Pessimistic	+55	+114	+212

The matter of climate change was the main topic of the Second World Climate Conference (Geneva, 1990). The conference agreed that the international consensus of scientific understanding of climate change points out that without actions to reduce emissions, global warming is predicted to reach 2 to 5°C over the next century, a rate of change unprecedented in the past 10,000 years. The warming is expected to be accompanied by a sea-level rise of 65±35 cm by the end of the next century. There remain uncertainties in the predictions, particularly in regard to the timing, magnitude, and regional patterns of climate change (18).

Regarding the Mediterranean, and knowing that one of the most pressing requirements with regard to research into the enhanced greenhouse effect is the need for regional scenarios of climate change, the Climatic Research Unit of the University of East Anglia had been commissioned by UNEP to produce a Mediterranean Basin scenario. Regional-scale changes in climate can only be estimated using general circulation models (GCMs) (19). For the Mediterranean Basin, the only GCM result that one can place confidence in is a general warming. The regional transient-response warming rate is likely to be similar to the global-mean rate, with no evidence of any marked differences between seasons. However, these greenhouse-related changes may well be masked by natural climatic variability for decades in the future. For precipitation changes, there is some evidence of increased autumn precipitation occurring in the northern part of the basin and of decreases in precipitation occurring in the southern part, but *it is impossible to estimate the timing or magnitude of these changes.*

All of the above suggestions must be taken as possibilities rather than probabilities. They are amenable to more detailed investigation using existing data and model results. Indeed, there is much that can be done to reduce uncertainties and to develop better methods for interpreting GCM results (19).

Although scientific information is available, reliability is not yet fully realized, since probabilities are still unresolved regarding regional climatic scenarios.

Problems of Assessing Long-Term Changes of the Nile Delta and Cause-and-Effect Relationships

Perhaps the most crucial problem of the northern Nile delta is the continuous erosion that will be accelerated by a sea-level rise as a consequence of climatic changes.

Rivers are the arteries of continents. They are the chief carriers of water, salt, organic matter, and mineral particles from land to the sea. From the beginning of humankind rivers have received special attention. Cultures that depend mainly on rivers as a water resource are particularly vulnerable to regional changes of rainfall pattern if the catchment area is small. Moreover, minor shifts of climatic zones will have dramatic consequences.

Most African rivers have highly seasonal flows. However, construction of dams (e.g., the case of the River Nile) and reservoirs has obliterated most of the seasonal flow patterns.

The Nile River, since Pliocene time, has formed its classic fan-shaped delta and has contributed large volumes of sediment to the adjacent Egyptian continental shelf.

Since the construction of the High Dam at Aswan, the Nile transports virtually no sediment to the sea, thereby causing coastline erosion. Although the effect of decreased sediment flux has been difficult to quantify, erosion rates along several sectors of the delta coast accelerated as a result of deprivation of silt nourishment.

Successive Intervention of the Nile in Egypt

Man's intervention in the flow of the Nile dates back to Pharaonic times. Modern intervention began with the construction of the "Delta Barrage" near Cairo in 1861. The barrage sluices were opened to let the floodwaters flow, but made possible the beginning of perineal irrigation instead of basin flooding. The practice was developed with the Low Aswan Dam constructed in three stages (1902, 1912, and 1933). The dam was also provided with sluices to let the flood waters flow with their sediment load. This dam was designed to generate hydroelectric power and to store water during the 3-month flood season with release at the end of the flood. This dam was intended to extend the flood's duration downstream without carrying any water storage over from year to year and, accordingly, without storing any significant amount of Nile silt.

The Aswan High Dam is therefore the second dam to be built in the vicinity of the Nile's first cataract at Aswan. Its construction began in 1960. The Nile was closed around a temporary structure in 1965, the dam was completed in 1967, and by 1970 all 12 turbines were in operation. The High Dam was designed to store the Nile flow (84 billion m^3), so there would be no excess beyond the actual needs of 55.5 billion m^3 as permitted under the Nile Waters Agreements of 1929 and 1959, and to ensure that minimum flow would be maintained regardless of whether there was a flood or a drought upstream.

The Nile, with a drainage area more than 2×10^6 km^2, once carried an average of 100×10^6 t year^{-1} of sediments to its delta in the Mediterranean Sea (20). As some downstream and marine problems became apparent as soon as the river Nile was dammed, such as progressive coastal erosion, the construction of the Aswan High Dam gave Egypt many good opportunities, both economic and social.

The effect of the effective influx of the sediment load has been difficult to quantify. One problem is the erosion of the delta since 1900 after more than 100 years of standing progradation. Still, the erosion of the Rosetta branch between 1909 and 1945 was only a few hundred meters, while between 1945 and 1972 (presumably most occurring after 1965) it was more than 1.5 km (21).

Analysis of aerial photographs taken in 1955 and 1983 for the unstable zone of the Nile delta revealed drastic shoreline changes during this period, where the outer margins of both promontories have been clearly eroded (22). Besides, analysis of a series of landsat images of the Nile delta for 1973 and 1978 shows that erosion has occurred all along the delta coast, except in a few areas. The promontories were by far the most eroded areas (23).

The relationship between the shoreline retreat along the Nile delta and sea-level trends, by the Bruun rule, indicates that the sea-level rise has, itself, a relatively minor effect on coastal erosion. The sea-level trend at the Nile delta coast is found to be only one of several effects on the shoreline retreat (10).

Continental Shelf of the Nile Delta: Deprivation of Sediments and Changes in the Bottom Configuration

Examination of long-term changes of mapped distribution patterns of shelf sediments as well as bathymetric charts indicate low sediment accumulation rates and erosion of several areas along the Nile delta shelf. Comparison of charts of 1919–1922 and 1986 shows that erosion of the bottom dominates the inner continental shelf with accretion at sinks such as embayments and the downslope of the inner shelf. A sharp increase in bottom erosion was noticed in front of the Rosetta and Damietta promontories, with bottom accretion at some major sinks. In general, the net erosion is higher than accretion (24).

The diminished Nile flow and sediment entrapment by the High Dam coupled with the effects of intensified irrigation, channelization, and land reclamation projects in the northern delta during the past two decades have markedly reduced fine sediment supplied from the coast to the continental margin seaward of the inner shelf (15). In consequence, the seafloor is not receiving new depositional input from the Nile. Thus, unless replenished, this area almost certainly will be eroded by bottom currents.

The coupling of erosion by bottom currents and reduced sediment input from the Nile is likely to induce depositional changes on the shelf that could be monitored. It is expected that (a) the configuration of sediment distribution patterns on the mid-shelf will be altered as surficial deposits are displaced toward the east, and (b) if present conditions are maintained the mid-shelf mud blanket probably will be further reduced. Also to be considered are the effects of recent depositional changes on the mid-shelf that, in turn, could affect sedimentation patterns on the inner shelf closer to the shore (15).

Implications of Scientific Uncertainty for the Environmental Impact Assessment and Integrated Coastal Zone Management

The coastal zone is of great economic and environmental significance. This significance is further increasing and has important implications for all who use and enjoy the coast.

Coastal resource use is varied and manifests itself in a number of ways, sometimes conflicting and sometimes complementary. The first is settlement at the coast; the second is the economic importance of coastal resources, and the third is that inadequate coastal zone management can generate significant international problems.

The coastal zone in most countries worldwide is currently under severe and increasing pressure. A number of factors are contributing to this situation: rapid urbanization of the coast; pollution from residential, commercial, and industrial activities; tourism development; resource allocation conflicts between and among resource users; and continuous development in hazards-prone areas.

Although scientific and technological knowledge have witnessed many advances and improvements, various approaches to marine and coastal environmental management and to the regulation of activities are widely practiced.

Marine and coastal resources and activities cannot be managed successfully if they are dealt with individually or in isolation. Decision makers are the first to feel the need and importance of their access to comprehensive and innovative tools, in the managerial process of the protection of the marine environment. This is the basis for formulating their decisions.

Successful coastal management is issue driven and achieved by resolving existing problems with a combination of science, policy, lawmaking, and administration. How programs evolve is highly dependent on the social, political, cultural, and economic circumstances in the country concerned, and thus each program is unique. However, the experience gained from past success and failures of others can be of great use to current practitioners (25).

Integrated coastal zone management (ICZM) is most simply understood as management of the coastal zone as a whole in relation to local, regional, national, and international goals. This means the integration of environmental protection goals into economic and technical decision-making processes, the management of the impact of agricultural runoff and industrial and urban waste disposal on coastal zone water quality, the coordination of tourism policies with nature conservation policies, and the coordination of pollution policies within different parts of a particular coastal zone.

The most widely practiced coastal zone management is the ICZM. Within the ICZM, or independently, is the EIA. However, these activities are mainly based on perceived certainties.

The scientific and technical aspects of the EIA process largely are based on *perceived certainties* regarding available data, information, and approaches (see Fig. 11.1). Since EIA is the process of analyzing the positive and negative effects of a proposed project, plan, or activity on the environment, when *uncertainties* are inherent in proposed activities (as they always are), state-of-the-art EIA procedures need to develop methods and techniques to factor consideration of the uncertainties into the procedures to a greater extent.

The sustainable use of resources can be seriously affected by man-made or natural events. Gradual changes, such as climate change with a corresponding rise in global sea level, will certainly have negative impact. Therefore, EIA procedures require the inclusion of natural and man-made hazards (*uncertain*) with identification of possible impacts.

The setup of methods to develop parameters or boundaries around our scientific information is needed in order to provide levels of confidence in the information. In this respect fewer *probabilities* and more *possibilities* are needed for studies concerning climatic changes in global and regional levels. Moreover, assessment of the impact of climatic changes should be approached more realistically. Although scientists are still unable to provide precise predictions and models for future climate change, our scientific knowledge of climatic changes, causes, and consequences has been considerably enhanced in less than a 10-year period.

Since the structure of EIA is founded on certainties, the question is where to accommodate uncertainties. Boundaries will be minimum between these two paradoxical approaches because of the anticipated availability and reliability of scientific information and data relevant to climate change. In this respect, the definition of realistic boundary conditions derived from global climate change, management policies, and field data is of particular significance. Without such an approach, EIA remains a restricted tool to assess and minimize the consequences of uncertainties to be encountered and, hence, to provide alternative actions.

Within the identification of the possible impacts, socioeconomic aspects have to be fully comprehended. Thus, the developing of models integrating physical and socioeconomic aspects of the projected impacts of climatic changes and the sea-level rise has to be achieved. I deem it necessary, therefore, to quantify the magnitude of this problem through the integration of different kinds of models and assessment studies to cope with the expected natural and/or man-made hazards.

Quantitative and qualitative models should be based on sensible assumptions and regional climate models, actual and anticipated socioeconomic structures, physical parameters of the coastal and marine environments, and uncertainties. These models will serve to define trends of evolution of the area and aid in the setup of an information system and a decision support system in marine and coastal resource management. Models afford policy options to formulate a response strategy and create a link between marine science and economy and decision makers.

It might also be useful to revise legislation and procedures of the EIA process in relation to scientific uncertainty. In this respect, projected hazards (either natural or man-made) have to be emphasized within the EIA.

Recommendations

Some of these recommendations (26) address information and research priorities concerning future strategies dealing with uncertainties.

It is imperative that uncertainties should be fully regarded within the process of EIA and ICZM.

Given the current inadequacies of existing data, models, and predictions concerning the potential impacts of climatic changes, specialized agencies should support continued research activities designed to address present areas of uncertainty in these predictions.

Our current ability to predict global, regional, and local climatic changes with any degree of certainty is inadequate. It is imperative that more accurate predictions of local climate and sea-level changes be developed if the uncertainty concerning the magnitudes and rate of potential impacts is to be reduced.

The inadequacies of current predictions concerning impacts necessitate improved data collection and management systems for effective national, regional, and global exchange data, and their use in model and scenario verification and environmental planning.

In order to detect predicted changes it will be necessary to establish long-term monitoring programs.

Studies of the frequency of extreme events are needed to assist in the prediction of the probabilities of occurrence and to detect whether or not the frequency and intensity of such events are changing in relation to mean climatic conditions.

There exists a clear need to determine the priorities for action at local, regional, and global levels, based on a sound assessment of vulnerability to climate change impacts and taking into consideration other sources of environmental change and degradation.

References

1. UNEP. An approach to environmental impact assessment for projects affecting the coastal and marine environment. UNEP Region Seas Rep Stud No 122, 1990.
2. Jeftic L, Milliman JD, Sestini G, eds. Climate change and the Mediterranean. UNEP Publ. London: Edward Arnold, 1992.
3. IPCC. Climate change, the IPCC impacts assessment, report prepared for IPCC Working Group II, 1990.
4. Broadus J, Milliman J, Edwards S, Aubrey D, Gable P. In: Titus JG, ed. Rising sea level and damming of rivers: possible effects in Egypt and Bangladesh. Effect of changes in stratospheric ozone and global climate, 4. Sea-level rise. UNEP/EPA: 1986:165–189.
5. El-Sayed M Kh. Implication of climate change for coastal areas along the Nile delta. Environ Prof 1991;13:59–65.
6. Sestini G. In: Jeftic L, Milliman JD, Sestini G, eds. Implications of climatic changes for the Nile delta. Climate change and the Mediterranean. UNEP Publ. London: Edward Arnold, 1992:535–601.
7. El-Sayed M Kh. Implication of sea level rise on Alexandria. In: Frassetto R, ed. Proc. First Int. Meetings Cities on Water Impact of Sea Level Rise on Cities and Regions, Venice, 1989, 1991:183–189.
8. Warne AG, Stanley DJ. Late Quaternary evolution of the northwest Nile delta and adjacent coast in the Alexandria region, Egypt. J Coast Res 1993;9:26–64.
9. El-Gindy AA, Eid F. Sea level variation in the Mediterranean. Rapp Comm Int Mer Medit 1988;31:196.
10. Frihy OE. Sea level rise and shoreline retreat of the Nile delta promontories. Natural Hazards, 1992;5:65–81.
11. El-Sayed M Kh. Sea level rise in Alexandria during the late Holocene: archaeological evidences. Rapp Comm Int Mer Medit 1988;31:108.
12. Kashef, AA. Salt-water intrusion in the Nile delta. Groundwater 1983;21:160–167.
13. Ayyad M, Le Floch E, eds. An ecological assessment of renewable resources for rural agricultural development in the western Mediterranean coastal region of Egypt: case study; El Omayed test area, 1983.

14. Kandil S, ed. University and industry: Alexandria case study. 1st Edition. Delta Center, Alexandria, 1993. (*in arabic*)

15. Stanley DJ. Subsidence in the northeastern Nile delta; rapid rates; possible causes and consequences. Science 1988;240:497–500.

16. El-Sayed M Kh. Socio-economic aspects of climate change impacts in Alexandria: prospective study. 7th Int. Symp. Environ. Pollut. and Impact on life in the Medit., France 1992.

17. UNEP. Implications of climatic changes in the Mediterranean. UNEP/MAP, Athens 1988: p 27.

18. Palutikof JP, Guo X, Wigley TML, Gregory JM. Regional changes in climate in the Mediterranean Basin due to global greenhouse gas warming. MAP Tech. Rep. Series No 66, UNEP, Athens, 1992.

19. Wigley TML. Future climate of the Mediterranean basin with particular emphasis on changes of precipitation. In: Jeftic L, Milliman JD, Sestini G eds. Climate change and the Mediterranean. UNEP Publ. London: Edward Arnold, 1992:15–44.

20. Milliman JD, Meade RH. World-wide delivery of river sediment to the oceans. J Geol 1983; 91:1–21.

21. Sestini G. Geomorphology of the Nile delta. In: Proceedings Seminar Sedimentology Nile Delta; UNESCO/UNDP/ASRT, Alexandria: 1976:12–14.

22. Frihy OE. Nile delta shoreline changes: aerial photographic study of a 28 year period. J Coast Res 1988;4:597–606.

23. Smith SE, Abdel Kader A. Coastal erosion along the Egyptian delta. J Coast Res 1988;4: 245–255.

24. Frihy OE, Nasr SM, Ahmed MH, El Raey M. Temporal shoreline and bottom changes of the inner continental shelf off the Nile delta, Egypt. J Coast Res 1991;7:465–475.

25. Coastal Resources Centre "CRC." Case studies of coastal management: experience from the United States. The University of Rhode Island, CRC Publ. No. 1001, 1991.

26. UNEP. (Prepared by M. Gerges and M El-Sayed) Potential impacts of expected climate change on coastal and near-shore environment. UNEP Regional Seas Reports and Studies No. 140, 1992.

Environmental Problem Solving in an Age of Electronic Communications: Toward an Integrated or Reductionist Model?

12

James Perry
Elizabeth Leigh Vanderklein

Introduction

The information superhighway will not solve our twenty-first-century environmental problems. In fact, many think it will exacerbate them. It is rare that a technological concept such as the Internet comes along which has such revolutionizing potential. However, it is even more rare that aspects of technology develop according to their earliest visions. In the years to come, there will be many isolated "success stories" in which cooperative and intersectoral planning is facilitated by the Internet and related communication technologies in a way impossible before they became so prevalent. However, we argue that without creative initiative on the part of managers and problem solvers, the Internet is more likely to support and entrench existing models of environmental management than to revolutionize them.

The Internet, on-line data bases, and the vast world of the cybernet are revolutionary in theory, but in practice are merely extensions of current print and communication media (1). As such, they yield increased efficiency and speed but do not significantly change the nature of communications. Consequently, we suggest that unless used creatively and purposefully, these technologies will simply accentuate existing impediments to environmental problem solving. Electronic information technologies are poised to cement the gridlock that plagues current environmental problem solving, skew the represented electorate, promote the polarization of concerned parties, promote a "sound-byte" approach to communication and decision making, expose people to more information than they can process, and mislead many to believe that quantity of information is a substitute for quality. Rather than facilitating carefully considered consensus decisions, these technologies may well increase our social and scientific uncertainty, thus forcing a situation in which environmental decisions are increasingly issue based, nonintegrated directives and substituting the mistake of impetuosity for the former frustration of outdated data and lengthy deliberations.

We present these arguments from three points of view. First, we present an overview of environmental decision-making processes and demonstrate how they have changed over the past 25 years. This section illustrates ways in which landmark, integrated environmental acts of the early 1970s have been continuously dissected and amended, creating the current situation in which environmental goals are increasingly obscured by fragmentation of authority, dominance of local politics, a reliance on only incremental changes in legislation, and the increasing role of self-interest groups. We argue that such single-issue, crisis management decision making must return to a more integrated format if environmental problem solving hopes to effect real change.

Second, we review the growth of the Internet itself and evaluate the communication character of its many components. Evaluating both trends in communication etiquette over the Internet (netiquette) and the human response to the availability of information, we illustrate how Internet-based communication is peculiarly suitable for use by activist, single-issue electronic communities, and how it gives the perception of information certainty while in fact increasing uncertainty. The combination of Internet's popularity, ease, communication character, and the existing trends in environmental decision making pose critical challenges to environmental problem solvers—challenges in which decision makers are likely to find themselves reacting to issues rather than planning for them.

We suggest, however, that our scenario is not predetermined. Through critical analysis and planning, many of the environmental pitfalls of these technologies can be avoided. Consequently, the third section of this chapter is devoted to a discussion of the rhetoric used in both environmental and technology writing. Through this, we attempt to clarify the messages that organizations and individuals involved in decision making will be receiving from the press, the government, and individuals. Understanding the language used to debate future scenarios will enable decision makers to clarify the ways in which communication technologies may have a real impact on their field. This perspective will enable managers to plan appropriate strategies for incorporating and responding to the challenges of electronic communication technologies.

Through these three sections we present a hopeful but cautionary tale for the next era of environmental decision making. The intent of this chapter is not to suggest firm courses of action that managers should take in the face of radical changes in communication technology, but to prompt individuals and agencies to address these issues in exercises of strategic planning for the future. Through a critical analysis of the advantages and disadvantages of these emerging technologies for environmental problem solving, agencies and individuals can mold their use of cyberworld communications in constructive and productive ways.

How Environmental Decisions Are Made

Two conflicting models are currently evident in environmental decision making and policy. The first is toward an increasingly sectoral and single-issue format in which government decision makers, the courts, and one or two particularly vocal citizen groups define the direction of environmental management (or create the gridlock that prevents changes in management). While few would argue that this process reflects ecologic principles or the concerns of the affected communities, it is often the most politically viable and is being furthered by social forces and institutional stagnation.

The second trend conforms to the sustainable development model and reflects principles of multisectoral integration, information exchange, and broad citizen participation. Though widely accepted in theory and practice, it has yet to be widely adopted because of political and logistical constraints centering around the effort that must go into achieving integration, information outreach, and participation. We present brief outlines of the social forces influencing the adherence to these two models. Following this section we analyze ways in which current information technologies are likely to influence adherence to either of these two models.

Factors Influencing the Sectoral Approach

Most analysts, conservative and liberal alike, agree that current U.S. environmental legislation is ripe for change. The bold experiments enacted in such legislation as the 1969 National Environmental Policy Act (NEPA) and the 1972 Clean Water Act Amendments have run a long and fruitful course, but are now mired in administrative and political quagmires (1–5). Fulfilling even the most basic integrated environmental planning missions of NEPA remains a distant goal (6, 7). The expansion of responsible agencies, competition among agencies, confusing rules and regulations, spiraling costs, and a reductionist approach to legislation have all eroded confidence in existing environmental management infrastructures (8–10).

In many ways, the constitutional framework of the United States is inefficient for solving environmental problems. The Constitution, with its system of checks and balances on authority, was never designed for a technological age and was certainly not designed to manage problems that are inherently multisectoral and require collaborative effort. Instead it was intended to limit power and protect individual liberty (11), properties that are in many ways antithetical to modern environmental policy needs. Federalism disperses authority by subdividing it among federal, state, and local institutions. In environmental management, however, this fragmentation has led to confusion and delay through its expanding complexity, duplication, and jurisdictional rivalries (5, 9). In addition, the short-term nature of the constitutionally mandated voting cycle (i.e.,

2–6 years), while a critical element of democracy is a disincentive to implementation of long-range and large-scale decision making. When lawmakers have only a few years in which to earn the favor of voters, they are inclined to push heavily for local interests and address problems through incremental change with short-range appeal instead of tackling more complex problems reflecting larger scales and longer time frames.

Two decades of environmental policy analysts have identified a long list of factors that contribute to current issue-based, sectoral environmental problem solving. Among the causes are the complexity of environmental issues, high costs, scientific uncertainty, and intense interest-group competition (12, 13). Others have suggested that the issue of leadership has primacy. In particular, these analysts feel that issue-driven environmental problem solving is driven by ineffectual leadership, especially the decreasing influence of Congress as a whole and the corresponding increasing influence of individuals, whether talk-show pundits or powerful legislators (5, 8–10). In addition to these factors, the adjudication of environmental law has contributed to a noncomprehensive approach. Much recently enacted environmental legislation is written nonspecifically. In turn, the power and importance of judicial interpretation of environmental law has increased substantially (8, 10, 14). Each of these factors contributes to the current confusion and gridlock, but each may safely be said to derive from four more intractable root causes, detailed below: (1) fragmented authority, (2) influence of political action groups, (3) localism, and (4) incrementalism (4,5).

Fragmented Authority

The problem of fragmented authority is that no one really has authority. For example, there are approximately 70 congressional committees and subcommittees responsible for water resources management; nearly 30 federal agencies have authority affecting some (frequently overlapping) aspect of the environment. Complicating this even further is the tendency of states to duplicate federal statutes and agencies, adding layers of state fragmentation upon the federal layers (5). Competition among committees and agencies as well as among national and state authorities results in protracted bargaining and compromise, diffusion of laws, and confusing and contradictory legislation. Inconsistencies are found even in fundamental concepts such as the evaluation of risk. For example, of the 34 laws regulating public exposure to hazardous or toxic substances, 7 laws require consideration only of health risks, 2 require consideration of the available technology, and 25 mandate some kind of risk assessment, balancing costs, and risks avoided (9).

Influence of Political Action Groups

Over the past 25 years, political action groups, from environmental groups to business organizations have grown in membership, power, and savvy. Many now employ permanent lobbyists in Washington and the significantly expanded opportunities for "standing to sue" have increasingly moved the environmental battleground into court venues. Many of these groups have been responsible for important advances in environmental legislation. For example, they have used judicial intervention to force attention to and action on mandated legal responsibilities of the government or polluting entities. To the extent that they promote creative and integrated policy, they can be vital and positive forces in environmental problem solving and environmental law (13, 15). Important non-government organization (NGO) coalitions and consensus reports have also furthered important changes in policy (13, 16).

Political action groups can also, however, impede progress in problem solving. In particular, standoffs between industry and environmental groups have been held responsible for incidents of environmental gridlock and the impossibility of consensus building (12). On a smaller scale, community opposition to projects such as landfills or other waste sites can derail progress in comprehensive environmental management (17). Small special-interest groups have also learned how to generate significant press for limited views, giving rise to notable distortions in nationwide interests (8, 18).

Localism

The influence of local needs has always dominated U.S. politics. While local representation is critical in any democracy, exclusive loyalty to local needs impedes environmental management through weakening attention to national or regional needs. Looking at each problem through a local lens causes the composition of atypical constituencies to affect the health and environment of the nation as a whole. In addition, it causes Congress to favor legislation with specific local payoff (e.g., construction grant programs) often at the expense of other vital, more far-ranging legislation. Thus, more agencies and local committees are set up with responsibility for narrower and narrower goals, without a broad view of how these goals affect each other. A classic vicious cycle, localism both derives from and promotes issue-based legislation and a reductionist approach to environmental management.

Incrementalism

Incrementalism is the primary mode through which legislation is changed in today's political climate. In order to satisfy special interests, competing agencies and committees, and maintain a favorable voting record, legislation is most often changed only in small steps. Incrementalism may be beneficial in some environmental programs. Hastily enacted legislation often reflects ill thought out relationships and may not engender public support. Consequently, the legislation is likely to fail, thereby wasting time, resources, and money, and diminishing the desire of parties to work toward consensus in later issues (12).

Incrementalism may be the "safer" alternative, but it is often an ineffectual one which inhibits policy innovation and creates significant time delays in resolving critical problems. It has been cited by many as a major obstacle to needed environmental policy reform and progress in solving environmental problems (5, 9, 12).

In all, the legislation of environmental programs has become mired in a crisis mentality in which the pollutant or environmental crisis of the year results in next year's major environmental law (5, 9). Changes in regulatory priorities, deadlines, agency responsibilities, and priority issues are frequent, additive, and inhibit progress on environmental agendas. Congress' own information basis for decision making remains terrifically convoluted and disorganized (19), aggravating both the reliance on public pressure and the tendency to make only the most cautious of decisions.

The Sustainable Development Paradigm

Environmental problems do not stop at community, state, regional, or international boundaries. Neither do they stop at resource boundaries. The quintessential example of this is acid rain, which results when air pollution precipitates downwind on vegetation, water bodies, and soil surfaces. Acid rain causes significant changes in water quality, dramatic declines in fish abundance and has been implicated in declines of vegetation health and rigor. In turn, changes in these resources affect local industries such as fishing and tourism. In the 1960s, Rachel Carson aptly publicized the spread of chemicals through the environment and educated people to the interconnections implicit in patterns of chemical use. Scientific studies have catalogued the impact of forestry activities on downstream water quality, of water quality on fisheries, of fisheries on local economic well-being, and so forth. These ecologic interconnections underlie the emphasis on holistic and biocentric environmental policy, which seeks to ensure ecosystem integrity and stability while still ensuring public use of natural resources (20–22).

The inescapable logic of this view is that unless environmental management is integrated, it will only treat symptoms and will not resolve environmental problems. Thus, throughout the same period during which U.S. environmental institutions have subdivided authority and responsibility, evidence has accumulated that the new paradigm for successful environmental decision making is one that involves integration across affected sectors, and that true integration can only be realized through more effective information management (i.e., education) and broad public participation (22–25). Support for this approach is evident in the intent, if not the actual practice, of many natural resource agencies. Expressions of this awareness can also be seen in the public and political sentiment for attention to increasingly long-range and large-scale (e.g., regional to global, multidecade) issues such as global warming and tropical deforestation.

Given the dominant character of U.S. environmental management today, this global view may either be construed as the ultimate irony or as a mask hiding the prevalence of local interests. Whichever reality it represents, for those who accept the view that environmental issues are among the most critical facing the nation, the trend in environmental decision making evident in the last 20 years is alarming. Instead of tackling environmental issues with a multisectoral, educational, participative approach, current environmental policy is being eroded through rewards to the squeaky-wheel issue of the day.

Recognition of the need for integration in environmental policy is not new. Early preservationists such as Aldo Leopold and John Muir and early conservationists such as Theodore Roosevelt and Gifford Pinchot all acknowledged the interconnectedness of the environment and the need for the environment to be treated as a whole. According to Bartlett (26), the terms "policy," "administration," and "environment" were first linked together in 1963. Since then, recognition of the importance of integrated management and public participation has come through many different channels, from efforts of local communities, state agencies, and foreign assistance efforts (5, 27, 28). Typical among these is a review by Mitchell (22) of the efforts of seven countries to achieve integrated water resources management. In his conclusion to the book, Mitchell notes that

> . . . legislation by itself does not seem sufficient to achieve integration. . . . much more attention needs to be directed to cultivating an organizational culture and attitudes that reinforce the benefits to be derived from integration. At the moment, most of the incentives encourage less rather than more integration. Development of skills in bargaining, negotiation and mediation will have to be central to improving the atmosphere for more effective integration.

Will Environmental Policy Change?

Although there is a crisis in confidence in environmental management, it is impossible to determine in what way environmental management is likely to change. Environmental problem solving evolves in response to changes in societal values, infrastructures, laws, and scientific understanding of issues. Many competing influences are currently shaping the tug of war over policy developments. For example, advances in science are encouraging policy makers to consider integrated and larger-scale solutions. Foreign assistance programs and some state-level agencies have also gravitated toward integrated programming, being influenced both by science and by their past failures (5, 29).

However, perhaps even more influences are countering these trends toward integrated management. For example, U.S. citizens have become frustrated with scientific principles they do not understand. Consequently, support has eroded for programs such as the Endangered Species Act, which reflects ecosystem-based principles. In addition, the whole weight of the U.S. constitutional structure and traditions of pluralism almost ensure that sectoral problem solving will continue to be the norm. The enormity and entrenchment of the environmental management infrastructure will resist significant change and the features detailed above (i.e., incrementalism, localism, fragmentation, and interest group influence) will stand as impediments to significant new directions (13, 21, 26).

The Internet as a Potential Player in Environmental Problem Solving

A new tool for players in the delicate balance between status quo and change has been introduced through the advent of the Internet and related communication technologies. These new programs facilitate information retrieval, speed communication, and increase the range of personal contacts. Consequently, they will have a significant role in the debate between these two environmental management paradigms. These new communication technologies are poised to change the way in which people, institutions and scientists communicate, prioritize, and implement decisions. In turn, they can influence the direction for change by bolstering the strength of different players. The remainder of this chapter is devoted to an analysis of the ways in which the Internet and its associated communication "character" is likely to exacerbate existing trends such as single-issue dominance in environmental decision making.

This chapter does not address the important political changes currently occurring in the United States, except to ask the reader to remember that environmental decision making occurs in relation to dominant social values, as well as in response to political changes. For example, the Reagan era undermined the strength of many of the U.S. environmental infrastructures but in response to these policy changes environmental action organizations such as the Environmental Defense Fund, the Nature Conservancy, the Audubon Society, and others experienced a surge in membership and support. Influences such as these will continue to vacillate over time and are important in understanding the ways in which decision making takes place and the priorities placed on the many components of environmental decisions. These influences are a pertinent backdrop to this chapter and play a very visible role in environmental problem solving. However, we suggest that other, less visible forces such as the Internet and related communication technologies will pose an increasingly dominant role in environmental problem solving.

Possible Relationships Between the Internet and the Environment

The advent of the electronic communications such as the Internet (also, the National Information Infrastructure (NII) or Global Information Infrastructure (GII)), on-line data bases, electronic mail, and their family of related technologies has sparked intense debate regarding the social implications of a nation, or world "on-line." Will these developments change the fabric of society or will they simply fade into the background as one more communication tool? Will the Internet per se be a common vehicle through which all people can communicate freely or will different systems and different services dissect society even further? And what does the debate have to do with the business of environmental decision making?

There are no clear answers to many of these questions. However, environmental decision making is a vital social process significantly influenced by media, public opinion, scientific knowledge, economics, and politics. In short, environmental problem solving is influenced by communications strategies. As such, these new communications technologies can change mechanisms of interaction and information retrieval critical to environmental management. Its impact on environmental decision making may be enormous. In those cases, for example, where the Internet facilitates improved communication among scientists, citizens, and decision makers, it can assist in lowering the degree of scientific and social uncertainty that is translated into policy, and thus have a powerful, positive and significant role in decision making. However, if and when using the Internet becomes a shortcut around careful information retrieval and interpretation, or promotes crisis management over strategic planning, then it stands as an equally powerful but negative influence on environmental decision making.

The Internet has already played important roles in environmental law and environmental action, mobilizing action to stop projects or bringing pressure to bear on companies responsible for environmental damage (15, 30, 31). People send queries over electronic mail and the Internet about toxic chemicals, organizing local communities, community problems with water quality, and effective strategies for organizing community action against a perceived problem. Additionally, organizations such as Greenpeace and Enviroline send information and action alerts through their own and other relevant on-line services, listservs, and bulletin boards. On a case-by-case basis, these may result in improved information exchange, improved local decision making and faster resolution of issues. However, this approach clearly reflects crisis management. Unless it is part of a larger proactive management plan, such issue-by-issue mobilization will only further the fragmented model of environmental management; the same approach which has already proved untenable.

Prototypes of the Internet have been in use for decades, but until recently it has been the tool of only a select few. Enormous growth in the past 3 to 5 years has made the Internet the norm for communication within and among university scientists and in many government agencies. However, the very recent expansion of the Internet into a commercial realm and its potential expansion into the majority of North American and Western European households has fueled a debate about predictions of Internet's impact. On one hand, technological utopians claim that the Internet is the tip of a large iceberg of new technological changes that will improve our lives, erase considerations of race and economic standing, and allow us all equal access to the world's mass of information. Countering them are Luddite philosophers who warn that the technology will bring "big brother"–like monitoring and through its video and audio technology, an end to creative thought (32–34).

The Internet: What It Is, Is Not and Might Be

The Internet is peculiarly undefinable. It is a commonly applied term for a large complex of "on-line" resources. Thus, as used here the term "Internet" is shorthand for everything from virtual libraries, with their on-line and searchable data bases of books, journals, journal articles, and government documents, to local freenets that allow citizens of cities and towns access to on-line information about everything from taxes to phone rates and local entertainment. In between these are a vast maze of resources: electronic mail (e-mail), bulletin boards and usenet groups, World Wide Web (WWW) pages and Gophers as well as ftp and telnet sites, "chat" lines in real time and new, daily emerging technologies. Although technically different, other technological developments are commonly considered as part and parcel of the Internet (e.g., technologies such as "intelligent communications" and videoconferencing).

Each interrelated arm of the Internet has its own information character, each appropriate to a different sort of information and use. For example, the explosion of WWW pages has been furthered by the development of software that allows video and audio images to be integrated with text; this icon-driven, point-and-click technology enables browsing through posted information. It is, however, a format suited for information outreach rather than interactive information exchange. For the latter, a heavy proportion of Internet use is relegated to e-mail and listservs that allow members to post comments, queries, and problems to a group of like-minded or like-trained citizens. New developments of chat lines in "real time" also fall under the category of exchange technologies where gophers, ftp, and telnet sites all fall under the category of information outreach.

It is difficult, if not impossible, to estimate the number of actual Internet users, host sites, and, maybe more importantly, the type of services most commonly used by the on-line community (35). Numbers published on-line via listservs such as CPSR-Global note growth of nearly 25% per quarter and project that in the United States alone the number of users will top 100 million by the year 2000. Current estimates of the number of U.S. users rests between 3.5 and 4 million (36). While early growth in the Internet market was largely in the United States, the rate of growth in international markets is now surpassing the rate in the United States. Although there is no accurate way to count users, current estimates range from 15 million users (37) to 35 million (38) scattered over 150 countries. One report estimates that the number of Internet users is expected to reach 180 million by the year 2000 (38).

Although impressive, these numbers may distort the utility and popularity of Internet services. Internet's novelty lies in the potential for two-way, or multiple-direction communication. Some use this Internet capacity to its fullest. Devoted followers, "surfers" of the Internet, they keep up with new services and news groups and monitor the progress of world events through the net. However, the vast majority of people log on only occasionally, and even fewer contribute to the vast array of bulletin boards, listservs, and chat lines. In practice, Internet as a tool, especially an environmental one often does little more than make public the views of a few regular participants.

Changes in the Internet

The Internet is not one system: it is composed of a web of more than 45,000 autonomous networks. Until recently, the dominant domains of Internet users were academic and government institutions (i.e., .edu or .gov suffixes on addresses). The use of these domains was facilitated by government subsidies and university fees that allowed users apparently free access. While these domains continue to increase in volume and use, enormous increases in the commercial and network domains are outstripping increases in the "traditional" educational and government domains. These developments present interesting and as yet unexplored implications for the future of the Internet (i.e., use is changing in character as it grows).

Most social analyses of the Internet have speculated that growth beyond academic and government circles would be severely restricted by financial realities and lack of pertinence. These predictions appear to have underestimated the creativity of businesses and marketers. The Internet is a chameleon. Because it is merely a conduit to vast sources of knowledge, its potential utility to different audiences is diverse. It appears that new potentials are only now being realized by the commercial sectors. Efforts by advertisers to use the medium and the new privatization of the Internet itself may vastly change the nature, scope, and composition of both on-line resources and the on-line community.

Other significant forces may also be changing the future of the Internet and related technologies. Government policy itself may have an enormous influence on who "controls" the Internet and who has access to its diverse resources. For example, the question of what constitutes electronic copyright could have drastic implications for restrictions on the net as well as, some claim, for personal freedom. Recent government proposals contain broad interpretations of what qualifies for copyright restrictions and would switch the focus of copyrights from control over distribution to control over reading, primarily because the advent of downloading any file via electronic access requires that readers make copies of the document to their machine.

Another area of policy concern is the privatization and commercialization of the Internet. Issues of regulation, competition, and monopoly control have all emerged in the debate over what the next stage will be for these electronic communication technologies. While some feel that the nature of the Internet is already defined by its current users, others feel that competition and commercialization of it could severely slant the resources toward entertainment and away from the information provisions. In addition to changing the nature of the Internet and related technologies, this might also further aggravate problems of societal access. The pressure to seek equality in information access comes when services are seen as socializing or political phenomena, but not when they are in the economic goods and services domain (39). While changes are bound to take place, their nature and their susceptibility to political winds are unknown: it is important to remember throughout the Internet debate how much is still up in the air and that changes in the structure of the web itself could affect the environmental scenarios outlined below.

The Environment Versus the Character of the Internet

As stated earlier, the emerging paradigm in environmental management is based on increases in three elements: integration, information, and participation. Here we address whether the growth of Internet communications will advance or retard emergence of that paradigm. We do not denigrate the positive social influence that the Internet (or its evolutionary successor) will certainly have in the future of environmental problem solving (15, 30, 40). Neither is it our purpose to promote the use of the Internet. Instead we acknowledge that some form of these electronic communications are here to stay and that they are likely to have real impacts on patterns of communications, research, education, activism, and policy formation. Consequently, a critical analysis of the potentials and realities of electronic communications may help policy makers, managers, and individuals plan critical strategies for the future.

Electronic information technologies have been adopted by proponents of sustainable development and environmental equity as the democratizing tool that will transcend barriers to participation, knowledge, and fundamental change (1, 15, 30). Within Agenda 21 of the United Nations Conference on Environment and Development (UNCED) is the endorsement of electronic information technologies and the claim that they can close the data gap and improve information availability. Much like Francis Bacon's famous "knowledge is power" quote, modern environmental activists propose the free flow of information facilitated by electronic communication technologies (e.g., the Internet) is an "effective safeguard against environmental harm" (15). These scenarios propose that the Internet will improve environmental decision making of the future because of the ways in which barriers of space and time are eliminated (compressed) (41). Specifically, they suggest that electronic communication technologies will facilitate environmental problem solving by bringing about broad changes in our ability to obtain data and engender public discourse. They will:

1. Increase the ability to maintain and gain access to timely data bases (i.e., increase knowledge). Through this change, one of the current impediments to environmental decision making (i.e., reliance on outdated data) can be reduced.
2. Democratize decision making (i.e., increase participation and integration). At its fullest potential, Internet allows much wider audience participation in discussions and periods of public comment on proposals. Thus, there is the potential for more people from broader interests to be represented in any given environmental discussion (42).
3. Increase the speed of resolution of given issues. The ability to obtain existing, disparate data over a period of weeks instead of months or years and the ability to solicit opinion quickly "on-line" can streamline the problem-solving process.

Internet's *potential* to accomplish these goals is not in question. However, in the following section we critique these three ideas and suggest that they are more utopian than realistic. Without critical rethinking of our environmental problem-solving needs, the Internet will be an avenue to streamlining status quo processes.

A Realistic Interpretation of Internet's Role in Environmental Problem Solving

In the preceding subsection we outline the three arenas through which the Internet can, potentially, revolutionize environmental problem solving. In the sections that follow we take a critical look at each of these and outline a more realistic interpretation of the likely effects of the Internet on environmental problem solving.

Timely Data Bases

Environmental decision making requires up-to-date data, but rarely requires, nor is furthered by, immediate access to data. Environmental decisions are made through integrating information from a variety of fields (e.g., biologic, ecologic, geographic, social, and economic information). Consequently, the skill in decision making comes in balancing these factors and that balance depends on understanding time trends and risk analyses. As a reflection of social values and accumulated scientific information, good decision making cannot happen overnight. This is not an argument for use of outdated data in decision making: where there are current and reliable data, it is imperative to incorporate them into the decision-making process. However, currency alone is not a selling point in scientific veracity: acting on current information that is inaccurate and uncor-roborated may be as damaging (and perhaps more so) than acting on older data. Even several years of data may inaccurately reflect long-term trends, reflecting, instead, an aberration in weather patterns for that period, inaccu-racies of new testing equipment, or a host of other invalidating measurements. Without patience and corroboration of measurements, science loses its analytic and predictive power.

Critiques of the legislative process have documented the impact of using outdated data sets in writing environmental legislation. In some instances, newer data would have changed the nature of regulations as Congress would have been more aware of changes in industrial technologies. In spite of the accuracy of those specific examples, at a larger scale there is a greater risk in being seduced by "newness" of immediately available information. In the book *Science Under Siege* the author proposes that public "scares" frequently override scientific evidence and force policy makers into rash policy formation (18). Oth-ers have demonstrated how hastily approved legislation runs significant risk of failure, wasted resources, and public ire (12). Thus, the argument that good decision making will be improved by access to new, current data is only partially substantiated. Some decisions will be improved, certainly, but there is also a risk that other decisions will be made impetuously, and that data interpretation will be left to those without the tacit knowledge to understand it (see below).

The Democratization of Knowledge?

The democratization of knowledge is a phrase often used in articles promoting the virtues of electronic communications (1, 15, 30). However, in order to fulfill this promise, Internet communications must meet a large and likely unattain-able set of goals. Among the most important of these for environmental man-agement is that the Internet must broaden not only people's contact with each other but the subjects they discuss. It must also be broadly accessible by a rep-resentative cross section of the populace. Perhaps most fundamentally, in order to achieve its democratizing effect, users must be able to navigate comfortably among the Internet's many services and understand the information they encounter. A tool is only as good as the skill of the user: if people use the Internet as no more than an on-line newspaper or as a faster, but substantively unchanged mail system, then it will become little more than a substitute for existing media and will not achieve its lauded potential.

Evidence from many avenues suggests that the Internet may fall far short of its democratizing potential simply because of the ways people choose to use it. For example, as detailed below, listservs, e-mail, and news groups are, as anticipated, broadening people's exposure to each other. However, dominating this spatial reduction is the unanticipated result that contacts are increasingly limited in context and subject area. Consequently, people are limiting, rather than expanding, their exposure to new arguments, positions, and ideas. In addition, the likelihood that the Internet will ever be fully accessible to a broad section of the population is increasingly unlikely given gender, financial, and educational constraints. While technological developments make software increasingly "user friendly" (with a whole new generation of user friendly software due out in 1995, such as Microsoft's "Bob"), people may not always understand the information to which they now have access, and the possibility for misinterpretation of important social or scientific information is enormous. Even more fundamentally, the enormity of the scope of information available on the Internet creates the dangerous assumption that quantity is a substitute for quality.

Narrower: Not Broader Contacts

As the Internet expands it becomes more and more a medium for narrowing, rather than broadening, contacts. The attraction of much of the Internet is, in fact, in finding a like-minded virtual community. In a recent book, John Pavlik (36) suggests that there are four critical dimensions that define the current and future uses of Internet and its associated electronic communications: (1) people's interest in specialized topics, (2) people's desire to send and receive, not just retrieve, information, (3) people's difficulty with identifying their information needs, and (4) concerns for privacy. The first three of these may explain the dominant popularity of e-mail in comparison to other Internet services. To date, e-mail gains the largest share of the Internet-use market, estimated at over half (32). The other large cuts of the Internet pie are taken up by bulletin boards, listservs, and usenets: those areas which enable the user to carry on a dialogue.

Certainly many people have found friends and colleagues over the net they would not have met otherwise. Many of these are people from other countries and other cultures. Net lore is full of stories of cross-cultural barriers breaking down in the face of speaking with virtual friends. However, throughout history people from other walks of life have met in unusual and new circumstances (wars, for example) and formed lasting friendships with seemingly unlikely companions.

However, it is only in the most utopian view that the Internet can really be viewed as a democratizing tool. The unique difficulty about links made through cyberspace is that, while they may include people from many walks of life, those contacts are usually made because of common, narrow interests and they are notably ephemeral. Environmental problem solving requires not just debate across ideological lines, but even more importantly it requires that those involved stick to the problem resolution process. Environmental problem solving is often a lengthy, emotional, and sometimes transforming process, and the continuing involvement of players is vital. Yet, Internet communities are notably short-lived. Because community members are faceless and frequently anonymous, there is little pressure to resolve complicated issues. More frequently if someone disagrees with a discussion line in a listserv or bulletin board, they simply sign off (43, 44). Environmental problem solving will never be served when those involved can simply fade away.

The narrowing of Internet circles is not restricted to casual users. The trend is also increasing in and among professional circles. While use of Internet is extensive and rapidly expanding, expansion is generally occurring along pre-existing lines. Thus, NGOs prefer, for reasons of efficiency and cost, to limit their interaction to forums for NGOs only. Some NGOs even maintain closed discussion groups. Government people interact primarily with other government people, and businesses with businesses. While this focus is not surprising, it is antithetical to the touted improvements of global communication enabled by the Internet. Enormous proliferation of discussion groups improves our communication reach, but promotes ever smaller arenas for discussion. The end result will be (or is already) a narrowing of interests and exchanges rather than an expansion of them. For example, to maintain a broad view of even one natural resource management discipline, water quality, it is theoretically necessary to belong to over 40 separate water discussion groups, whose foci range from groundwater flow to Mideast issues to global change impacts. Because of both the number of groups and the volume of discussion in each group, staying abreast of each becomes prohibitively time consuming. The end result is a choice to join only one or two groups or to sign off completely.

An important consequence of narrowing focus is a change in on-line language. Recent interchanges on several listservs highlight this problem. People assume they are speaking to like-minded colleagues so their use of this publicly accessible medium becomes rife with shorthand, "off-the-record" comments. Occasionally these are forwarded to news groups with opposing viewpoints, resulting in angry, defensive exchanges. The trend is as apparent on "scientific" listservs such as Ecolog-L as it is in the more sociopolitical ones such as Enviroline and Econet. Thus, this search for ideological kin appears to lead to decreased rather than increased dialogue. The direction in the long run is creation of more firmly entrenched positions and less willingness to compromise: "after all, I know through my usenet that there are thousands of folk who think like I do." Rather than democratizing the realm of ideas, problems such as increasingly narrow contexts, their ephemeral nature, and the associated linguistic shorthand used by members of Internet communities appears to be supporting the confirmation of our own beliefs, decreasing debate, and perhaps decreasing even the interest in or willingness to participate in debate. Thus the Internet has extraordinary potential to create more entrenched positions as like-minded groups reach more members with less need to invoke balanced analyses or language. Combating this trend will place even more pressure on environmental managers and decision makers to bring groups face to face and work through differences of opinion and interpretation.

Uneven Access

As indicated, another critical dimension in the democratization issue is access. To be truly democratizing the Internet must either represent everybody or at least represent an even cross section of the populace (1, 15, 39). Neither has yet been achieved, nor is it likely that either will ever be achieved (45). Uneven access has many causes, from inherent problems involved in staying abreast of quickly evolving technology to more fundamental societal questions of politics, unequal distribution of resources and differences in communication styles and interests across gender, educational, economic, and demographic lines.

Despite the enormous expansion of the Internet, many segments of society are not currently enjoying the fruits of Internet or its private access lines such as Compuserve, America-On-Line, or Prodigy. Internet's initial popularity was in part because it was subsidized, and thus apparently free to affiliates of academic, government, or organizations whose employers subsidize the service. The unemployed, self-employed, or those employed by most small businesses rarely have access to Internet, or must pay for the service on home computers. Members of disadvantaged communities not only have no access to Internet through their employers, they rarely have a computer or awareness of what it could bring to their communities. Libraries, it has been proposed, can fill this gap by providing community computers and on-line hookups, but if people do not frequent the library the information is still effectively denied to them. While some of these access considerations are doubtless due to the fact that the technology is still relatively new, the universal access that is theoretically possible is highly unlikely because of financial and knowledge constraints in poor, or even many middle-income households and communities. Globally, the fact that users now communicate with others in distant countries can create the illusion that these discussions reflect true global concerns (15). Those involved in such discussions still represent a small and selective minority of the global population who share similar educational and economic experiences and, most often, speak English.

Uneven access is also apparent in the on-line community itself. The speed of new developments in both hardware and software creates logistic difficulties and harsh financial choices for computer users. Within 6 months or a year, a substantial investment in computer hardware may be "outdated" by the latest software. Each new software upgrade requires more and more memory and faster and faster chips. Upgrading memory is possible but relatively expensive and many, if not most, computer users use older, slower, less sophisticated programs but are interacting in an environment designed for the very latest version of every program. This results in memory overloads, system failures, and, overall, a reduced ability to utilize the newest technologies. The implication for managers or decision makers interested in reaching a broad segment of the public is that even though a user might be on-line and reachable through e-mail or a discussion group, providing information through a new technology service may not reach as broad an audience as intended. While utilizing new technology can be elegant and can reach new audiences, relying on it as an outreach system may, ultimately, limit the spread of information.

User demographics also illustrate uneven use patterns across gender, educational, and economic lines. In the environmental sector, this uneven use may have significant implications for sharing and receiving information. For example somewhere between 80 and 90% of all Internet users are male (43, 46). Researchers have also identified that characteristically male patterns of language and debate predominate in Internet-based discussions, having the effect of further reducing the involvement of women in on-line discussions (43). Even more gender gaps have been discovered in the way men and women use computers, as a whole, and the Internet, specifically. For example, women are more "task oriented" on computers as a whole, but when they do log onto the Internet, women tend to seek out companionship rather than specific information. In contrast, men are more likely to use computers as a "toy" and a source of specific information. While these tendencies are broad and will differ substantially across user groups and access services, they may still have important implications. They are especially pertinent to environmental managers since years of research confirms that women and men respond to environmental problems, threats, and challenges differently. In general, it appears that women are more likely to be involved in local issues and more likely than men to respond to concerns of potential harm (to themselves, family, or the biosphere) from environmental conditions (47). If managers are gleaning responses to an environmental issue from those on-line, it may present a skewed view of citizen opinions on an issue. Similarly, if managers, educators, or policy makers use on-line services as an environmental information tool, they may largely miss an enormous (and receptive) segment of society (i.e., women).

Similar biases are apparent in other areas: education is a better predictor of new technology access than is income alone, and occupational status is more important in telecommunications and computer use than any home or family applications. Further, having a white-collar job or attending school is a high correlate of new technology adoption (48). On top of technology access, alone, the typical on-line family earns nearly 80% more than the typical American family, is more likely to be college educated, and is younger than the average (49). Disregarding for a moment the gender issue, these demographics do conform reasonably well with those of the most environmentally active of the U.S. population (9). Consequently, (male) computer users, particularly Internet users, already reflect the most informed, environmentally active cross section of the population. Interestingly, environmental information services have been slow to develop on the Internet. Among many thousands of cyberworld resources, only a small percentage relate to the environment (although this percentage represents a large number of services) and many important environmental services such as EcoNet require extra user fees.

The implication for environmental managers may be twofold: unless user demographics change dramatically, heavy reliance on Internet communications will largely miss target audiences of women, the less educated, and the manufacturing and noninformational services sectors of the economy. Also, the male bias of Internet users can be expected to underrepresent society's environmental concerns as a whole. Internet users, however, are vocal and informed, and in their own use of the system are likely to skew interest toward specific local issues (or issues of the minority on-line community). In either case, it is the broad picture that will suffer. This information suggests that environmental managers and decision makers will need to know and correct for the composition of the on-line community if they rely on it as a tool for surveying public opinion on an issue. Even more importantly, outreach and education will need to focus on those communities not well represented on-line. That focus is not well developed today.

User Knowledge

The Internet's potential is often untapped because of user frustration, confusion, or use at a level that does not match the technological potential. Democratization of decision making rests not just with the dissemination of information, but the gaining of knowledge. One difficulty with scenarios equating Internet use with democratization is that it equates information alone with knowledge. This equation is simply unrealistic (15, 41, 50). The ability to understand information is not implicit in the ability to download it. Neither is the fact that information is somewhere in cyberspace a guarantee that it will be accessible to users. Even user-friendly program browsers such as Mosaic for WWW often obscures pathways that contain extensive information. Information is often housed in subpathways of access sites whose utility is not immediately obvious (e.g., sunsite), or relevant information is housed in many different areas and user search interest is exhausted before the search is exhaustive.

Three issues are relevant to understanding the problem of user knowledge. The first is the problem of tacit knowledge. How does a new user learn the tricks of computer use and navigation? Even more importantly, how does on-line information look to people with different training and is there a significant opportunity for misinterpretation when people are receiving information without having appropriate background to interpret that information? The second issue is that of the overwhelming volume of available on-line information and the ways in which people respond in the face of too much information. Lastly, there are important and hard-to-answer questions with respect to ingrained biases in either data bases or in the way people search for new information.

The Problem of Tacit Knowledge and Context   A major area of concern for the future and expansion of these new communications technologies and an even more substantive criticism of the democratization concept is the problem of tacit knowledge (1, 8, 50). Tenner (32) classifies tacit knowledge as "skills and ideas that may not be recorded in written form but that arise from person-to-person learning and experience." It is important in any discussion of the Internet for several reasons. At one level, tacit knowledge is the way in which people learn to understand how to use electronic communications technologies. To quote Tenner (32): "Learning any game or skill requires immersion in a group of people who already have the skill." Despite utopian visions that the Internet, on-line data bases and other resources will be accessible to all, there is a very real problem of training if the medium is going to meet anything close to its user potential (51).

A second, more contentious, issue for environmental problem solving is the interpretation of data to be found on the Internet, bulletin boards, listservs, or on-line data bases. It is tacit knowledge that distinguishes trained professionals in one field from those in another, or from laypersons: it is the context brought to a review of data that is critical (1, 50). To an environmental manager looking at GIS data retrieved over the Internet, it may be acutely obvious that patterns of geology form the matrix for a regional water-quality problem. However, the same data base would reveal far different information to someone without that environmental tacit knowledge. Consequently, while the Internet facilitates discussion and sharing of knowledge among professionals, if the same data are available to those without the critical tacit knowledge to interpret them, the information could lead to vastly different (and inaccurate) interpretations of a given situation.

Pfaffenberger (1) proposes a different possibility: that interpretational limitations are sufficient enough to act as a barrier to untrained information seekers. In his scenario, without tacit knowledge in a given field, information will remain effectively unavailable. Thus, Pfaffenberger feels the danger of inaccurate interpretations is low, but also discounts the idea that information technologies will democratize information.

The Problem of Too Much Information   While information access is wonderful in theory, in reality too much information can be a serious impediment to learning, leading to the collapse of meaning (40) and diminished attention spans (15). In the face of such overwhelming choices as exist on the Internet, most people conduct only the most superficial of information surveys and they tend to be distracted by tangential subjects that cross the screen. Information overload also increases users' reliance on digests and compilations of information (52). The ease of these compilations may in the short run increase productivity and information retrieval. However, in the long run they may impoverish the information retrieval process since users reduce their own browsing of primary and related secondary sources, lose relevant threads and give each subject less time.

In the face of too much information, we are swayed by volume and over-whelmed with all there is to know. Suber (50) suggests students may be "duped" by the gamelike quality of on-line searching into believing they under-stand the information they have retrieved. The overall effect of too much infor-mation, then, may be (1) the inclination to include even less information in the decision-making process, (2) the inclination to rely more heavily on inter-pretations of others, and yet (3) the tendency to feel more confident in our newly found "knowledge." The result may well be that our decisions become weaker instead of stronger.

As an example, Internet access to on-line libraries vastly increases our abil-ity to conduct bibliographic searches. However, several trends in end-user searching illustrate that this field is not developing according to its potential. The first is that people trained in on-line searches exhibit early enthusiasm and repeated searching, but after a year only a small percentage continue to use on-line searches (1). Second, most searches are accomplished via one or two data bases despite the existence of hundreds; people have limited time, patience, and enthusiasm for what cannot be found quickly (1). Third, searches commonly reveal more articles than one can effectively digest. The clear trend is to rely heavily on abstracts instead of reading a complete article, chapter, or book. Yet abstracts do not enable one to analyze the hypotheses, methods, or conclusions of a paper and do not enable the synthesis and comprehension that comes from reading full articles. Suber (50) expresses a concern that students who limit their searches to what is on-line will "truncate the field" and that the roots of modern day fields will over time be lost.

The Issue of Bias   Proposing that the "virtual" environment will be democratiz-ing rests on the assumption that societal biases will be eliminated in virtual communities. Many testaments have been given to the fact that being nameless and faceless while on-line enables people to be more open and communicative than they would be in face-to-face communications. However, research suggests that the mere lack of verbal or audio clues is not sufficient to eliminate people's bias. Simple awareness of age, gender, or affiliation with different organizations is enough to establish status (e.g., biases) in on-line relationships (53). Distinct digital personas, to which people respond in specific ways, can develop through just the use of names, signatures, and characterizations (54). Consequently, proposing that facelessness of communication, alone, will eliminate barriers

denies inherent social realities: people search out not only similarities as discussed earlier, but differences as well. The embryonic on-line community may
develop a different system of biases and status hierarchies than the society
around it, but status and bias will remain a part of that community.

Bias in the world of the Internet is also possible from many different fronts,
each with its own peculiar implications for environmental decision making. For
example, bias may occur in the

- Ways in which people choose to search Internet resources, hence in
  the information they will eventually find
- Development of the original data bases: the journals catalogued for
  any given data base, the maintainers' list of key words and searchable
  phrases
- Format of data bases; some easily accessible, others not so

Increased Speed of Decision Making: Reduced Quality and Increased Immediacy?

One of the largest threats the Internet poses to environmental decision making
may be the flip side of just those things that are touted as the key to a new
democracy. Internet custom has developed quickly to promote messages and
communications that are simple, direct, and easily digestible. The media people
have sound bites, the Internet has info-bytes (44). Long postings to moderated
groups are frequently edited down to stay within these info-byte guidelines,
and net-etiquette calls for keeping postings and e-mail messages to single subjects. Rather than intertwine subjects, the pressure to reduce and simplify sets
up a preference for the shortest, most direct communiqué. Thus, the Internet's
greatest strength—its link with so many other people—comprises one of its
greatest limitations: the volume of information and the number of potential
communicators makes complex debate a rare use of the Internet.

As a result of the info-byte syndrome, successful utilization of the technology confers a clear advantage to action-oriented messages. It also conveys
an advantage to the "new" issue of the day, while chronic problems are relegated to slower technologies of print and paper (15). The Internet is uniquely
suitable for specific information, whether for references or identification of
problems, but it is also uniquely suitable for relaying short, media-laden appeals
for action. Consequently, the Internet has tremendous potential for mobilizing
citizen action. It is already being used in this way by many groups, from Greenpeace to Amnesty International, and has already been used successfully in many
campaigns to give local issues global exposure in order to bring about greater
international pressure on local decision makers (31). While many would pro-
claim this a positive (democratizing) force in environmental decision making,
it may be a dual-edged sword.

As indicated throughout this chapter, it does not follow that the Internet has equally extraordinary potential for presenting balanced education on issues. While political action groups may organize campaigns effectively through Internet resources, without complementary efforts to enter balanced viewpoints into a discussion the Internet will promote the crisis mentality of environmental problem solving, increase polarization, and increase issue entrenchment. Decision makers will be increasingly confronted with the need to make quick decisions, a pressure that leaves little room for balanced consideration. A management process that is forced to be reactive to crises rather than proactive will never achieve an integrated end.

Technology Rhetoric and Environmental Problem Solving

We are increasingly deluged with portrayals of the "information society" of the future: of dramatic changes in education, entertainment, and social organization (37, 41, 55, 56). To social and technology historians, these arguments invoke a strong sense of déjà vu. Only a decade ago, cable communications were touted as the democratizing medium of the future (57). Philosophers of the times predicted similar grand and tragic results from the advent of the horseless carriage, telephones, and video technology. None of the contemporary predictions were able to see the secondary effects of these technologies that have had such enormous impacts (e.g., the highway system which grew out of the "horseless carriage"). More important than exact predictions is the process of debate over what changes technology may bring: the writings themselves reveal strengths and weaknesses in our social processes that are more revealing than their predictions. A careful review of these writings helps to separate the inherent character of the technology from the assumptions of those analyzing it.

A study of the rhetoric of technology writing reveals some important challenges ahead to environmental decision makers. On one hand are the scenarios that lull us into complacency: that the technology itself has a transforming character and all our decisions of the future will be cooperative, democratic, and participatory. On the other hand are dire scenarios predicting greater central control over decisions, greater control over information (or, more accurately, over the information most people receive through news digests) and thereby less public involvement in the decisions which affect them. How do decision makers navigate among these predictions?

We conclude this chapter with a critique of the two most dominant genres about technological change facing our society, utopian and antiutopian scenarios, and look at the conventions inherent in these portrayals. As all agencies, companies, and individuals confront new technological frontiers, it is vital to retain a critical eye pertaining to the messages about these technologies. Hidden in the rhetorical discourse of many of these essays are the kernels of truth for environmental managers and policy professionals as to how to effectively incorporate these technologies.

Technology Writing: Hidden Messages and Hidden Assumptions

Social commentary, whether ecologic, technological, or sociological most often divides itself between utopian and antiutopian views, with utopian scenarios the primary genre in North American technology writing (e.g., Louis Rosetto's [55] portrayal in *USA Today* of "Techno wonders") (58). Interestingly, it probably is equally fair to say that antiutopian scenarios are the dominant genre of ecologic writing (59). In effect these scenarios are two sides of the same coin, with the utopian scenario creating an idealized view of what technology promises and the antiutopian scenario imposing the vagaries of human greed, power, and ignorance on the expression of the technological potentials (31, 58). Singly, each scenario creates a nearly untenable view of the future, but taken together they provide great insight to how technologies and their human users interact.

Both utopian and antiutopian scenarios are written authoritatively and detail how aspects of computerization (or other technologies) will create or aggravate dominant social problems (58, 60). They differ primarily in their consideration of the ways in which humans will interact with a technology. Utopian technology writing, for example, considers the future in a human vacuum. In these views, technology transcends human nature, and the new society is one in which people always cooperate to make things work and conditions of education, empowerment, and ability are maximized. In contrast, antiutopian treatises place significant emphasis on the (negative) ways in which technology will crystallize failings in human nature and human society. Antiutopian scenarios focus on the "tragic" of the technology: the 1984, big-brother scenarios, or cataloguing the end of creative thought. Both scenarios are exaggerations of the way people interact with technology (41), but their arguments are persuasive and seductive. Managers and decision makers in the next decades must be sensitive to the manipulative appeal and both the strengths and weaknesses of these dominant genres.

In order to create their portrayals of technological determinism, technology writers and philosophers take clear (if sometimes inadvertent) positions with regard to important social processes such as conflict, skill training, and social change (1, 58). Each of these three areas, detailed below, is clearly relevant to environmental decision makers, and the fact that they stand out as untenable under either utopian or antiutopian genres highlights an important opportunity for environmental managers.

Portrayals of Social Conflict

Technological utopian analysts portray a world free of substantial conflict, whereas technological antiutopians usually portray certain fundamental conflicts (e.g., rich versus poor) as inevitable and unalterable (58). Neither of these scenarios adequately reflects the past, present, or likely future conditions for environmental management. Certainly the utopian view where conflict is negligible is exceedingly unlikely: environmental issues are quintessential examples of social conflicts. However, neither is the antiutopian scenario a likely one for the future. Societies do evolve, and environmental policy is peculiarly devoted to balancing conflicts among segments of society. While neither of these genres adequately models the nature of social conflicts from the perspective of environmental managers, the lack of a good model, alone, presents a valuable opportunity. It illustrates that the way managers choose to address (or ignore) conflicts in the future may influence the effect of environmental management on Internet usage, and vice versa.

Skill Acquisition and the Transfer of Knowledge

The problem of skill acquisition and the transfer of knowledge is central to all questions of the adoption, use, and misuse of technology. It is a pivotal question in evaluating whether communication technologies will affect environmental decision making: simply having access to the Internet does not imply that the resources available through the net will be well or thoroughly utilized. Do the technological potentials create a false sense of security? Are people even capable of utilizing the vast resources? If used sloppily, is information gained from the Internet any better or any worse than information gained through face-to-face and hands-on research?

Utopian and antiutopian scenarios adopt very different perspectives on skill acquisition. In utopian writing, the question of skill acquisition is glossed over: people have whatever skills they need to use systems adequately. In a parallel context, utopians may recognize that new technologies cause new problems, but firmly believe that additional technology will solve the problem (e.g., Rosetto [55]). Conversely, in antiutopian views people either have the appropriate skills, and they use them to manipulate others in a negative way or the author foresees general confusion about the use of technologies (58). Clearly, however, skill acquisition is neither as black nor as white as these would suppose. As illustrated throughout this chapter, the use of electronic information technologies will certainly expand. As it does, some will use the technologies effectively and some will not; some will use them for altruistic good and others for socially miscreant ends. Environmental managers will face the far more difficult questions of degree: is the public obtaining enough balanced information to make informed decisions. In turn, are agencies receiving enough balanced perspective on citizen values and goals to make informed decisions?

Imperatives of Social Change

Perhaps the most transparent aspect of both the utopian and antiutopian genres is their certainty that a specific social change is predetermined by the technology. However, environmental change is almost defined by contingencies. Environmental effects occur simultaneously on many different temporal and spatial scales, and affect both the structure and function of plant and animal communities. Thus, there are many likely environmental and social changes and technology alone does not predetermine the direction of change. It can, however, as demonstrated in this chapter, concretize changes that occur because of parallel influences. The pertinent question for environmental problem solvers is whether communication technologies will complement existing trends in environmental management, and thus institutionalize them or whether they can be used creatively to build a new paradigm.

Making Management Happen

We have outlined likely scenarios for the development of the Internet, related technologies, and environmental problem solving. However, we stand outside the traditions of utopian or antiutopian theorists in recognizing that no future scenario is predetermined. Information technology users have significant power to direct the course of these technologies. Consequently, environmental managers, problem solvers, activists, and concerned citizens may all take these cautions to heart in guiding their own use of these technologies. This section does not propose concrete solutions, but hopes to suggest that through initiative, creativity, and planning, the concerns addressed within this chapter may be, if not avoided entirely, controlled and turned into advantages.

Throughout this chapter we have outlined areas of deep concern in how these technologies will affect the future of environmental management. These concerns tend to fall into several categories: (1) areas in which Internet rhetoric eclipses reality, (2) areas in which Internet use is tending to conform to existing patterns (e.g., is conservative) despite the appearance of radical differences, (3) areas in which Internet may be creating new communication dynamics.

Mixing Healthy Skepticism with Initiative

In the first category falls examples such as the claim that everyone will soon be on-line, that user-friendly software will enable everyone access to the full range of the cybernet and that improved speed will improve decision making. We recommend caution and healthy skepticism toward these claims. Those involved in environmental problem solving may best be advised to take these claims with a grain of salt and to use these technologies knowing that they still represent a limited cross section of the population.

The second category, those in which Internet use appears surprisingly conservative, includes such areas as on-line search patterns, the struggle with too much knowledge, and equating information with knowledge. These areas may respond well to creative problem solving. Numerous opportunities exist to try to develop Internet's potential in these areas. The Internet can assist in ongoing processes such as complementing data searches, increasing the scope of public notices and improving the process of public comment. The operative words, though, are those that stress using the Internet as part and parcel of other strategies (e.g., *complementing* data searches): where it is used to link people with each other and with new information, but where these links are coordinated with other media and personal interaction.

Example successful models of Internet use come from school programs. In these, students follow the work of researchers or explorers through e-mail, but support this exposure with class projects, speakers, and other multimedia efforts. The same multimedia efforts may prove successful for environmental managers. For example, managers could work with local freenets to establish environmental information lines but simultaneously work with schools, community groups, and other organizations to integrate the flow of information and dialogue throughout a community.

Other options include offering environmental courses that are run partially through Internet and partially in person or in small discussion groups; organizing Internet watcher groups that help monitor the discussions, concerns, and level of knowledge of citizen groups; assisting NGOs in the development of their environmental educational resources (on-line or in other media). These specific examples are not important, but they illustrate that the ability to take the initiative in utilizing electronic communications is critical. They also illustrate that the Internet may be best used when it is a complement to face-to-face and multimedia efforts.

The third area outlined is that in which Internet may create new dimensions to existing problems. These would include the probability of increased pressure from diverse political action groups and the possibility that widely available data may be interpreted poorly by those without the tacit knowledge to understand them. These areas have their pros and cons, as described in this chapter, but environmental problem solvers must at least be cognizant of the possibilities and must plan ahead in order to address these changes. Ironically, the best way to divert these possible effects would likely be through effective coordination meetings (i.e., regular personal contact) with representatives from political action groups (including businesses and industry), scientists, extension personnel, and state and federal natural resource managers.

Internet Intermediaries

One of the largest constraints experienced by any Internet user is the time required to stay abreast of issues. Simply because of the volume of information available on-line, and because of the steepness of each individual user's learning curve, it is unrealistic to expect that individual managers will substantively change their decision-making practices due to the advent of electronic communications. Many people in positions of public responsibility rarely use their e-mail because of the large volume of unrelated messages routinely received.

However, writing off Internet resources on the basis of this marginalization argument is unfair to the resource and shortsighted. A more realistic approach, being explored intentionally in some cases and by default in others is to have an "Internet librarian" (intermediary) available. Much as current librarians do, this person would be able to stay current with the growth in on-line resources and technologies, know the information resources most likely to be useful to those within an agency or business, and provide timely advice to those needing information.

Adopting the Principles of Other Programs to Use On-line

Another avenue deserving exploration is the use of computer-mediated problem solving. The success of these programs has been well documented under controlled circumstances (22–24). It might be possible to expand the audience, modify the "classroom environment," and incorporate conflict management techniques into guided Internet-based conferences or public meetings. Importantly, the success of these models rests on facilitation, continuing participation, and guided exercises, precisely the three aspects missing from most Internet contacts. However, there is certainly no functional barrier to using the Internet in this way, and there is great potential for managers involved in environmental policy setting to incorporate the principles of these workshops and software programs into a guided Internet discussion.

Conclusions

We have shown that environmental decision making has changed in the past 25 years, growing away from a comprehensive, integrated approach toward one dominated by sectoral and single-issue legislation. Despite this trend, the preferred paradigm for effective environmental legislation remains one focusing on multisectoral integration, broad public participation and effective education (knowledge).

We have also demonstrated that the ways in which the Internet, on-line data bases, and other communication technologies are being used are, overall, far different from their democratizing, revolutionizing potential. Will the Internet accentuate the existing trend toward single-issue decision making? Does Internet have the potential to tip the scales toward any single party involved in the decision-making process? Does Internet's very size serve as an impediment to knowledge gathering? We suggest that the answer to each of these questions is yes, if environmental managers and problem solvers do not become actively involved in shaping the ways these new technologies are used.

Many of these questions posed in this chapter are not unique to the Internet, but reflect the way information has been gathered and shared for centuries.

Illiteracy has kept knowledge in the hands of the elite for millennia: access to public libraries is an even newer phenomenon; even with card catalogues and only a few libraries at their disposal, people rarely exhausted the range of information sources available. Information has always been shared most extensively with like groups; that is the whole point of journals, magazines, and professional societies.

What is unique about the advent of the Internet is that the technology has raised large, and probably unrealistic, expectations about our improved abilities to communicate and retrieve information. For example, information technology proponents claim that the technology itself is transforming and will transcend educational boundaries. However, inherent learning styles have still been shown to have a fundamental influence on how users searched for information (and what information they found) (61). Again, the importance of learning styles is not new: what is significant is the expectation that technology will eliminate these inherent patterns. It is the gap between those technologically derived expectations and the very conventional ways we use those tools that may create the most important impediments to good decision making. Management agencies and users must recognize this gap and plan carefully to optimize their use of the technology.

Another important, possible, impediment to decision making is related to the ways in which Internet is suited to increasing public pressure and public mobilization around issues. While at one level this is a positive scenario enabling improved access to environmental decision making and improved public participation, it also clearly runs the risk of relegating environmental decision making to the issue of the day (i.e., decision made at the mercy of the best-mobilized, most well represented on-line community). As reflected in Torsten Hagerstrad's (1991) model of information diffusion (31), as information diffuses it becomes distorted and ephemeral. Thus, the Internet may promote wider information dissemination, but the information being transferred becomes increasingly shallow and, thus, poorly suited for equitable environmental decision making. Successful models of environmental decision making depend on depth of understanding, getting parties to communicate, and arriving at a consensus with regard to environmental risks and economic potentials. Often this is a lengthy process involving numerous opportunities for face-to-face contact and full expression of frustrations and feelings.

As an information-gathering tool, the Internet is certainly useful in support of the decision-making processes. However, the difficulty with seeing the Internet as the great democratizer and enabler of decisions is in equating its information retrieval facility with transfer of knowledge, promotion of understanding, and encouragement of cooperative decision making. Internet will impede decision making if it is seen as a panacea: it may complement it if our expectations are technology-realistic.

Acknowledgments

We are grateful to all our faceless electronic colleagues who contributed to the ideas and development of this chapter. Dr. Chet Grycz, Mrs. Barbara Rhodes, Dr. Rich Liroff, Chris Stuart, Dr. Barbara Knuth, and Jean Bonhotal reviewed drafts of the manuscript and provided helpful comments. Supported in part by the Minnesota Experiment Station under project 42-24 of the MacIntirre Stennis Cooperative Forestry Act. Paper 21, xxxx of the Scientific Journal Series, Minnesota Experiment Station.

References

1. Pfaffenberger B. Democratizing information: on-line databases and the rise of end-user searching. Boston: G.K. Hall, 1990.
2. Blaug E. Use of the environmental assessment by federal agencies in NEPA implementation. Environ Prof 1993;15:57–65.
3. Boggs J. Procedural vs. substantive in NEPA law: cutting the gordian knot. Environ Prof 1993;15:25–34.
4. Dickerson W, Montgomery J. Substantive scientific and technical guidance for NEPA analysis: pitfalls in the real world. Environ Prof 1993;15:7–11.
5. Rabe BG. Fragmentation and integration in state environmental management. Washington, DC: The Conservation Foundation, 1986.
6. Ensimger JT, McLean RB. Reasons and strategies for more effective NEPA implementation. Environ Prof 1993;15:45–56.
7. Malik M, Bartlett R. Formal guidance for the use of science in EIA: analysis of agency procedures for implementing NEPA. Environ Prof 1993;15:34–45.
8. Howell D. Scientific literacy and environmental policy: the missing prerequisite for sound decision making. New York: Quorum Books, 1992.
9. Rosenbaum W. Environmental Politics and policy. Second edition. Washington, DC: Congressional Quarterly Press, 1991.
10. Smith Z. The environmental policy paradox. New York: Prentice Hall, 1992.
11. Kraft ME, Vig NJ. Environmental policy from the seventies to the nineties: continuity and change. In: Vig NJ, Kraft ME, eds. Environmental policy in the 1990s. Washington, DC: Congressional Quarterly Press, 1990:3–32.
12. Kraft ME. Environmental gridlock: searching for consensus in Congress. In: Vig NJ, Kraft ME, eds. Environmental Policy in the 1990s. Washington, DC: Congressional Quarterly Press, 1990:103–124.
13. Ingram HM, Mann DE. Interest groups and environmental policy. In: Lester JP, ed. Environmental politics and policy: theories and evidence. Durham: Duke University Press, 1992:135–157.
14. Wenner LM. The courts and environmental policy. In: Lester JP, ed. Environmental politics and policy: theories and evidence. Durham: Duke University Press, 1992:238–260.
15. Hall BW. Information technology and global learning for sustainable development: promise and problems. Altern Social Transform Hum Govern 1994;19(1):99–132.
16. Mitchell RC. Public opinion and the green lobby: poised for the 1990s? In: Vig NJ, Kraft ME, eds. Environmental policy in the 1990s. Washington, DC: Congressional Quarterly Press, 1990:81–102.
17. Mazmanian D, Marell D. The NIMBY syndrome: facility siting and the failure of democratic discourse. In: Vig NJ, Kraft ME, eds. Environmental policy in the 1990s. Washington, DC: Congressional Quarterly Press, 1990:125–144.
18. Fumento M. Science under siege: balancing technology and the environment. New York: William Morrow, 1993.
19. Chartrand RL, Ketcham KC. Opportunities for the use of information resources and

advanced technologies in Congress: a study for the joint committee on the organization of Congress. Inform Soc 1994;10:181–222.

20. Perry J. Water quality in the 21st century: Proactive management at the ecosystem level, landscape and decade scale. Conf Proc: Spanish Limnol Soc, 1993.

21. Dryzek JS, Lester JP. Alternative views of the environmental problematic. In: Lester JP, ed. Environmental politics and policy: theories and evidence. Durham: Duke University Press, 1992:314–330.

22. Mitchell B, ed. Integrated water management: International experiences and perspectives. London: Bellhaven Press, 1990.

23. Marshall P, Atkinson D, Ee L, McKay J. Computer-aided group problem solving: an overview of successful methodologies. Environ Prof 1993;15:159–163.

24. Friend J. The strategic choice approach in environmental policy-making. The Environ Prof 1993;15:164–175.

25. Bonnicksen T. The impact process: a computer-aided group decision-making procedure for resolving complex issues. Environ Prof 1993;15:186–193.

26. Bartlett R. Comprehensive environmental decision making: can it work? In: Vig NJ, Kraft ME, eds. Environmental policy in the 1990s. Washington, DC: Congressional Quarterly Press, 1990:235–256.

27. von Droste B. Introduction: sustainable development for all. Nature Resour 1992;28(1):2–4.

28. Hanson AJ, Regallet G. Communicating for sustainable development. Nature Resour 1992;28(1):35–43.

29. Perry JP, Vanderklein EL, Strecansky B. Environmental management in Slovakia: teaching people a new way to think. Environ Prof 1995: in press.

30. Bonine JE. Internet and environmental law. Internet Soc News 1992;1(1):26–27.

31. Barker ML and Soyez D. Think locally, act globally? the transnationalization of Canadian resource-use conflicts. Environment 1994;36(5):12–36.

32. Tenner E. Learning from the net. Wilson Quart Summer 1994:18–28.

33. Anderson C. The rocky road to a data highway. Science 1993;260:1064–1065.

34. Maddox T. The cultural consequences of the information superhighway. Wilson Quart Summer 1994:29–35.

35. Schwartz M, Quarterman J. The changing global Internet service infrastructure. Internet Res 1993;3(1):8–25.

36. Pavlik J. Citizen access, involvement and freedom of expression in an electronic environment. In: Williams F, Pavlik J, eds. The people's right to know: media democracy and the information highway. Hillsdale, NY: Lawrence Erlbaum, 1994:139–162.

37. Gomery D. 1994. In search of the cybermarket. Wilson Quart Summer 1994:9–17.

38. "Edupage" 11/29/94. Edupage@ivory.educom.edu.

39. Doctor RD. Seeking equity in the national information infrastructure. Internet Res 1994;4(3):9–22.

40. Clarke R. Electronic support for the practice of research. Inform Soc 1994;10:25–42.

41. Webster F. What information society? Inform Soc 1994;10:1–23.

42. Nilan M. Speculations on the impact of global electronic networks on human cognition and human organization. Internet Res 1993;3(1):47–56.

43. Wylie M. Women and Netiquette. Digital media perspective: a Seybold report. December 13, 1994.

44. Postman N. Technopoly: the surrender of culture to technology. New York: Alfred K. Knopf, 1992.

45. Deetz S. Representation of interest and the new communication technologies: issues in democracy and policy. In: Medhurst M, Gonzalez A, Peterson TR, eds. Communication and the culture of technology. Pullman: Washington State University Press, 1990:43–62.

46. Tomasaitis N. Why women don't log on. ComputerLife Feb 1995:139.

47. Stern PC, Dietz T, Kalof L. Value orientations, gender and environmental concern. Environ Behavior 1993;25(3):322–348.

48. MacEvoy B. Personal communication. 1995

49. *Investors Business Daily.* January 11, 1995:A4.

50. Suber P. The database paradox: unlimited information and the false blessing of objectivity. Library Hi Tech 1992;10(4):51–57.

51. Hart CA. A learning organization perspective on training: critical success factors for Internet implementation. Internet Res 1994;4(3):36–44.

52. Koebrick A. Communities, sustainable development and the national information infrastructure. Paper commissioned for the Bauman Foundation at the request of the Office of Management and Budget. Available from koeba@ruby.ils.unc.edu. 1994.

53. Kilger M. The digital individual. Inform Soc 1994;10:93–99.

54. Clarke R. The digital persona and its application to data surveillance. Inform Soc 1994;10: 77–92.

55. Rosetto L. Oh, what techno wonders. *USA Today* September 15, 1994:11A.

56. Max DT. The end of the book? *Atlantic Monthly* 1994;274(3):61–71.

57. Hiskes AL, Hiskes RP. Science, technology and policy decisions. Boulder/London: Westview Press, 1986.

58. Kling R. Reading "all about" computerization: how genre conventions shape nonfiction social analyses. Inform Soc 1994;10:147–172.

59. Killingsworth M, Palmer J. Ecospeak: rhetoric and environmental politics in America. Illinois: Southern Illinois University Press, 1992.

60. Segal HP. Technological utopianism in American culture. Chicago: The University of Chicago Press, 1985.

61. Wood F, Ford N, Walsh C. The effect of postings information on searching behavior. J Inform Sci 1994;20(1):29–40.

The Implications of Scientific

Uncertainty for Environmental Law

13

Donald A. Brown
Patrick Zaepfel

Introduction

Other chapters explain the pervasive nature of scientific uncertainty in predicting the effects of human actions on human health and the environment. This chapter will describe how the law has treated scientific uncertainty in environmental controversies. The chapter demonstrates that the law approaches scientific uncertainty in environmental matters quite differently depending on (1) the type of legal action, (2) whether scientific evidence is offered in the form of an expert opinion, (3) whether the legal action was initiated as an appeal from a government decision, and (4) the specific statute that is the focus of the legal action.

This analysis demonstrates that the law relating to scientific uncertainty in environmental matters is inconsistent and often incoherent. Because the law assumes that scientific procedures can separate fact from error, the law deals very poorly with matters in which there is pervasive scientific uncertainty about the consequences of human action. Moreover, it is often argued that for human actions that create serious but uncertain threats to human health or the environment, the law should authorize governments or persons to take or require others to take preventative action. This view is often referred to as the precautionary principle. This chapter will describe how much of current law on scientific uncertainty in environmental matters is antithetical to the precautionary principle.

This chapter begins with a review of the precautionary principle. Next, it analyzes the law of admissibility of scientific evidence in court proceedings. Then there is an analysis of the law of expert testimony. This is followed by a review of the law of scientific uncertainty in government decisions. Next, the chapter concludes with a description of the law relating to scientific uncertainty in the preparation of environmental impact statements.

The Precautionary Principle

One of the international agreements reached at the United Nations Conference on Environment and Development, generally referred to as the Earth Summit, in Rio de Janeiro in June of 1992 was the Rio Declaration on Environment and Development. The Rio Declaration is considered to be "soft law," which means that it is not directly enforceable, but it is expected to act as a normative goal that nations should follow in adopting more specific laws and regulations in the future. Principle 15 of the Rio Declaration states (1):

> In order to protect the environment, the precautionary approach shall be widely applied by States according to their capabilities. Where there are threats of serious or irreversible damage, lack of full scientific certainty shall not be used as a reason for postponing cost-effective measures to prevent environmental damage.

The precautionary principle establishes the norm that nations should take steps to protect the environment before potentially serious harmful effects of proposed or existing actions are fully proven. The precautionary principle departs from many traditional approaches of law that presume that no harm has occurred until a party can demonstrate damage and causation. For this reason, the precautionary principle represents a fundamental shift in many legal principles that might otherwise apply to environmental controversies. Most importantly, under a precautionary approach to environmental law, the burden of proof is shifted to the party who seeks to undertake potentially harmful activity (2). Therefore, if governments follow this principle, they will act so that the benefit of doubt is given to human health and the environment rather than the polluting activities (3).

If laws are to be enacted to implement the precautionary principle in environmental matters, they will authorize action that err on the side of protecting human health and the environment. However, the duty to err on the side of human health and environmental protection under the precautionary principle is in proportion to what is at stake; it is not absolute. If a proposed action could possibly seriously damage ecosystems, then the duty is stronger to err on the side of protection. The greater the threat, the stronger is the duty to take protective action.

Hans Jonas argues that in predicting long range consequences of human actions where there are serious possible consequences, humans should be guided by the "heuristic of fear" in predicting consequences (4). That is, humans should give preference to the bad over good predictions. Particularly where the use of technology could cause serious irreversible consequences, where the stakes are high, decision makers should give more weight to prognosis of doom than of bliss. According to Jonas, the philosophical reason for the duty to give more weight to the prediction of harm is premised on the notion that present generations do not have a right to gamble with the interests of other generations nor to act so that life on Earth is jeopardized. How does existing law on scientific uncertainty measure up to the precautionary principle? This question will be examined in succeeding sections.

Scientific Evidence in Legal Proceedings

Because environmental decisions must be made in the face of pervasive scientific uncertainty, legal rules on the use of scientific evidence in court proceedings often determine when environmental laws may be enforced or implemented. If rules of evidence restrict the use of scientific evidence in court proceedings to that which is highly certain, enforcing or implementing an environmental law may be impossible in matters where certain scientific evidence is theoretically or practically unavailable. Therefore, rules on the use of scientific evidence in legal proceedings must be recognized as important normative choices about when environmental laws may be enforced. A review of law on the use of scientific evidence in court proceedings in the United States reveals that rules differ, depending on the type of proceeding or how the evidence is presented in the proceeding.

Legal rules often are designed to implement societal norms. For instance, in criminal proceedings, the prosecution must prove that the accused is guilty beyond a reasonable doubt. In contrast, in civil proceedings the plaintiff typically has the burden of proving his or her case by a mere preponderance of the evidence. Thus, the burden of proof in civil cases is usually understood to be much less rigorous than that required in criminal proceedings. In criminal proceedings, the law implements the social norm that is preferable to let a guilty person go free than imprison an innocent person. In civil disputes, the law assumes that the person with the better proof should prevail and therefore allows a party to prevail even when that party's evidence is inconclusive.

Similarly, scientists follow rules designed to implement the societal norm that science should find "truth." Because of this norm, in scientific research, scientists often follow, by convention, the 95% confidence rule (5). This rule means that scientists should not conclude that they have proven causation unless the evidence describes a response between cause and effect at a 95% confidence level. When following this rule, scientists are deciding to be 95% certain that a particular hypothesis is reliable (5). The relatively high level of certainty implied by the 95% confidence rule implements the social norm that science should not draw false conclusions or speculate.

As the following analysis will demonstrate, the law dealing with the use of scientific evidence in civil proceedings often confuses these criminal, civil, and scientific norms and is rarely consistent with the norm implied by the precautionary principle.

Admissibility of Scientific Evidence in Tort Law and Other Court Actions to Restrain Threatened Action

A "tort" is a wrongful act that causes a recoverable damage. In the United States, legal actions to recover environmental damages or to obtain a court order to prevent harm from pollution may be brought under a variety of tort theories, including public and private nuisance, negligence, and trespass. Persons who bring tort actions must prove their case through the introduction of scientific evidence that meets certain minimum standards of reliability in order to be admitted. Similarly, when government goes to court to restrain persons who are about to cause environmental harm under a variety of environmental laws, courts will require that scientific evidence meet these same standards of reliability. In civil proceedings in federal courts in the United States, courts follow the federal rules of civil procedure on the admissibility of scientific evidence.

In 1993, the United States Supreme Court changed the standard of admissibility of scientific evidence in all civil proceedings in *Daubert v. Merrell Dow Pharmaceuticals* (6). In *Daubert,* the U.S. Supreme Court rejected a 70-year-old test for admissibility of scientific evidence that had been established by the case of *Frye v. United States* (7). *Frye* was a criminal case in which the evidentiary question before the court was whether the results of systolic blood pressure test, a precursor of the modern lie detector test, should be admitted. The *Frye* test allowed into evidence only that scientific evidence that had gained "general acceptance" in the particular scientific field in which it belongs. Because the systolic blood pressure test had not been accepted by the scientific community, the court in *Frye* held the test to be inadmissible.

The *Frye* test was eventually applied to civil as well as criminal cases in the United States. Therefore, courts prohibited the introduction of scientific evidence in civil proceedings unless the scientific evidence had reached high levels of acceptance by the scientific community. Under *Frye,* causal evidence of damage to the environment or human health could not be admitted in proceedings unless high levels of certainty of causation had been acknowledged by the relevant scientific community. As a result, persons who had a reasonable basis for a belief that they may had been harmed by exposure to chemicals could usually not recover damages if the evidence did not establish causation with a high level of certainty. For example, if someone with cancer could prove that they had been exposed to chemicals that caused cancer in animals, they would nevertheless be unable to recover damages without reliable epidemiologic proof of causation of cancer in humans under *Frye.* Yet, epidemiologic evidence of causation of cancer in humans is sometimes practically impossible to obtain due to ethical limitations on testing humans and the need for a very large sample size to establish statistical significance for substances that do not cause injury for every exposure. Therefore, a rule that excludes animal studies in disputes about causes of cancer in humans makes recovery difficult if not impossible for substances that may actually cause cancer.

A stringent rule, like *Frye,* that limits scientific evidence to scientifically certain evidence arguably makes sense in criminal cases where the law assumes that a strong presumption of innocence is in society's best interest. However, such a rule is more problematic in civil proceedings where the appropriate evidentiary norm has traditionally been that the party should prevail that has the stronger evidence. Even though *Frye* is more obviously appropriate in criminal, rather than civil, litigation, courts have applied *Frye* to civil proceedings because they have been concerned that scientific evidence that is not reliable could potentially confuse a jury. However, in seeking to limit scientific evidence to that which is reliable, the *Frye* court followed the scientific norm designed to limit scientific statements to those that are truthful. Moreover, the *Frye* rule is clearly inconsistent with the precautionary principle because it prevents legal action in environmental cases where the scientific evidence is uncertain.

Although *Frye* is no longer the federal rule on the admissibility of scientific evidence, it is still the test of admissibility in some states, for example, Pennsylvania and California. For example, in *Commonwealth v. Crews,* the Pennsylvania Supreme Court refused to admit DNA evidence following *Frye* (8).

In *Daubert,* the United States Supreme Court liberalized the *Frye* test of admissibility of scientific evidence. The court announced the following four-pronged analysis to assist the court in determining whether the evidence is reliable enough to be admitted:

1. Is the scientific method used by the expert to derive an opinion capable of being tested? That is, is the method capable of being shown to be false? If the method is not capable of being shown to be false, then the method is not scientific and, hence, not admissible. Examples of common environmental methods of analysis that cannot be verified are many, including environmental models.

2. Has the scientific method been subjected to peer review publication? Publication only strengthens admissibility; nonpublication does not impart inadmissibility. Publication provides public scrutiny of the method, increasing the likelihood that substantive flaws in the method will be detected. Yet, many analytical tests and assumptions used in environmental matters have never been subjected to peer review. For instance, in risk assessment, analysts often make assumptions about the toxicity of untested chemicals by drawing analogies from other chemicals that have been tested for toxicologic properties.

3. Does the method have a known error rate, or do outside standards exist that can monitor the method? Obviously, a low error rate will encourage admissibility. Many analytical tests and assumptions used in environmental matters, such as environmental models, have no known error rate.

4. Does the method have general acceptance in the relevant scientific community? Acceptance by the scientific community is not needed for admissibility, as it was under *Frye*. A method, however, with only minimal support in the general scientific community properly may be viewed with skepticism.

Courts that have recently applied *Daubert* have made it clear that scientific evidence must still be shown to be reliable as a precondition for admission, even though the test for reliability is more relaxed than *Frye*. For instance, in *Conde v. Velsicol*, the court refused to admit evidence that chlordane sprayed on a home foundation to eradicate termites caused illness (9). Similarly, in *Re: Joint Eastern & Southern Dist. Asbestos Litigation,* the court excluded evidence showing some relationship between the plaintiff's colorectal cancer and asbestos exposure (10). Yet, some courts have shown more flexibility to admit scientific evidence after *Daubert*. For instance, in *Hopkins v. Dow Corning*, the court allowed an expert to testify on the toxicity of silicone breast implants where the conclusions about toxicity were based in part on immune system disorders found in animal studies (11).

Although *Daubert* may have liberalized the rule on the admissibility of scientific evidence in civil proceedings, recent interpretations of *Daubert* still prohibit evidence that merely establish possibilities. That is, scientific evidence that establishes a reasonable basis for concern about harm, but relies in part on speculation, is probably still not admissible after *Daubert*. If this is so, the rules on the admissibility of scientific evidence in civil proceedings in the United States are inconsistent with the Rio Declaration's precautionary principle. Although such a rule is arguably defensible under the norm that those that have not actually caused damage should not be forced to pay damages for which they are not culpable, the current legal test of admissibility of scientific evidence makes recovery virtually impossible for possible but unproven threats. For instance, legal actions that stem from exposure to substances where there is reasonable basis for concern, but where conclusive proof does not exist, will probably be unsuccessful even after *Daubert*.

Expert Testimony

Assuming that a civil plaintiff has overcome any problems of admissibility of scientific evidence, the plaintiff must prove his or her case by a preponderance of the evidence. In cases involving scientific evidence, experts are usually called upon to give opinions about the meaning of the scientific evidence. In such cases, expert opinions will be allowed if the expert can testify that the opinion can be given with a ''reasonable degree of scientific certainty'' (12).

Therefore, even if all scientific evidence passes a *Daubert* test and is, therefore, admissible, testimony by an expert relying on scientific evidence will be admitted only if the expert can testify that his or her opinion is given with a reasonable degree of scientific certainty. Courts often interpret this requirement for reasonable scientific certainty so that it is consistent with the preponderance of the evidence test applicable in civil litigation. As a result, expert opinions are admitted only if the expert can testify that the conclusions embedded in his or her opinion are "more probable than not." If experts cannot establish that the evidence on which they rely allows them to testify that their conclusions are "more probable than not" then the opinions are usually not admissible. Thus, an expert that attempts to show that harm is merely possible will not be allowed to testify. For this reason, scientific evidence that establishes the possibility of harm, but does not establish that the harm is more probable than not, will not be admitted through expert testimony. Therefore, expert testimony about potential environmental harm or danger to human health may not be admissible in any civil proceeding unless the evidence supports the harm or damage will occur by a preponderance of the evidence. Thus, the rules on the admissibility of expert testimony may be inconsistent with the precautionary principle.

Administrative Action

Laws dealing with environmental matters avoid many of the admissibility problems encountered in other civil actions by giving governments certain powers to take legal action. These powers include the power to take action if the government determines that an activity creates a "threat" to human health or the environment and the authority to adopt enforceable standards. Through a grant of power to government to take action where a "threat" of environmental damage exists, the government avoids the problem of showing causation of damage or harm in fact. For instance, the Comprehensive Environmental Responsibility and Liability Act (usually referred to as Superfund) authorizes injunctive action upon a showing of "a threat of imminent and substantial endangerment." In one case, the court held that "the United States need not prove an actual imminent and substantial endangerment, but may obtain relief on proof that the danger may exist (13).

Similarly, if the government has the power to enforce standards directly, the standards are presumed to be a definition of environmental harm, thus avoiding evidentiary problems associated with showing actual harm. For instance, under the Clean Water Act, states in the United States set water-quality standards through rule making; and environmental harm is presumed if government can show that an action caused a violation of these water-quality standards.

Another way in which environmental law avoids the admissibility problems of uncertain scientific evidence is to limit judicial review of administrative actions to the administrative record created by the administrative agency. For instance, in legal disputes about cleanup remedies under the Superfund law, it was held that "[i]n considering any objections to the Environmental Protection

Agency's cleanup of spilled waste oil contaminated with PCBs, the court may only look to the administrative record" (14). In cases where judicial review is from the administrative record, the court does not call witnesses nor admit evidence, but simply reviews the record of public comment about the proposed action prepared by the administrative agency. Because there are no witnesses or evidence heard in such court proceedings, there are no problems of admissibility of scientific evidence. In record review matters, the agency's actions are afforded deference and must be upheld if they are based on relevant factors and are not a clear error of judgment (15). In one record review case, the court concluded (16):

> Where the environmental protection statute is precautionary in nature, evidence is difficult to come by, uncertain, or conflicting because it is on the frontier of scientific knowledge, regulations are designed to protect public health, and the decision is that of an expert administrator, the court will not demand step-by-step proof of cause and effect; but the administrator may apply his expertise to draw conclusions from suspected, but not completely substantiated relationships between facts, from trends among facts, from theoretical projections from imperfect data, from probative preliminary data not yet certifiable as fact, and the like.

Much of the body of environmental law in the United States authorizes the government to take precautionary action in writing regulations and making case-specific decisions under these regulations. Although under these laws administrators may decide questions of uncertainty in favor of environmental protection, most laws do not require that decision makers resolve questions of uncertainty in this way. Therefore, if an administrator believes for political or other reasons that he or she should not regulate until the scientific basis for regulation attains high levels of certainty, environmental laws will not force a precautionary regulatory action.

Because of pervasive scientific uncertainty in environmental problems, technical experts within government have often refused to act out of fear that they will enrage legislators who will have them fired if it is discovered that they have imposed insupportable costs upon constituents. This reluctance is also consistent with most scientific training. The scientist is trained to be very conservative in asserting cause-and-effect relationships. Many traditionally trained scientists will not act quickly if there is uncertainty about the cause of an environmental problem. If a position, once taken, is later discredited by subsequent scientific research, the technical person who made conclusions about the cause-and-effect relationship may suffer peer sanctions for being associated with a faulty scientific hypothesis. That is, they will be accused of being "bad scientists." Because scientists are taught to be silent in the absence of proof, scientific norms may be inconsistent with public policy norms. Therefore, some administrators may be inclined to take no protective action until matters of scientific uncertainty are resolved. This inclination derives from the confusion created by the conflict between the norms implied by the precautionary principle and the scientific norms of normal scientific research.

Although many do, not all environmental laws clearly prescribe a precautionary approach. Some environmental laws require a finding of environmental harm as a factual prerequisite for protective regulatory action. For instance, the law that regulates pesticides in the United States, the Federal Insecticide, Fungicide, and Rodenticide Act (FIFRA), requires that the administrator of Environmental Protection Agency (EPA) for pesticides in use at the time FIFRA was passed, determine that a pesticide causes an unreasonable risk to human health or the environment. This determination must be made before the administrator may take action to limit the use of the pesticide (17). Although under FIFRA the administrator need only find "risk" rather than actual harm to make the determination of risk, he or she must rely on "validated tests" or other "significant evidence" (18). "Validated tests" are defined to be tests that are "consistent with accepted scientific procedures" (19). "Significant evidence" is defined to be "factually significant information . . ." (19). Therefore, EPA's determination of risk of harm must be based on valid scientific tests, a standard that is likely to be interpreted as one that requires normal scientific proof. Because of the need to find evidence of harm before taking action, FIFRA has often been criticized for putting the burden of proof on the government rather than on the manufacturer of the pesticides. As a result, EPA has been often been severely criticized for the slowness of reviewing the thousands of chemicals that were in commerce when FIFRA was passed.

When environmental science is uncertain about the environmental consequences of human action, insisting on high levels of scientific proof before government action may be taken is a prescriptive rule that puts the burden of proof on government decision makers and protects the status quo. Such a rule may prevent protective government action where there is a reasonable basis for concern, but where science is uncertain about the consequences of certain human activities. The standard of proof that should be required of regulatory action is an ethical question, not a scientific one. If society lets scientific standards dominate legal institutions, it is making ethical choices that may be inconsistent with the precautionary principle. Although insisting on rigorous certainty may make sense in criminal cases where society wants to implement presumptions of innocence, more flexible standards of admissibility are appropriate in matters where government is expected to act according to the precautionary principle.

Those wishing to challenge government action for either its use of less than certain evidence or its failure to act on evidence of harm need to show that the government action is arbitrary or capricious, contrary to law, or based on insubstantial evidence in the administrative record (20). Because the burden of proof is on the challenger, both those who wish to challenge government's overreliance on speculative evidence and those who desire to challenge for failure to act where there is some evidence of potential harm will have great difficulty in showing that the government acted "arbitrarily and capriciously." This is so because, generally speaking, courts tend to defer to the agency's decisions in scientific disputes litigated from administrative records (21). However, courts will sometimes overturn agency actions if they feel that the agency has failed to demonstrate in the administrative record that an adequate scientific basis exists for its decision. For instance, courts have upheld appeals of EPA decisions when (1) EPA failed to do adequate testing of a substance before listing it as a hazardous waste; (2) EPA made a decision about the toxicity of a substance at a site contaminated with hazardous substances without determining whether the substance was present in highly toxic or low toxic form (22, 23).

Cases where government actions are overturned for failure to provide adequate scientific basis for action are more common than cases that are successful challenges to government's failure to act in the face of uncertain evidence of harm. That is, courts sometimes will overturn government action if the government found potential harm while relying on an inadequate scientific basis for the determination, but courts rarely, if ever, require government to act after determining that uncertain threats exist. (See discussion of Environmental Impact Statements.) Courts tend to look on false positive (an erroneous conclusion that an activity is harmful) with greater scrutiny than false negative (an erroneous conclusion that actionable harm exists when evidence of harm is uncertain). However, in both cases, courts tend to defer to government expertise.

Mathematical Models and Environmental Decisions

Mathematical models are frequently used in developing environmental regulations and in making day-to-day environmental decisions. In the United States, Clean Air Act implementation relies on the use of models in permitting decisions, enforcement, and long-range planning. The Superfund program and the permitting process under the Resource Conservation and Recovery Act use models to determine groundwater flow, predict transport of hazardous pollutants, assess risk, rank hazardous sites, and determine natural resource damages. Clean Water Act implementation relies on models to set effluent limitations for discharges into surface waters. The ability to predict environmental impacts under the National Environmental Policy Act also depends on the use of a variety of different types of models.

These models may never achieve levels of certainty achieved in other scientific endeavors because (1) ecologic systems are open systems rather than the closed systems described by the models. Because ecologic systems are open, the models fail to deal with unmeasurable and underdetermined parameters as well as certain cause-and-effect relationships (24); (2) ecologic models cannot be verified or validated in ways that other scientific processes can be tested (24); and (3) models usually must make simplifying assumptions for theoretical and practical reasons.

Courts have traditionally deferred to agency expertise in challenges to mathematical models (25). For instance in *Edison Electric Institute v. EPA,* the court upheld EPA's decision not to account for biodegradation of chloroform in a subsurface groundwater model even when EPA admitted problems with the model (26). In upholding the model, the court deferred to the agency's substantive expertise to weigh limitations of the models used by EPA.

Although most courts have given the benefit of the doubt to the government in its use of models, judicial review of computer models in environmental decision making has been uneven (25). For instance, in *South Terminal v. EPA,* petitioners successfully challenged EPA's use of a model used in the Metropolitan Boston Air Quality Transportation Control Plan (27). In this case, the court found that EPA relied on insufficient evidence in constructing and applying an air pollution model. Similarly, in *Cleveland Electric v. EPA,* petitioners successfully challenged the coefficients used to represent assumed weather conditions in air pollution models (28). In *Chemical Manufacturing Assn. v. EPA,* the court overturned an EPA model used for a chemical's air dispersion after the court concluded that the petitioners had successfully demonstrated that the model inappropriately assumed that the chemical would be dispersed as a gas rather than as an aerosol (29).

How courts review models is often a function of the judge's expectation about the degree to which government decisions should be based on sound science. If judges do not understand the inevitable imprecision entailed by the use of any environmental model, they are likely to decide cases in a way that undermined the precautionary principle.

When courts defer to government expertise, the reasons given by the court usually have to do with the greater expertise of the administrative agency, rather than a recognition that an agency is applying the appropriate societal norm to uncertain scientific evidence. Some mathematical models are often described in multiple pages of complex mathematical formulae, beyond comprehension to all except the most mathematically literate. When science is this impenetrable to a judge, the decision is most likely to defer to agency judgments. To deal with such complexity, some have argued for science courts staffed by technically competent judges. Yet, increasing the technical literacy of the judges will not guarantee that the court is applying the appropriate societal norm to resolving the controversy in dispute.

The Duty of the Government to Speculate About Environmental Impacts in Environmental Impact Statements

Principle 17 of the Rio Declaration states that (1)

> Environmental impact assessment, as a national instrument, shall be under-
> taken for proposed activities that are likely to have a significant adverse
> impact on the environment and are subject to a decision of a competent
> national authority.

What does the law require of preparers of environmental impact statements (EISs) when they are faced with possible but uncertain environmental impacts? In the United States, EISs are required by the National Environmental Policy Act (NEPA) for government actions that have potential significant impacts on the environment. In addition, environmental impact assessment requirements similar to NEPA have been adopted by over 30 countries (30).

Under NEPA, courts take a "hard look" at the adequacy of the EIS and require that the EIS represent a "full disclosure" of the environmental impacts of the proposed action (30). To enforce the "hard look," courts distinguish between a "substantive" and a "procedural" challenge to an EIS (30). The duty to reject a project on environmental grounds is generally referred to as NEPA's "substantive" duty, while the duty to prepare an adequate EIS is a "procedural" duty.

The United States Supreme Court found in *Strycker's Bay Neighborhood Bay v. Karlen* that NEPA did not impose a substantive requirement that elevated environmental concerns over other concerns (31). Courts have, however, consistently held that NEPA requires that EIS analyses are based on a "good faith effort" of the agency to fully identify environmental impacts (30). NEPA is thus understood to create procedural duties to carefully examine potential environmental impacts, but no substantive duty to refrain from taking actions that have adverse environmental impacts. Thus, successful challenges to EISs under NEPA have almost always followed from failure to adequately examine potential impacts rather than from unwillingness to mitigate adverse environmental impacts. An EIS must fully explain its inquiry, analysis, and reasoning to survive a procedural challenge, but once having described the potential impacts, there is no duty to avoid environmentally damaging behavior (30).

Examples of recent cases that found violations of NEPA for failure to carefully examine potential environmental impacts of proposed federal actions are the following:

1. In *Public Service Co. v. Andrus,* the court found that the failure of an EIS to study the potential impacts of all nuclear activities at a national engineering laboratory, including existing and ongoing programs for shipment, receipt, processing, and storage of spent nuclear fuel in an EIS violated NEPA (32).

2. In *Portland Audubon Soc. v. Babbit,* the court found a decision of the Bureau of Land Management not to supplement the EIS when scientific evidence available to the Secretary of Interior raised significant new information relevant to effect of logging on spotted owl violated NEPA (33).

3. In *Seattle Audubon v. Mosely,* the court found that NEPA requires that an agency charged with preparing an EIS take a "hard look" at environmental consequences of a project, and that the EIS disclose risks, present alternatives, and respond with reasoned analysis to opinions of reputable scientists concerning hazards (34). The failure of the United States Forest Service to prepare an EIS that included a full-scale statistical viability study of northern spotted owl, without an explanation of why it was not necessary or feasible to do the study, violated NEPA.

However, recent cases that have found no violation of NEPA for failure of EIS to adequately study impacts include the following:

1. In *Conservation Law Found. v. Federal Highway Admin.,* the court held that NEPA does not require an EIS to analyze cumulative impacts where two previous EISs alerted public to arguably cumulative impacts involving the project (35).

2. In *Salmon River Concerned Citizens v. Robertson,* the court found that NEPA does not require it to decide whether an EIS is based on the best scientific methodology available, nor does it require the court to resolve disagreements among various scientists as to methodology (36). Therefore the court found no violation of NEPA by a United States Forest Service EIS on a proposed use of herbicides where there was a reasonably thorough discussion of the significant aspects of probable impacts, even where there was disagreement among experts about the extent of the impacts. The court upheld the Forest Service's analysis in deference to the agency's expertise and because the EIS contained a reasonably thorough discussion of significant aspects of the probable environmental consequences.

3. In *Northern Crawfish v. Fed. Highway Admin,* the court held that NEPA did not require that the Federal Highway Administration consider the impact of a highway on a park recently acquired by city where EIS adequately considered cumulative impacts and indirect impacts (37).

Congress did not address the problem of scientific uncertainty in identifying potential impacts when it passed NEPA. In 1978, the United States Council on Environmental Quality (CEQ), an agency created by NEPA to coordinate EIS preparation, adopted regulations that addressed the problem of scientific uncertainty (38). These regulations provided that if scientific uncertainty exists that can be cured by further research, the agency must do or commission the research. If the necessary research is exorbitantly expensive or beyond the state of the art the agency had to make it clear that uncertainty existed and include a "worst-case analysis" in its EIS (39). When the full extent of environmental impacts from an agency action was uncertain or unknown, an agency was under a duty to discuss the worst possible consequences and the probability of their occurrence. Federal agencies were reluctant to comply with the worst-case analysis regulation and asserted that the regulation involved excessive speculation and exceeded the rule of reason that defined the parameters of NEPA analysis.

One of the major issues regarding the worst-case regulation was whether the "scientific uncertainty" or "gaps in the relevant information" had to pertain to a reasonably foreseeable significant adverse environmental impact. In *Sierra Club v. Sigler,* the court rejected the government's contention that a threshold of reasonable foreseeability had to be met before the worst-case analysis regulation was invoked (40). The court held that the worst-case regulation was triggered by scientific uncertainty or gaps in the information that were "essential" or "important" to the federal agency's decision. The court determined that the probability of a worst-case outcome needed to be identified in the EIS, but that a high probability of occurrence was not a trigger to preparation of the worst-case analysis. Similarly in *Southern Oregon Citizens Against Toxic Sprays, Inc. v. Clark,* the court rejected the federal government's contention that the worst-case analysis was required only for reasonably foreseeable environmental impacts (41).

The CEQ was particularly critical of these judicial decisions that required federal agencies to discuss disastrous impacts when the federal agencies believed that there was no credible scientific evidence to support the probable occurrence of such impacts. The worst-case analysis rule received much criticism for delaying decision making and forcing the agency to speculate about low-probability events. In response, the rule subsequently was amended by CEQ in 1986 (42).

The new rule stated that if information is not available, the federal agency must make reasonable efforts, in light of overall costs and the state of the science, to obtain missing information that, in its judgment, is important to evaluating significant adverse but reasonably foreseeable impacts on the human environment (38). If the costs of obtaining the information are exorbitant or the means of obtaining it unknown, agencies had to

1. State that the information is incomplete or unavailable.
2. State the relevance of this information to evaluating reasonably foreseeable significant environmental impacts.
3. Summarize credible scientific evidence relevant to evaluating impacts.
4. Evaluate these impacts based on theoretical approaches to research methods generally accepted in the scientific community. Reasonably foreseeable impacts shall include low-probability catastrophic impacts if the analysis of the impact is supported by credible scientific evidence, is not based on pure conjecture, and is within the rule of reason (30).

Thus, the amended regulation stated that incomplete or unavailable information necessary to fully describe impacts must pertain to a reasonably foreseeable significant adverse environmental impact rather than all possible impacts. This position was advocated by the federal agencies and rejected by the federal courts in the litigation surrounding the worst-case regulation. The amended regulation appeared to be a victory for the federal agencies after their main judicial defeats. Nevertheless, because the amended regulation defined reasonably foreseeable as "impacts which have catastrophic consequences, even if their probability of occurrence is low," the amended regulation still required an analysis of low-probability events if they had serious potential consequences. As a result, some commentators saw the new rule as essentially the same as the old worst-case analysis rule determined by the courts (42). Yet, the new rule, while retaining the old obligation to describe low-probability events, attempts to limit speculation by requiring that analysis be based on "credible scientific evidence." However, one commentator argues that since credible scientific evidence allows "theoretical approaches or research methods that are generally accepted in the scientific community," the new rule is not much of a change from the original court-created one, despite the rule's attempt to limit speculation (43). However, the new rule may be interpreted by the courts as a limitation on an EIS preparer's obligations to identify serious possible impacts where scientific theory is weak because low-probability events must only be described where there is "credible scientific evidence."

If the new rule is interpreted to not require identification of catastrophic but low-probability events where one must rely in part on speculation, the new rule is inconsistent with the precautionary principle. This is so because the precautionary principle requires that nations not take actions that have "possible" serious adverse environmental impacts without regard to the ability to accurately predict the probability of the event's occurrence. The new rule implicitly presumes that an action is not harmful unless credible scientific evidence supports that it is harmful because it does not require identification of harmful effects unless there is credible scientific evidence of the harm. Under a precautionary approach to environmental law, the burden of proof is shifted to the party who seeks to undertake potentially harmful activity to prove that it is not harmful. Because the new rule does not require that an EIS identify possible harm unless there is credible evidence that identifies that harm, the rule gives the benefit of the doubt to the *status quo*. The new rule is therefore inconsistent with the precautionary principle.

Conclusions

The above analyses demonstrate that the law on scientific evidence often deals inadequately with human actions that may have serious but highly uncertain adverse impacts on human health and the environment. In matters of civil litigation, only evidence that reaches relatively high levels of certainty is admissible, while experts may not speculate about uncertain potential impacts. Although courts will defer to administrative agencies that make protective decisions in the face of uncertainty when laws authorize precautionary action, some laws create rigid scientific hurdles that must be overcome before the agency may take protective action. If, for political reasons, an agency follows rigid scientific norms under laws that authorize a precautionary approach, challengers will not be able to force government to take precautionary protective action because courts will defer to agency expertise especially where science is uncertain. Moreover, when governments prepare EISs there is no duty to identify possible serious impacts if there is no credible scientific basis for concluding that these impacts may actually occur.

Environmental law has failed to deal adequately with scientific uncertainty and as a result fails to implement the precautionary principle adopted at the Earth Summit. If the world is to implement the precautionary principle and fulfill its responsibilities to future generations, explicit direction must be provided by the laws that govern people's behavior. An analysis of the American statutes and case law demonstrates that this is simply not occurring. To implement better the goals of the Earth Summit, lawyers, policy makers, and scientists must begin to work cooperatively to create legal mechanisms that provide for scientific uncertainty.

References

1. United Nations. The Rio Declaration on Environment and Development, 1992.
2. Weintraub B. Science, international regulation, and the precautionary principle; setting standards and defining terms. NYU Environ Law J 1992;1:172.
3. Roht-Arriaza N. Precaution, participation, and the greening of international trade law. Oregon J Environ Law Litig 1992;7:57.
4. Jonas H. The imperative of responsibility. Chicago: University of Chicago Press, 1984.
5. Cranor CF. Regulating toxic substances: a philosophy of science and law. Oxford: Oxford University Press, 1993.
6. *Daubert v. Merrell Dow Pharmaceuticals, Inc,* 113 S.Ct. 2786, 1993.
7. *Frye v. United States,* 293 F. 1013 (D.C. Cir. 1923).
8. *Commonwealth v. Crews,* 640 A. 2d 395, 1994.
9. *Conde v. Velsicol,* 24 F. 3rd 809 (6th Cir, 1994).
10. *In Re: Joint Eastern & Southern Dist. Asbestos Litigation,* 827 F. Supp. 1014, 1993.
11. *Hopkins v. Dow Corning,* 33 F. 3rd 1166 (9th Cir. 1994).
12. *In Re: Paoli Litigation,* 35 F. 3rd 717 (3rd Cir. 1994).
13. *U.S. v. Conservation Chemical,* 619 F. Supp. 192, 1985.
14. *U.S. v. Mexico Feed & Seed,* 729 F. Supp. 1255, 1990.
15. *Citizens to Preserve Overton Park, Inc,* 401 U.S. 402, 1971.
16. *Ethyl Corporation v. EPA,* 541 F. 2d 1, 1976.
17. 7 U.S.C. sec. 136–136y.
18. 40 C.F.R. sec. 154.3(i).
19. 40 C.F.R. sec. 154.3(e).
20. *U.S. v. Akzo Coatings of America,* 719 F. Supp. 579, 1989.
21. *Browning-Ferris Industries of South Jersey, Inc. v. EPA,* 31 ERC 1088, 1990.

22. *American Mining Congress v. EPA*, 907 F. 2d 1188, 1990.
23. *National Gypsum Co. v. EPA*, 968 F. 2d 40, 1992.
24. Oreskes N, Shrader-Frechette K, Belitz K. Verification, validation, and confirmation of numerical models in earth sciences. Science 1994;263:641–643.
25. Case CD. Problems in judicial review arising from the use of computer models and other quantitative methodologies in environmental decisionmaking. Environ Affs Law Rev 1982; 10:251–281.
26. *Cincinnati Gas & Electric Co v. EPA*, 578 F. 2d 660, 1978.
27. *South Terminal v. EPA*, 504 F. 2d 646, 1974.
28. *Cleveland Electric v. EPA*, 572 F. 2d 1150 (6th Cir., 1978).
29. *Chemical Manufacturing Assn. v. EPA*, 28 F3d 1259, 1994.
30. Mandelker DR. NEPA, law and litigation. New York: Clark Boardman, 1992.
31. *Strycker's Bay Neighborhood Bay v. Karlen*, 444 U.S. 223, 1980.
32. *Public Service Co. v Andrus*, 825 F. Supp. 1483, 1993.
33. *Portland Audubon Soc. v. Babbit*, 998 F.2d 705 (9th Cir, 1993).
34. *Seattle Audubon v. Mosely*, 798 F. Supp. 1484, 1992.
35. *Conservation Law Found. v. Federal Highway Admin.*, 827 F. Supp. 871, 1993.
36. *Salmon River Concerned Citizens v. Robertson*, 32 F. 3rd 1346, 1994.
37. *Northern Crawfish v. Fed. Highway Admin*, 858 F. Supp. 1503, 1994.
38. Council on Environmental Quality. Regulations for preparing environmental impact statements, 40 CFR Section 1502.22, 1988.
39. Reeve M. Scientific uncertainty and the National Environmental Policy Act—the Council of Environmental Quality's Regulation, 40 CFR Section 1502.22. Wash Law Rev 1984;60: 87.
40. *Sierra Club v. Sigler*, 695 F.2d 957 (5th Cir. 1983).
41. *Southern Oregon Citizens Against Toxic Sprays, Inc. v. Clark*, 720 F.2d 1475, 1478 (9th Cir. 1983), cert. denied, 469 U.S. 1028, 1984.
42. Fitzgerald E. The rise and fall of the worst case analysis. Dayton Law Rev 1992;18:1.
43. Tutchton J. Robert v. Methow Valley Citizens Council and the new 'worst case analysis' regulation. Environ Law 1989;8:287.

Science Assumptions and Misplaced Certainty in Natural Resources and Environmental Problem Solving

14

Lynton K. Caldwell

The intended end product of scientific research applied to a major environmental problem is usually a policy decision. Environmental problems are characteristically "solved" at two levels—scientific and political. Because major environmental problems are broadly inclusive geographically and socially, their solutions (implied or intended) usually are expressed through government. Some environmental problem solving can be expressed through findings presumed to be essentially scientific. These environmental problems and findings may, but often do not, directly imply policy decisions (e.g., problems in toxonomy, entomology, geodesy, paleontology). Solutions to environmental problems involving relationships between human society and its ambient environment are seldom "strictly scientific"—human culture, and all that it implies, influences the way in which the problem is defined and science is used.

In considering the influence of scientific uncertainty or unwarranted assumptions on the process of public-policy making it is important to understand the relationships between scientific and political levels of problem solving. They are seldom clearly separable—interactive influences are characteristic. Although an environmental problem, especially if large and complex, may have a scientific answer, the ultimate solution for human society is political, expressed as policy. Multidisciplinary teams are usually needed to define and propose solutions for large and complex environmental problems. Because there may be alternative answers to an environmental problem (as in environmental impact assessment), policy solutions may require a choice involving nonscientific considerations. To the extent that the science solution is uncertain or disputed, the latitude of the policy maker to act, or postpone action, is broadened.

Intermediate between the scientific and political levels of problem solving is the policy analyst who may, or may not, be conversant with science and scientists. Even at the highest levels of science opinion and advice, as in the Presidential Office of Science and Technology Policy, proposed science-based

problem solutions must compete with other considerations (financial, economic, military, bureaucratic, and ideological). Uncertainty in environmental
problem-solving may occur at all levels or stages in the problem-solving process.
Because of the other-than-scientific considerations, an environmental policy
solution can seldom be finalized at the science-research level. If, as is usually
the case, the problem solution must occur at the policy level, the process of
policy analysis between scientific findings and policy decisions is likely to be
critical for the applied solution.

At this stage, unwarranted certainty and misconception of the actual problem can lead to ineffective policy solutions. Worse, it may lead to unforeseen
and unwanted consequences. The so-called paradigmatic or ideational context
in which a problem is perceived influences—may even determine—the way in
which the problem is defined. For example, is the problem of "world hunger"
too little food, inadequate distribution, or too many mouths? Unexamined
assumptions about the "problem" may lead to different answers based on spurious certainty in the validity of the problem as defined. The argument of this
chapter is that environmental problem solving requires careful examination of
underlying assumptions. A specious certainty about a perceived problem—especially shaped by cultural and political biases—may be a greater threat to problem
solution than is scientific uncertainty.

Policy-related environmental problem-solving characteristically responds to
a perceived social demand for preventive or remedial action affecting human
behavior in relation to the environment. A problematic situation may arise out
of the impact of a society upon the environment or the impact of natural forces
on society. In either case the object of ultimate inquiry is societal. Accordingly,
public institutions and government are likely to become involved. If so, environmental problem solving is incorporated in the processes of public-policy
making in which human assumptions and values beyond the purview of science
are encountered. Depending on the criteria applied, a societal demand may not
correspond to a genuine social need. Thus, quite apart from uncertainties in
scientific hypotheses and methodologies, a large area of uncertainty may exist
in assessing the significance and salience of human attitudes, beliefs, and behaviors in the creation and solution of environmental problems.

To some extent, assumptions of unwarranted certainty are consequences of
the social-political environment in which scientific investigation occurs. To be
socially credible and politically acceptable, some assumption of certainty is
required in findings alleged to be scientific. These allegations in support of official policy may exceed confirmed scientific evidence or even contradict the
probabilistic findings of "good" science. Perversions of science have been notable in defense of politically initiated natural resources policies, in standards for
acceptable residue of pesticides and herbicides in food and water, and in economically inspired attacks upon environmental policies (e.g., for protection of

the ozone layer and in prevention of global climate change) (1–10). Where politically preferred policies are at stake, efforts to downplay scientific uncertainty and to assume certainty where it does not exist may be expected. Our political-legal culture has little tolerance for admitted doubt or uncertainty in formulation of public policies. Scientific work is carried on within this social context and cannot wholly escape its influence. Too often, certainty is tacitly assumed on matters that unbiased prudent judgment would regard as uncertain.

Failure of science to result in more effective policies is not necessarily the fault of research methods per se which, appropriate to particular aspects of a problem, fall short of addressing its full dimensions. Political circumstances and financial support may require an essentially linear response to a multilinear problem. This is partly because people tend to have a limited span of attention, and a concern focused primarily upon "here and now" issues. These issues perceived to be salient are likely to be narrowly conceived and become candidates for political attention. Other factors are involved, especially the intentions and receptivity of policy makers. These may engender conceptual and methodological hazards to the validity and utility of science-based problem analysis focused on public (i.e., political) issues. Among these, one of the more common and fundamental is the error of identifying and defining a research problem in conformity with the perceived pressures and preferences involved in a policy choice. Accordingly the focus on the actual problem may be misplaced. Certainty regarding problem definition and the reliability of methods applied toward its solution ought not be assumed when underlying assumptions are uncritically taken as "given" and their validity insufficiently examined. Scientific methods more often yield probability rather than certainty. An unqualified and untested assumption of certainty in problem definition risks error compounded by uncertainties inherent in scientific methods.

I do not suggest that for every uncertainty that requires policy choice there is a researchable problem awaiting a definition. The problematic character of an issue or situation often derives from human values and normative assumptions. The natural phenomenon may not be problematic; it is people's perception of the situation in relation to expectations and values that appears to present a problem requiring solution. Floods, earthquakes, hurricanes, deforestation, or soil erosion per se do not of themselves present problems for policy research. They become scientific problems when scientists seek to understand them and their implications. They may become policy problems when people seek to reduce their vulnerability (often self-imposed). In the latter case problem identification also addresses cultural or psychological phenomena, shaped by assumptions, desires, and values.

In the sciences, problem definition normally occurs within parameters established by consensus among qualified investigators. And this consensus embodies assumptions and evaluations constituting the paradigm in which a situation is seen as problematic. There are circumstances in which problem definition by any criterion is elusive. For socially sensitive or divisive issues affecting the welfare or survival of social groups or institutions, there may be a narrow range of possible political responses. Yet lasting success in coping with problematic circumstances may depend upon their being realistically defined, regardless of untested assumptions derived from cultural values or political preferences. Reality as perceived by scientists may differ significantly from perceptions held by the general public, by policy makers, and by funders of research. But, of course, scientists as an occupational "genus" do not necessarily share the same perceptions or assumptions. Problem issues such as control of human reproduction, sustainability of social systems, genetics of intelligence, and requisites for biodiversity exemplify policy issues in which differing assumptions lead to differing definitions of scientific problems.

Some Types of Science-Based Problem Solving

The vulnerability of policy-related environmental research to conceptual bias differs significantly with the type of problem addressed. For policy problems that are essentially technical (and they are relatively few in number), scientific analysis may be largely (but not always) free from preconceived bias. Even here, however, notably in relation to agriculture and energy production, some scientists and engineers have shown a conceptual prejudice (prejudgment as implicit certainty) in their approach to policy choices. This is notably evident in assumptions that agricultural biotechnology will indefinitely counteract world hunger, or that fossil fuels and atomic nuclei are the only sources of energy worth serious consideration.

The following discussion, however, focuses on those nontechnical types of policy-relevant research in which a valid identification of the "actual" problem and its ramifications is essential to coping with it in the course of policy formation and execution. Although intended as decision informing, this research need not be done merely to accommodate the immediate perceived needs and preferences of decision makers or their clients. Beyond this, it may seek reliable answers and alternatives in the identification and explication of basic social and environmental problems requiring popular education and understanding (11–17).

There is nothing methodologically unique about policy-informing research focused on problems relating to natural resources and environment. For physical phenomena in nature there are science-based tests of truth (more often of probability) that narrow the scope of problem definition, and make policy-related findings less flexible than in research relating almost wholly to interpersonal human relationships. Policy issues relating to natural resources and environment are sometimes distorted by biases of personal advantage or of professional or political acceptability. To these susceptibilities, more often observed in politicians and public officials, scientists are not wholly immune.

The validity test for science-based problem-related policy research is not its short-run political acceptability. Trends and long-run consequences should be indicated to the extent that the evidence permits. Yet, its findings to be reliable must be applicable in form and content to the political decision process, although not skewed to serve a predetermined political preference. The utility of this research depends on its providing information that would enable the political decision process to cope as effectively as possible with the real problem. There is more than one kind of policy research, but the kind discussed here engages in "speaking truth to power" (18, 19).

Focus and Process in Policy-Related Research

Occasions for conflict between traditional politics, conventional economics, and scientific knowledge have increased with enactment of legislation to protect health and environment from hazards inherent in "unecologic" aspects of human behavior. The action-forcing provisions (Section 102) of the National Environmental Policy Act, and mandatory principles and standards in other U.S. statutes, along with enhanced opportunity for public scrutiny under the Freedom of Information Act, have brought environmental and resources policy making from behind the closed doors of legislative committee rooms and government bureaus into the public forum and before the courts of law. Particular forms of policy-related problem identification have been prescribed by law, including environmental impact analysis, cost-benefit analysis, technology assessment, and risk assessment. An effect of these processes has been to enlarge and to diversify the conceptual approaches and analytic methodologies available to policy researchers. And they increase the occasions in which policy analysis (e.g., relationship between research findings and policy options) can be invoked, and explicate the systematic basis of alternative problem solutions.

Typically, government policy making relating to natural resources and environment should reckon with not merely one, but with many outcomes from a single policy decision. The diversity, complexity, and often the variability of environment-resource problems over periods of time, impose hard-to-satisfy requirements for the validation of research hypotheses and assumptions. And, of course, policy-relevant research cannot rely wholly on exact science. To suppose that science could be made exact by covering it with a gloss of pseudoscientific research technique, giving a superficial appearance of statistical rigor, would only mislead the unsophisticated or precommitted.

Expansion of knowledge on any complex topic often reveals and may even create previously unperceived problems. The complex, multidimensional character of a great many resource and environmental issues presents the researcher with difficulties in specifying research methodology and in reliance on any single test of so-called validity. Regardless of intentions, solutions to the environmental problems of human societies are likely to lead to multiple and sometimes unintended outcomes. As Cook and Campbell observe, ". . . it is more important to note the multiple outcome issue and to know how to deal with it than it is to be able to classify it neatly under one of our validity labels" (20). Focusing policy-oriented research on the actual problem creating situation is the first step toward the productive joining of concept and method. Premature conceptualizing, misinterpreting, or underestimating the problem are fundamental errors in problem-focused research; therefore attention should be given to error detection and to methods to prevent omissions or correct deviations of focus from what should be the object of research.

Ideally, administration of policy related problem-solving research may be regarded as a cybernetic process, steering an investigation toward resolution of a policy problem. At risk of oversimplification, problem-defining research may be categorized by focus on what is or was (historical/descriptive), what should be (prescriptive/advocacy)—or what may be (possible/probable futures), including forecasting of the probable consequences of alternative choices. Each of these categories may be the object of legitimate and methodologically sound research employing scientific concepts and methods, but not excluding other forms of inquiry (e.g., logic and history). Each category may have a fundamentally different influence upon the selection of a research approach.

The first category, historical/descriptive research, is intended to reveal or explain the dynamics of a given policy and to explore its origin and evolution. It is a form of inquiry common to the study of anthropology, art, religion, politics, and law. Interpretations of research findings with policy implications appear in the writings of physical and biologic scientists and engineers (e.g., Walter Bradford Cannon, Reneé Dubos, J. Robert Oppenheimer, Garrett Hardin, Jay Forrester, etc.). Much of this writing is more in the nature of personal perspectives or of reporting investigations or findings having problematic implications for humanity rather than of proposing applied science policy solutions.

The second category, prescriptive/advocacy research, defends a conclusion, or in extreme cases a preconceived policy, and is common to practical politics and litigation, and has also characterized some publicized disputes among scientists. Characteristically it claims to have found the right or best answer to the problem addressed. Susceptibility to policy advocacy is in no way peculiar to the social sciences, as controversies among physical scientists and biologists over policies regarding recombinant DNA, DDT, nuclear energy, world food production, and the protection of endangered species amply illustrate.

The third category, or decision-informing predictive research, differs from the foregoing approaches in two important ways. Research financed by grant or contract leading to application risks reflecting a predetermined policy preference, or the bias of a sponsor, or predilections within a science peer group. It may not attempt to analyze all feasible alternative policy choices and their probable consequences. A characteristic objective of this research is not only policy-relevance, but policy-applicability; that is, its findings are presented in the form of propositions upon which decisions can be made. The effectiveness of the policy may depend on the validity, reliability and persuasiveness of the science-based research and the extent of political and public receptivity.

Although each of the foregoing research approaches presents distinctive choices of research design, sources, uses of sources, and styles of presentation, distinctions among them are not always clear-cut, and the motives of researchers may be mixed. Nevertheless, a large amount of policy-relevant research can be classed as being primarily one of the three. It is obviously important to the quality of the research product that the researcher know which purpose he is pursuing. This knowledge is no less important to recipients of the research product, but readers or users often lack the discriminating habits of mind to perceive how differences in approach can lead to differences in conclusions.

Explanatory Function of Environmental Problem Solving

There are several reasons for enlisting scientific research on problems of policy relating to natural resources and environment. First, in a society threatened by incipient environmental crises and seemingly intractable problems of resource supply and management, it is a kind of research for which reliable knowledge and understanding are needed. Second, where there is an intent to inform the decision process comprehensively in relation to the scope of a problem, a special responsibility for problem definition rests on the science researcher. This requires asking the "right" questions and avoiding unsubstantiated assumptions. Problem solving in these areas characteristically requires factual data that ideally should set the dimensions and direction of research, and that are not redefinable for the convenience of either researchers or official policy makers. Third, in order to be adequately decision informing, the research must present decisionable alternatives if there are rational choices. It must be informing to all who are significantly involved in the decision process, and this may range from a chief executive to an entire organization or electorate, and increasingly to the collectivity of nations. Environment and resource policies characteristically affect large numbers of people, and alternatives may affect different groups of people differently. Especially in a democratic society, informing the public and the policy makers of the nature and significance of an environmental problem involves an explanation of causes and consequences and the implication of alternative solutions.

I do not discount the value of research that only describes or partially explains an environmental policy problem, provided that its limitations are acknowledged and understood. Especially when they are serial and cumulative, such inquiries, limited historically or geographically, may contribute insights that significantly increase understanding of relationships between human behavior and many aspects of the Earth's environment. Such investigations may indirectly influence policy making, although this may not have been their purpose. Research of this type in archaeology and history, for example, may uncover hitherto unperceived circumstances in which analogous problems for present-day policy may be latent (for example, in the consequences of environmental mismanagement or of population growth exceeding a sustainable resource base) (21). Unfortunately, the policy relevance of this research is often overlooked until a problem, hitherto unperceived, becomes visible and a policy decision is necessary. Relevant research done in the past may be overlooked or discounted because it is assumed to be out-of-date or to have no contemporary application.

Scientific research relative to natural resources and environment has on occasion uncovered policy "time-bombs," previously unrecognized by technician researchers who, working (for example) as agronomists, geologists, or ecologists, have not foreseen the implications of their findings. Some environmental effects believed to be foreseen, however, may not in actuality prove to be valid. For instance, in the Minnesota Reserve Mining controversy, knowledge that taconite tailings discharged into Lake Superior were similar to asbestos fibers had little policy significance until linked to possible asbestos-induced illnesses. This information exploded into a major public health issue, changing the character of the controversy (22–24). Research on the impact of pesticides on ecosystems, and of the effects of CO_2 and SO_2 emissions on the soil quality, and water and atmosphere, revealed harmful consequences noted by scientists before the need for a policy response were generally perceived. Environmental consequences may be unobtrusive, cumulative, and delayed. By the time that risk becomes generally credible, remedial action may no longer be possible. Scientific assumptions regarding global warming and ozone layer depletion may not obtain general credibility because the assumptions cannot be "proved" to the satisfaction of skeptics.

Future-oriented scientific research (however significant its implications) may do little more for the public decision process than to document the latency of a problem which neither the public generally nor its political representatives regard as immanent. Problems revealed by scientific inquiry are seldom the problems of primary interest to politicians, personally, or in relation to their political responsibilities as they understand them. A political "problem" is often a surface manifestation of a more fundamental socially divisive disorder that politicians prefer to avoid. Thus, the public and its representatives settle for a superficial definition of a problem that may be addressed by acceptable, though ineffectual, remedies—"we're doing something about it"—even though the basic problem remains ill-defined and unaddressed.

If there is an unequivocal measurable problem confronting public decision-makers for which scientific advice is sought, why should opinions differ as to its character? This question is as old as it is familiar, partly epitomized by the parable of the blind men and the elephant. Even in a relatively exact and rational science such as physics, differences of opinion rise over the definition or significance of particular phenomena. But the achievements of physics have come about through efforts to resolve the differences via better understanding of the phenomena. In most respects physics does not offer applicable models for policy-relevant research, yet it does provide principles against which some policy-relevant applied research strategies can be compared—if only to better understand the circumstances that complicate problem definition.

Mislocating the Problem: The Parallax Fallacy

Methods of detecting error in assumptions are important in any area of scientific inquiry. Indeed, an essential function of research has been described as ''a search for error'' (25). The need for this search is acute in those aspects of policy wherein conventional ''common sense,'' political folklore, and misunderstanding of science tend to dominate decision making. Natural resource and environmental issues frequently fall into this category. A compelling reason for building strategies for error detection into any research effort is ''conceptual parallax.'' A relatively slight and apparently inconsequential difference in viewpoint or perspective can, over time and distance, displace the perception of the object of attention. A camera view finder that is a slight distance removed from the position of the camera lens may cause the viewer to perceive the object of his focus in a different position than it occupies in reality. The photographer may have to make adjustments in focus to compensate for parallax. Failure to do so may result in a photograph that misses or partially cuts off the intended object of the photograph. So it is in conceptualizing phenomena in the external world. To the extent that the object in the mind's eye is displaced from the object in reality, a failure of focus occurs.

Parallax—an apparent displacement of an object due to an observer's position—is, of course, a physical science concept and relates imperfectly to the subject of this chapter, except in one significant analogy. If the observer of a problem (scientist or investigator) assumes that the object (i.e., problem) that he perceives is where it is actually located, the assumption, if erroneous, will lead to conclusions that are, at least in part, erroneous. The ''parallax fallacy'' is an assumed certainty in seeing a problem in a perspective convenient or conventional to the observer. Because of the complex variabilities encountered, uncertainties in environmental problem solving are probably more numerous than in many (but obviously not all) physical science and engineering applications. Difficulty in finding the right methodological approach to solving a problem is compounded when the problem toward which science is applied is misconceived.

Misplacing the object of policy compounds other possible errors. Statistical rigor may mislead when applied precisely to data that is not relevant to the "real" problem (26). An example of the misleading use of statistical methods was published in *Science* under the title "Resources, Population, Environment: An Oversupply of False Bad News" (27). By comparing selected statistical evidence with generalized and journalistic overstatements of environmental hazards, the author purported to demonstrate that concern over population growth, resource depletion, and environmental deterioration was foolish and unfounded. One could concede the numerical accuracy of the tabulations and nevertheless show that none of the statistical "facts" proved the "bad news" to be false. An OP-ED commentary in *The New York Times* noted that (28)

> In the 19th century Lord Kelvin, the unassailable physicist of his day, using mathematics and the new laws of thermodynamics, determined that the Sun and the Earth were a mere 100 million years old, an eye blink in cosmological time. He was wrong, of course. His math was lovely, but his data were incomplete.

The commentator points out that, whereas the math published in "cautious scientific journals" may be unassailable, the conclusions that follow are not necessarily so. Such errors may also be consequences of misconceiving the problem or misreading the evidence. They may also result from insufficient regard for the limitations of mathematics (29). "The foundations of mathematics have serious flaws that may imperil all the sciences" (29). (For an earlier explanation of the parallax quasi-analogy see Ref. 30.)

The prevalence of scientific indeterminacy or uncertainty may therefore be compounded by conceptual error. Indeterminacy is a condition that in some form characterizes every field of knowledge (9, 10, 31). It is obviously greater in the more complex sciences—those relating to human social behavior. One might assume therefore that in those sciences there would be the greatest need for error detection. This may be true, but possibilities for remedial action are limited by the extent to which the causes of human error or misbehavior can be identified and significantly reduced, causes of which we have some, but as yet insufficient, knowledge. For example, a factor that poses a threat of undetermined risk to society is the vulnerability of nuclear reactors to human error or malevolence. What is the potential environmental risk of a plutonium-yielding breeder reactor? Obviously technical safety measures are insufficient to remove the chance of disastrous, unpredictable human intervention. Behavioral science is far from providing the understanding of human behavior that might lead to an applied science of behavioral prediction and control. Behavioral analysis may be able to identify risks and needed precautions for planners and policy makers, but its findings are possibilities, not certainties (32–37). Moreover, scientifically reliable solutions applied to a behavioral problem may be socially unacceptable for nonscientific reasons alleged to be ethical, moral, legal, or politically inexpedient.

Uncertainty unacknowledged is a factor that handicaps efforts to discover whether error has occurred. In political controversy over public policies, the limits to certainty are often cited as evidence that there can be no "right" answers to questions of choice. But "right" in this context has significance only if differing answers really relate to the same questions. When policy problems are differently defined, it is logical to assume that there may be at least as many apparently "right" (or wrong) solutions as there are perceived problems. The important question then becomes which definition corresponds most accurately to the *actual* circumstances that have caused the problem, regardless of how the problem may subsequently have been perceived or defined. The answer here must be sought through empirical research; there is no substitute for what in the language of remote sensing is termed "ground truth." Abstract findings must be tested by comparison with physical (not merely statistical) actuality.

Closely relating to the all too frequent indeterminacy of factual data is the bounded rationality of investigators and their clients. The capacity of the human brain to comprehend complexity and magnitudes is surely not infinite. There is also a practical problem of defining policy-relevant issues in concept and language comprehensible to intended recipients. Efforts to simplify issues or recommendations for easier comprehension may have the effect of misstating the issue. Human perception is inseparable from conditioned human mentality, and although the reach of the human imagination may be wide, it is nevertheless bounded in practical matters of human choice (38–40).

Unfortunately, as the reasoned errors of eminent scientists and scholars have demonstrated, error may coexist with intelligence, and educated incapacity has often blocked consideration of novel concepts or hypotheses. Educated incapacity is, of course, associated with conceptual constraints imposed by a researcher's commitment to a particular paradigm (i.e., the way in which his perception of the problem area has been structured by profession or by convention) (41, 42). A common cause of conceptual constraint is expressed by Miles's law that "where you stand on an issue depends on where you sit" (43). Perceived self-interest may lead to internationally misplaced definitions of policy problems, notably in defining resource and environment issues in agriculture, forestry, flood control, coastal erosion, energy, and mineral supply. Policy options in these areas have traditionally been defined by engineers, economists, resource developers, bureaucrats, lawyers, and politicians, many of whom have vested interests in solutions to an environmental problem as defined. Conceptual reductionism, moreover, has often simplified complex problematic questions to technical how-to-fix-it answers.

Another obstacle to problem definition to lies in the difficulty in foreseeing the effects of complex interrelationships interacting over time. Jay Forrester, among others, believes that human intuition, generally, is poorly adapted to perceive the convergence of apparently unrelated trends and that the behavior of social systems is often found to be counterintuitive (44). Computer simulation and modeling may partially compensate for this conceptual deficiency. But the availability of a compensatory technology does not promise its proper use. For the computer to counteract perceptual error, inputs and programming must be right, and they come from outside the computer.

The hypothesis of people as rational self-interested calculators has had advocates among some social scientists, notably among political economists. But its assumption is generally regarded as unrealistic and inadequate by those social and political scientists whose approach to the study of behavior has been strongly influenced by psychology and biology (32–37, 45). The balance struck by policy analysts between assumptions of human rationality or irrationality and the weight assigned to information as a factor in popular attitudes and official judgments will tend to affect the direction and hence the findings of their research (46).

Aside from insufficient, unverifiable, or erroneous information, a factor making for uncertainty in the reliability of science-based knowledge in environmental problem solving is found in the place where the problem is defined and the solution undertaken. The place may be a scientific laboratory or field investigation, or it may be a governmental or private corporate agency. Environmental problems isolated and researched in a laboratory may not be entitled to the degree of certainty that might be warranted if undertaken in situ in nature. Laboratory tests of the behavior of water pollutants may not equate to the fate of the same pollutants discharged into the actual environment at the end of a pipeline. Strict control over data may not always enhance certainty in understanding the problem *where it actually occurs* (e.g., in an uncontrolled environment). The pertinent locale of the problem may be displaced from nature to laboratory in an effort to obtain more rigorous control over the data. The search for certainty in scientific methodology may be as great a hazard to environmental problem solving as are unperceived uncertainties in definition of the problem.

Adapting Research Methods to Policy Problems

Initial perceptions of policy problems, by decision makers and researchers alike, tend to shape the formal definition of the problem and the methods chosen to investigate it. Thus the parallax phenomenon, if it occurs, carries risk of two types of error: error of misplaced concept or definition and errors of method. External invalidity, caused by misconceived concepts, is likely to induce internal invalidity in the collection and processing of data. This consequence is likely when uncertainties, which the preferred research methods are unable to handle, are ruled out of the investigation.

Conceptual and methodological errors may be compounded if the research project is defined improperly to accommodate a preferred research method. In the more flagrant cases, which are not uncommon in social and behavioral research, the investigators select or define the research problem to fit the methodology in which they are most competent and with which they are most comfortable (47, 48). Pursuit of an esthetically coherent solution may not be the right route to solving an inherently incoherent (i.e., messy) problem.

Methodological error, involving questions relating to the internal and external validity of research design and treatment of data, is possibly the most extensively considered topic in the literature of social and policy research methodology. Not unique to resource and environment questions, but of great frequency and significance, are external factors of the magnitude and dynamics (e.g., relative stability) of the researched phenomena. Facts regarding the relative sizes and numbers of things, and directions and relative rates of change, are often indispensable data for problem definition. Interactive relationships revealed by the data may be as significant or important to scientific analysis as are apparent linear trends. Among many possible examples are the ramifying interactive effects of the diminishing availability of critical minerals or energy sources, or the cumulating effects of exotic compounds such as dioxins, DDT, or PCBs in the environment. Appreciation of the multiple causes and effects of global warming caused by the "greenhouse" phenomena is apparent in many current science policy studies. But makers of national policies have great difficulty in dealing with multiple effects that their constituents do not easily comprehend especially when consequences are foreseen in the future, but not apparent in the present.

Dynamic models, physical or mathematical, may be useful for comprehension of the interactive changes within complex phenomena. The ability to manipulate data under conditions controlled through computers can sometimes be an aid to prediction and the identification of mistakes in method, in concept, or in policy in action (49–52). Measuring and counting are essential to the discovery and verification of scientific and technical knowledge, but applied mathematical technique is not itself science. There is a danger in enthusiasm for mathematical expression of relationships, especially through models, that concern with technique will supersede concern for the problem researched (38, 53–55). When the method becomes in effect the message, definition of the actual problem risks the error of parallax—of mislocating the intended object of research (56, 57). (Phillips concludes with a paraphrase of a line by Orwell. With respect to the elaboration of sociologic method, he writes: "It gave an appearance of solidarity to pure wind" (50, p. 179). Testing models is a legitimate function of research, but does not ipso facto lead to more accurate problem definition.

To the extent that it can be done, identification of a hierarchy of relevance is a promising method for determining the priority and sequencing of research tasks (58). A policy issue may not be confined to a single problem; more often a multiproblematic situation confronts the decision maker and the policy researcher. The order and emphasis to be accorded these respective interrelated problems is one of the more critical determinations that policy researchers must make. Here, informed subjective judgment may become a factor, possibly assisted by the opinion of peers as in Delphi assessments.

Of the many facts and interrelationships in any set of complex issues, not all are of equal intrinsic significance. Interrelationships are usually more difficult to define than are isolated facts, but facts and relationships change with time and perspective. Thus, a schematic ordering of interrelationships is valid only insofar as it reflects the dynamics of the problematic situation. Static models will seldom prove indefinitely reliable. If, as a matter of practical necessity, a model is temporarily fixed at any given point in time, the relationships it describes should be subject, if need be, to periodic retesting and revision or, if this is not feasible, to the best estimate of the probability and direction of significant change. Environmental impact assessment affords examples of the risks inherent in finalizing conclusions reached at a particular point in time without recognition of uncertainties impinging on an ecosystem or inherent in its dynamics that could invalidate conclusions drawn at a particular point in time.

Not of least significance in the conceptual structuring of the problem are its empty places—indicators of data missing or not understood. It is, of course, important to any decision process to know what information, normally discounted or overlooked, soft or missing, might actually affect assumptions regarding the problem as defined or presumed solved. The validity of findings is questionable when certain pertinent aspects of a problem or alternative solutions are put out of bounds by sponsors of the research (e.g., government agencies). The public-policy analyst, however scientific in intent, is to some degree incorporated in a political process. This is especially evident in those forms of evaluation analysis structured by public-policy considerations; examples include environmental impact analysis, technology assessment, risk assessment, and cost-benefit analysis. These processes include the inputs of fact, value, and judgment brought to a decision-ready statement of alternatives by researchers and policy analysts. Decision makers, whether executive, legislative, judicial, the general citizenry, or some combination among them, are presumably presented with a set of propositions from which rational choices could be made. There may be no practicable way to discover the full range or significance of predetermined concepts or whether in advance of trial, these concepts may in some measure incorporate error.

The foregoing considerations underscore the importance of understanding the problem situation as thoroughly as possible, including its institutional setting, before policy-focused analysis begins, although further understanding may be gained in the process of investigation. Elementary as this may seem, it calls for more patience and more realistic expectations regarding the design and strategy for problem solving than may be appreciated by research sponsors. The parties concerned are usually eager to get on with the research task. This attitude need not be bad provided that the problem-defining task is not slighted. But it often happens that the policy problem cannot be fully understood at the outset; a substantial amount of experience may be required before all critical scientific uncertainties and political and financial constraint on investigation of the environmental policy issue are clarified.

Flexibility in research strategy and a studied tentativeness in problem definition are qualities recommended to investigators in the early stages of policy-relevant research intended for implementation. And because the policy problem may not readily yield to analysis, and because the problem may turn out to be somewhat different than first perceived, it is unwise to lock the pre-research effort into a preconceived model or specific research methodology. Nevertheless, peer groups reviewing an application for research funding may decline to approve a proposal unless the investigator will state precisely what methodology will be employed.

Once the policy problem appears to be adequately understood, commitment to a particular methodology may be entirely appropriate. Environmental problems vary greatly in scope and complexity and some are amenable, and may be amenable only, to an extensive menu of research techniques. Some environmental problems, however, can be resolved largely upon the basis of statistical information or factual data regarding circumstances in an essentially isolated environment. Examples might include the effect of fluorocarbon aerosols upon the upper atmosphere or the relative efficiency of energy transmission. Such problem isolation, however, often omits consideration of other factors that affect the policy decision, notably political, social, economic, ethical, and legal aspects.

The ultimate object of environmental policy research is usually qualitative (59, 60). Quantitative methods are employed to establish facts and avoid or expose error, but it would be presumptuous to assume that the basis of policy decisions can be reduced to mathematical equations. Fashions in research methodology tend to influence investigators, particularly beginners; and in recent years quantification and statistical methods have often been employed where their utility, although indispensable, is also at best marginal. Precision is worth pursuing, but is not a substitute for or guarantor of relevance.

Qualitative analyses and quantifications should not be regarded as opposite; they may in fact be apposite or complementary. Qualitative policy research is concerned with conditions or outcomes that may be preference-ranked according to the values and expectations of the public and policy makers. There is no methodology specific to or peculiar to this type of research. All forms of investigative methodology should be available to the researcher. Determination of the methodological mix is the sequential research task once the conceptual phase has been completed. But these phases may be mutually interactive because the concept of the problem may be refined and corrected in the course of choosing as well as in applying research methods.

Implementation of Problem-Focused Applied Research

In the context of this discussion, policy choice is the ultimate target of environmental and resources research, a choice made through a decision process in which the researcher plays a part (11–17, 61–63). The initial process of public-policy making may occur at any level or division of government and may include the public electorate, but it is only through administration that policies are implemented. Thus it follows that although incremental inputs to policy occur at all political or hierarchical levels, it is administrative decision making and its applications in practice that are critical outcomes of problem-solving policy research. Environmental and resources policies tend to generate continuing issues. Science is relied upon to find answers that work, but even when an immediate issue may be "finally settled," it is likely to leave residual problems or generate new ones that interact with other aspects of policy. To the extent that science can assist reliable forecasting, it may increase the probability that policies will be chosen with a view to their sustainability and reduce the likelihood of counterproductive results.

Analytic studies of policy and decision-making processes including institutional histories, and case studies may help to indicate the kinds of factors that must be included in defining the policy problem in its larger and developing dimensions and identifying its uncertainties. These factors must be considered, so far as possible, in the context of the actual problems at hand or emerging. If problem-oriented research is to yield results that justify its efforts, investigators need to avoid unwarranted assumptions about how their findings will be used. A figure of speech called "the law of unintended consequences" should be kept in mind. Insofar as is reasonably possible, research findings should be stated so as to minimize possible misuse. Organization and decision theory may be woven into the substantive treatment of the research problem or segregated as explanatory background for research findings. Appropriate communication of research findings and implications is an essential phase of implementation, especially when uncertainties must be addressed. Scientists are not always the most effective communicators of the significance of their findings. Abstract or theory-laden discourse is seldom appropriate to the text of a policy research report that should not necessarily be treated as if it were a scientific research paper even though scientific methods were used.

Assessing Expert Opinion

An essential requirement of decision-relevant problem-resolving research is that its results be stated in the form of alternatives upon which decisions can be made by persons who are not experts in all substantive details. Neither the public nor its political representatives are ordinarily in position to make scientifically rational choices amongst incommensurable claims of specialists. How are decision makers to ascertain the validity and relative importance of findings in fields in which they have no expertise? Researchers must also cope with this difficulty because they must often utilize findings from disciplines other than their own, and must evaluate disagreements among the experts.

Differences in expert opinion cannot always be reconciled, and it may not be clear that one opinion is more nearly right than others. On many policy questions it is important to decision makers to appreciate differences and disagreements over the state of knowledge and informed opinion. To understand these disagreements it is necessary to examine (or if necessary to infer) their underlying assumptions and the weight of evidence cited in their support. An emerging problem may not be ripe for policy decision or action, or it may require a judgment limited to further study (or containment for the time being) of an assumed hazard.

One of the requisites for skillful decision-relevant analysis is the appropriate and effective use of the "right" specialists. But confirmed experts sometimes appear to be reluctant to draw policy-relevant conclusions from their own specialties. This may reflect their own uncertainties or their apprehension regarding the uses to which their knowledge might be put. But they may also assume a professional certainty based on their expertise. Specialists often tend to be skeptical of the competence of generalizers or of anyone who would attempt to relate findings in *their* specialties to other fields of knowledge. And so the investigator must learn enough about the substantive aspects of an issue to utilize the assistance of specialized experts and to ascertain, if possible, unacknowledged areas of uncertainty or bias in expert opinion.

Reliance upon expert advice, unchecked by persons having broad practical experience in the subject area, risks error in problem definition or research method. The multidimensional character of most resource and environment policy issues makes a procedure such as "scoping" (e.g., in environmental impact analysis under the National Environmental Policy Act) a practical way to inform policy makers of the full dimensions of their problem (64). The concept of scoping was intended by the council and perceived by most commentators as a means for the early identification of what are and what are not the important issues deserving of study in the EIS. See Section 1502.25, as well as 1501.7, which states: "There shall be an early and open process of determining the scope of the issues to be addressed and for identifying the significant issues related to the proposed action. This process shall be termed 'scoping.' " (Fed.

Reg. 55993.) For application of the problem identification and scoping concept to environmental research, see (65). As March has observed, "Policy questions seem to confound ideas about scientific specialization that we have developed around the canons of more basic research" (66). Basic research, in itself alone, is unlikely to be decision informing.

Analysts (collectively or individually) ought not attempt to preempt the role of the policy decision maker, but, as noted, should be able to appreciate the circumstances that influence the administrator's (or legislator's) behavior (3, 67, 68). The integrity of the investigator may be put at risk when research in business or government is expected to support predetermined policies. Contracted research is especially vulnerable to distortion in support of the sponsor's preferred outcomes. For example, scientific research relating to oil spills at sea, to the health effects of cigarette smoking, and to the environmental effects of various herbicides and pesticides has been vulnerable to distortion induced by self-interested providers of funding. Decision-relevant research ideally should identify alternatives, interrelationships, and probable consequences to a degree that lays the available choices before the decision maker in a clear and comprehensive panorama. This counsel of the ideal may seldom be attained. But it is a goal toward which reliable decision-relevant oriented research should be directed.

Achieving Synthesis

More than three decades ago, Paul H. Appleby, then Dean of the Maxwell School of Citizenship and Public Affairs and an experienced public administrator, in a personal communication to the author (6 May 1956) described the decision-maker's task for which useful policy research has relevance

> Specialist after specialist pursues analysis; who pursues synthesis, or even pursues analysis with any sensible orientation to the larger function of synthesis? It is the synthesis which involves all the heavy burdens of practitioners, and these burdens are heaviest when the social action is most complex and most complexly environed. Synthesis becomes more and more important as one goes up the hierarchy, and more and more important as one moves from the relatively specialized fields of private administration to public administration.

An argument for a synthesizing approach specifically to resource-environment decisions has been developed by Sears (69) and by Michael Bradley (70).

Analysis directed toward the more inclusive function of synthesis is the essence of decision-informing policy-related problem solving research. Analysis probes into and desegregates a policy issue that is more often a set of interrelating problematic issues. Synthesis reassembles the elements of these issues into propositions or alternatives capable of being acted upon by decision makers.

But it is in reductionist analysis, not in synthesis, wherein the strength of science is the greater.

The need for breadth of knowledge and skill in synthesis has been recognized even though it is seldom honored. A former Secretary of Commerce observed that we need "... something akin to the Renaissance Man who is accomplished at all disciplines, who is capable of a multidimensioned view of the world." But noting that "we don't seem to have many Renaissance men at work on the environmental problem," he concluded that we must "pull together a multidisciplinary team that can launch a Renaissance Man's passionate search for solutions that will match our passionate sense of the problem" (71). But Secretary Peterson did not suggest how to make the multidisciplinary team function coherently. A few broad-gauged specialists might pull themselves together in a common focus, but their ability to do so would depend upon their possessing the capacity for synthesis that distinguished the prototypical "Renaissance Man"—an unlikely prospect in a society committed to specialization.

In practice, at least three levels of synthesis may be identified. The first is conceptual synthesis, and it occurs when the diverse and often disparate elements of a problem situation are pulled together intuitively, then tested and integrated to form a coherent research design. Following analysis of the issue and identification of its causes, dynamics, relationships, and consequences, a second level of synthesis is reached. It characteristically is a delineation of the findings of scientific research. A third level of synthesis is reached when the decision makers, whoever they may be, evaluate and consolidate the research findings in deciding on a course of action.

At this point, if they are to cope with the problem successfully, the designated decision makers must synthesize a policy from the information and conjectures at hand including factors beyond the purview of science. This synthesis may, therefore, include inputs in addition to those derived from scientific research; but to the extent that unavoidable factors beyond scientific evidence cannot be reconciled in a coherent problem-solution, a reexamination of the problem is indicated. Not all problems of public policy are amenable to solution at any given time. An extended interval of time may be necessary for the reconciliation of initially incompatible conclusions. Unfortunately this reconciliation may come too late to avoid disastrous environmental consequences such as ozone layer depletion or the effects of global warming.

Too often the objectives of policy-related research have been conceived and pursued in linear terms and assumed certainties although the objects themselves have been enmeshed in a matrix of environmental interrelationships that could be made explicit only in multiple terms. Viewed in narrow linear terms of the policy objectives sought, maximum-yield agriculture, nuclear energy, sustained yield forest management, and air pollution control could be accounted successes. However, from multiplex perspectives linear success often generates new problems or difficulties. These "new" problems may reveal uncertainties that the linear strategies had either failed to see or had discounted as unmanageable, or irrelevant, or unacceptable. Techniques have now been developed to reduce the tendency to oversimplify complex problems in the push toward achieving an uncomplicated policy commitment. Impact analysis and technology assessment (often the same thing) provide correctives to linear thinking, but offer no assurance that their caveats will modify policy preferences.

Until recently the distinction between environmental and natural resources problems was largely unrecognized. Environmental relationships are inherent among all the elements that have formed the planet. Resources have been present only since humans have identified them as useful. Environmental policy today confronts problems of ecologic irreversibility and global universality that affect the survival of human society and all more highly developed forms of life. Environmental policies confront natural "laws" or interactions that cannot be changed by human volition, and for which human adaptation is required. Resource policies, except as they encounter inexorable limits in the environment, may be more easily modified or reversed—frequently with the assistance of technology. The processes of the natural world cannot be repealed by technology, which through manipulation of natural laws may enlarge human possibilities and adaptations as in biotechnology, medicine, engineering, or the exploration of outer space. This is collaboration with, not conquest of, nature.

It is essential to clarity of results that policy researchers determine whether their focus is upon resource issues that are essentially economic or upon environmental issues that encompass a great number of values, and to avoid treating one class of issues with criteria and tools appropriate primarily to the other, yet not overlooking their interrelationships. This means that for many resource-environment-related issues at least two forms of analysis may be required. Resource issues almost always have environmental implications, and nearly all environmental policies affect resources directly or indirectly and entail costs of some kind. Ultimately these differing conceptual approaches to a policy issue involving resources *and* environment may need to be combined and synthesized in decisionable alternatives. Science may provide denominators common to complexly related problems, thereby helping to discover elements basic to their comprehension that might otherwise be overlooked, or misconstrued, or conducive to suboptional solutions.

In pluralistic democratic societies, a preferred method of resolving economic-environmental disputes has been through a balancing of equities, a qualitative exercise often with a pseudoquantitative gloss that in itself is outside the scope of normal science. It is usually undertaken by analysts or policy makers, seldom by scientists. A plausible quantitative method for balancing equities has been cost-benefit analysis, an appropriate technique in principle, when properly employed, for there are many kinds of costs and benefits. If so far as possible, all significant real costs and benefits are taken into account, this technique may contribute to a synthesis contributory to rational decision making. But in the environmental aspect of a policy issue there are limits to the evaluation of proposals using only quantitative methods. Reductionist and economistic tendencies in modern society have too often narrowed cost-benefit analysis to an exercise in econometrics. The National Environmental Policy Act has a provision in Section 102(2)(b) to counteract this tendency through the development of methods and procedures "which will insure that presently unquantified environmental amenities and values may be given appropriate consideration in decisionmaking along with economic and technical considerations" (72).

To the extent that environmental problems relating to natural resources are posed in *wholly* economic terms (i.e., economism), they risk "rational" answers to the "wrong" questions. Economism is a set of mind that postulates economic values (narrowly defined) as fundamental to all others and employs economic criteria as the primary measure of the worth of all human activities. In the words of the theologian-philosopher Nicholas Brdyaev: "Combined with an uncritical acceptance of innovating technology, it has facilitated the creation of a new artificial environment incompatible with the needs of the whole man even as it threatens the natural environment in which man evolved" (73). Yet environmental policies that ignore fundamental economic relationships, particularly between costs and benefits, are likely to be inoperative. I do not believe that policy researchers have as yet discovered a generally applicable way to analyze problems from both an economic and environmental perspective and to bring these and other relevant matters into a synthesis approximating the true set of circumstances (74, 75). I recognize that this conclusion skirts the psychological and philosophical problems of "knowing", "truth" and "reality." But policy making per se is neither science nor philosophy, although it may be influenced by both.

The Human Element in Resource-Environment Research

To obtain a cross-disciplinary dialectic of analysis and synthesis that valid policy research requires, the following conditions appear necessary to avoid unwarranted assumptions that misdirect effort.

The first is that the organization of applied policy research be structured to facilitate cross-disciplinary inputs relevant and necessary to the discovery so far as possible of the nature of the policy problem, and alternative ways of dealing with it. This implies precautions to prevent any particular discipline, methodology, or paradigm from distorting problem definition and the research process. It is at least plausible to suspect that many policy research programs or institutions today have unrecognized biases and tacit agendas. Some are frankly unidisciplinary in orientation or methodology—economics, law, and narrowly defined systems analysis providing examples.

The second condition is that provisions for problem research allow greater recognition for skill in the art of synthesis than presently prevails. Bringing together a team of disciplinary specialists does not in itself ensure interdisciplinary or cross-disciplinary perspectives on policy issues. Nor does absence of a disciplinary bias necessarily imply a talent for synthesis. Our educational system and academic and professional life rewards specialists and analysts, but seldom synthesizers. To avoid disciplinary bias, we often place lawyers, statisticians, or systems experts in charge of policy research on the assumption that their expertise relates to logic and method and is neutral (i.e., impartial) as to substance.

A third condition is an organizational or institutional arrangement to facilitate cross-disciplinary problem analysis. Lacking an institutional facility and incentive system appropriate to the type of policy analysis which resource and environmental problems require, we are not likely to produce the capabilities that can bring coherence to multidisciplinary research. Arnold J. Meltsner, professor of public policy at Berkeley, has declared that "The social and institutional context for establishing policy analysis as a valued and permanent activity does not exist" (76). (Also see References 77 and 78.) In the absence of a popular sense of need, institutional arrangements for valid policy analysis leading toward synthesis are not likely to be developed. The National Environmental Policy Act (Section 204) provides ample authority for analysis and forecasting, but the Council on Environmental Quality has been unable to implement these provisions.

A fourth condition is allowance for counterintuitive and irrational tendencies in human behavior. Surely the greatest area of scientific uncertainty encountered by environmental problem solvers is the domain of social and behavioral science. Scientific uncertainties may appear to be more easily identified, defined, and assessed if the human element is not present. This is never the case, if the presence of the investigator applies as, for example, in the Heisenberg uncertainty principle. Scientific uncertainties in the investigation of nonhuman-related environmental problems, such as the causes of continental drift, recurrent ice ages, and extinction of dinosaurs, are at least identifiable, and the known measure of our ignorance is a strong antidote to unwarranted certainty. These physical science investigations are seldom described as "environmental." They are not "problems" within the political context of contemporary human society.

The greater number of problems explicitly described as environmental are problematic because of their relationship to human society. Human behavior is very often the cause or consequence of environmental problems that become social problems and hence problems for politics and policy making. It seems safe to say that most environmental problems of general public concern today are primarily people problems. They carry with them all of the nonscientific factors that characterize human society. They involve relationships and processes where the predictive power of science is weakest and where the known dimensions of uncertainty (scientific or nonscientific) are not only very great, but are often dismissed in pursuit of predetermined policy preferences.

A cursory review of some major, so-called environmental problems will illustrate the difficulty of applying science to problems that are basically human in character and are dynamically interrelating in the actual world. A very large area of interrelated ecologic-sociologic problems encompasses issues of population, economic growth, and environment. Inextricably linked in fact, they have hitherto been generally perceived as separate and until recently have seldom been treated as related. The legacy of the human past has led to an uncritical bias favoring generalized growth, notably of populations and their economies. Population growth tends to drive economic expansion that draws upon all natural resources—land, water, forests, minerals, wildlife—and that produces residuals that in one form or another are returned to the environment, often as "pollutants."

To untangle the scientific certainties and uncertainties in cases involving far-reaching interrelationships, and often strong human emotions, requires a degree of rationality not common in human affairs. Scientists of high repute are not by that fact immune to unwarranted assumptions of certainty on many of these complex issues. For example, some demographers have declared that the Earth could support several times its present population. They have done so, however, on optimistic statistical estimates of the ability of the environment to produce the food and water necessary to sustain the numbers. The ecologic, cultural, and political costs (and limitations) of the "heroic" measures necessary to sustain a world population of 20 billion humans at more than a minimal level of existence is discounted. When popular demands for resources for food, fuel, minerals or lumber exceed supply we speak of shortages, rarely of excessive demand resulting from too many consumers.

Technology is often alleged to make the improbable possible. Then there have been other scientists who see no problem in the amount of food available—the problem lying in its inequitable distribution. A related case of pseudoscientific certainty has been common among scientists in agriculture and nutrition in response to the alarms of economists and politicians about so-called food shortages. Agronomists and agricultural chemists have given humanity the double-edged miracle of the Green Revolution. Population growth has been taken as given, and so the problem is defined as too little food, not too many mouths. Plant geneticists focused on increasing yields rarely express uncertainty about possible environmental costs. Certain of the efficacy of means for sustaining population growth, the Green Revolutionists have promoted a no-win cycle of continuing need for more food. Collateral ecologic uncertainties and social costs have been underestimated or ignored (79–81). The probabilistic nature of science and the admission of uncertainties in issues relating to energy policy has led to a discounting by economists and politicians of science-based propositions in which there is substantial certainty of trends and consequences but uncertainty regarding timing and distribution of effects. The estimated duration of petroleum reserves and the advent of the actual observable impacts of global climate change are cases in point. Here, honest scientific uncertainty is utilized by status quo vested interests to question scientific reliability. And anticipated energy shortages are rarely related to the encouragement of growth that drives energy demand. Nor is there much public attention given to the linkages between population growth, material consumption, uses of energy, maximum yield agriculture, and environmental degradation. Questions of scientific certainty and uncertainty are lost in the megaproblem of an irrational human demand for unlimited energy and endless material affluence.

Conclusion

Science-based problem-solving analysis has too often been misdirected because unwarranted assurance was placed on preferred strategies of investigation, while uncertainties were either denied or put out of bounds as unresearchable. Implied certainty in the identification and resolution of policy problems has probably led to more flawed policy decisions than have the uncertainties inherent in science. Researchers and their sponsors have too often failed to appreciate the uncertainties inherent both in the definition of a problem and in the methods applied to its solution. In the complex cosmos in which humanity is immersed, some element of uncertainty is irreducible. But given time, teamwork, and all relevant information, on some issues a close approach to sufficient certainty may sometimes be approximated. If clarification and best-estimated solutions to problems confronting policy makers are sincerely sought, approximations qualified by estimated margins for error may be sufficient for arriving at reasonable and sustainable policy decisions.

This discussion of concepts and methods of science applied to problems of resource and environmental policy analysis has emphasized the task of problem definition. Its argument is that perception of policy issues is vulnerable for many reasons to the error of parallax—for example, to misplaced or biased assumptions regarding policies that affect natural resources and environment. These areas of inquiry were selected not as unique, but in which the risks of misdefining the issues were high, and in which large numbers of people could be significantly affected by policy outcomes. The risk and frequency of misdirection in social science research is surely greater than in the physical sciences, its biases being more numerous and its data less demonstrably valid. There is less excuse for misdirected efforts in those fields of science in which assumptions may be tested by demonstrably valid evidence.

This reasoning leads to the conclusion that problem-focused, multidisciplinary teamwork, pulled together and into common effort by an individual or small group with synthesizing capabilities, is needed to assist decision makers, whoever they might be. But modern society has not yet seriously addressed the need to set up institutional arrangements to cope with problems that it has created for itself. We therefore arrive at the unhappy conclusion that policy-relevant research often falls short of providing adequate policy guidance because the problems as popularly perceived do not adequately comprehend the circumstances that are creating the problems. The types of intellectual capability and institutional arrangements needed to cope with modern societies' burgeoning difficulties have yet to be developed. The circumstances and especially the vulnerabilities of present society indicate the need for higher priority and increased investment in policy-relevant research. The requisites for valid policy research may be obtainable over time, but our society and its political leadership have not yet asked that they be provided.

Under ideal circumstances (rarely encountered), the public, politicians, and scientists share a common understanding of the limits of scientific certainty and an appreciation of the risks involved in any decision. The weighing of real risks, costs, and benefits by mutually accepted methods and standards could lead to more achievable or sustainable solutions. For many reasons, opinions expressed by scientists on a particular issue may differ, sometimes absolutely but more often on details. For this reason public understanding of the weight of evidence and of relative risk is necessary in any effort to deal with resource or environmental problems in which science is invoked. Application of science to resource and environmental problems undertaken either in or out of the public forum will almost always have policy implications. Rational environmental problem solving (i.e., solutions that are effective) is not likely to be attained unless the parties involved share a common understanding of assumptions that are defensible by reference to empirical evidence.

References

1. Morgan MG, Hension M. Uncertainty: a guide to dealing with uncertainty in quantitative risk and policy analysis. Cambridge, UK: Cambridge University Press, 1990.
2. Funtowicz S, Ravetz JR. Uncertainty and quality in science for policy. Dordrecht, Netherlands: Kluwer, 1990.
3. Salter L. Mandated science: science and scientists in the making of standards. Dordrecht, Netherlands: Kluwer, 1988.
4. Jasanoff S. The fifth branch: science advisors as policymakers. Cambridge, MA: Harvard University Press, 1990:241–250.
5. Brooks LA, Van Deveer S, eds. Compelling knowledge: from science to certainty in public decisions. In: Saving the seas: science, values, and international governance.
6. Majone G. Evidence, argument and persuasion in the policy process. New Haven, CT: Yale University Press, 1989.
7. Schiff AL. Fire and water: scientific heresy in the forest service. Cambridge, MA: Harvard University Press, 1962.
8. De Marchi B. Uncertainty in environmental emergencies: a diagnostic tool. J Contingen Crisis Mgt (submitted).

9. Borch RH. The economics of uncertainty. Princeton, NJ: Princeton University Press, 1968.

10. Tversky A, Kahneman D. Judgment under uncertainty: heuristic and biases. Science 1974; 185:1124–1131.

10a. Price WC, Chissick S, eds. The uncertainty principle and the foundations of quantum mechanics: a fifty years' survery (a tribute to Prof. Wernner Heisenberg). New York: Wiley, 1977.

11. Heineman RA, Bluhm WT, Peterson SA, Kearny EN. The world of the policy analyst: rationality, values and politics. Chatham, NJ: Chatham House, 1990.

12. Jones CO. An introduction to the study of public policy. Second edition. North Scituate, MA: Duxbury Press, 1977.

13. Nagel SS, ed. Improving policy analysis. Beverly Hills, CA: Sage, 1980.

14. Simeon R. Studying public policy. Can J Pol Sci 1976;9:548–580.

15. Stokey E, Zeckhauser R. A primer for policy analysis. New York: Norton, 1978.

16. Walker WE. Public policy analysis: a partnership between analysts and policymakers. Santa Monica, CA: Rand Corporation, 1978 [Rand Paper Series, P-6074].

17. Rhodes RAW. Public administration and policy analysis. Westmead, UK: Saxon House, Teakfield, 1979.

18. Wildavsky A. Speaking truth to power. Boston: Little, Brown, 1979.

19. Collingridge D, Reeve C. Science speaks to power. New York: St Martin's Press, 1986.

20. Cook TD, Campbell DT. The design and conduct of quasi-experiments and true experiments in field settings. In: Handbook of industrial and organizational theory. Chicago: Rand McNally, 1980:223–326.

21. Weiskel TC, Gray RA. Environmental decline and public policy: pattern, trend, and prospect. Ann Arbor, MI: Pierian Press, 1992. (contains extensive annotated bibliography)

22. Bartlett RV. The reserve mining controversy: science, technology and environmental quality (with introduction and concluding observations by Lynton K Caldwell). Bloomington, IN: Indiana University Press, 1980.

23. O'Brien DM. What process is due? Courts and science policy disputes. New York: Russell Sage Foundation, 1987.

24. Merritt GJ. The reserve mining case—20 years later. Focus Int Joint Commission Activities 1994;19, issue 3 (Nov–Dec):1–4.

25. Landau M. Proper domain of policy analysis. Am J Pol Sci 21 May:423–427.

26. Morgenstern O. On the accuracy of economic observation. Second edition revised. Princeton, NJ: Princeton University Press, 1963.

27. Simon JL. Resources, population, environment: an oversupply of false bad news. Science 1980;208:1431–1437.

28. Teresi R. Physics is history. *The New York Times* OP-ED, June 11, 1994.

29. Barrow J. Pi in the sky. Oxford, UK: Oxford University Press, 1992.

30. Hunter E, ed. A problem of parallax: concepts and methods in policy analysis relating to resources and environment. In: Proc Conf Frontiers in Applied Political Science. Tempe, AZ: Arizona State University, Department of Political Science, 1980:139–160.

31. Platt JR. Man and the indeterminacies. In: The step to man. New York: Wiley, 141–155.

32. Shrader-Frechette K. Unsafe at any depth: geological methods, subjective judgments and nuclear waste disposal. In: Gould CC, Cohen RS, eds. Artifacts representations and social practice: essays for Marx Wartofsky. Dordrecht Netherlands: Kluwer, 1994:501–524.

33. Kates RW. Risk assessment of environmental hazard. New York: Wiley, 1978.

34. Lowrence WW. Of acceptable risk: science and the determination of safety. Los Altos, CA: William Kaufman, 1976.

35. Okrent D. Comment on societal risk. Science 1980;208:372–375.

36. Slovic P, Fischhoff B, Lichtenstein S. Rating the risks. Environment 1979;21:14–20, 36–39.

37. Perrow C. Normal accidents: living with high risk technologies. New York: Basic Books, 1984.

38. Simon HA. Models of man: social and rational. New York: Wiley, 1957.

39. Bartlett RV. Rationality in administrative behavior: Simon, science and public administration. Pub Admin Quart 1988;12:301–318.

40. Bartlett RV. Ecological rationality: reason and environmental policy. Environ Ethics 1986; 8:221–239.

41. Ball T. From paradigms to research programs: toward a post-Kuhnian political science. Am J Pol Sci 1976;20:151–177.

42. Dryzek JS. Rational ecology environment and political economy. Oxford, UK: Basil Blackwell, 1987.

43. Miles RE Jr. The origin and meaning of Mile's law. Pub Admin Rev 1978;38:399–403.

44. Forrester JW. Counter intuitive behavior of social systems. Technol Rev 1971;73:53–68.

45. Wahlke JC. Pre-behavioralism in political science. Presidential Address, American Political Science Association, 1978. Am Pol Sci Rev 1979;73:9–31.

46. Caldwell LK. The contextual basis for environmental decisionmaking: assumptions are predeterminants of choice. Environ Prof 1987;9:302–308.

47. Charlesworth JC, ed. Mathematics and the social sciences: the utility and inutility of mathematics in the study of economics, political science and sociology. A symposium, sponsored by the American Academy of Political and Social Science. Philadelphia: The American Academy of Political and Social Science, 1963.

48. Spulber N, Horowitz I. Quantitative economic policy and planning. New York: Norton, 1976.

49. Travis LE. Observing how humans make mistakes to discover how to get computers to do likewise. Santa Monica, CA: System Development Corp. [S.P. Series 776], 1962.

50. Morgan T. The theory of error in centrally-directed economic systems. Quart J Econ 1964; 78:395–499.

51. Landau M. On the concept of a self-correcting organization. Pub Admin Rev 1973;33:533–552.

52. Metlay D. Error correction in bureaucracy. Unpublished Ph.D. dissertation, University of California at Berkeley, 1978.

53. Greenberger M, Crenson MA, Crissey BL. Models in the policy process: public decision making in the computer era. New York: Russell Sage (Basic Books, Distributor), 1976.

54. Richardson J. Models of reality: shaping thought and action. Mt. Airy, MD: Lomond, 1984.

55. Harris H, ed. Scientific models and man: the Herbert Spencer lectures delivered at the University of Oxford, 1976. Oxford, UK: Clarendon Press, 1979.

56. Phillips DL. Abandoning method: sociological studies in methodology. San Francisco: Jossey Bass, 1973 (and items in the appended bibliography).

57. Coopman TC. Measurement without theory. Rev Econ Statist 1947;29:161–172.

58. Caldwell LK. An agenda for research. In: Moll KD, ed. Research to anticipate environmental impacts of changing resource usage: proceedings of the Stanford Research Institute symposium. Menlo Park, CA: Stanford Research Institute, 1975:89–118.

59. Patton MQ. Qualitative evaluation methods. Beverly Hills, CA: Sage, 1980.

60. Patton MQ. Utilization-focused evaluation. Beverly Hills, CA: Sage, 1978.

61. Lindblom C. The policy making process. Englewood Cliffs, NJ: Prentice-Hall, 1968.

62. Ripley RB, Franklin GA. Congress, the bureaucracy and public policy. Homewood, IL: Dorsey, 1980.

63. Woll P. Public policy. Cambridge, MA: Winthrop, 1974.

64. Council on Environmental Quality. National environmental policy act: implementation of procedural provisions: final regulations. Federal Register 42, No. 230 (Wednesday, November 29, 1978, Part IV), 55982.

65. Holling CS, ed. Adaptive environmental assessment and management. Laxenberg, Austria: International Institute for Applied Systems Analysis, 1978.

66. March JG. Science, politics and Mrs. Gruenberg. The National Research Council: Current issues and studies. Washington, DC: National Academy of Sciences, 1979:28.

67. Bryson L. Notes on a theory of advice. In: Freedom and authority in our time: twelfth symposium on science, philosophy and religion. New York: Harper, 1953:27–44.

68. SIGMA XI. Science and public policy linking users and producers. Research Triangle Park, NC: SIGMA XI, 1993.

69. Sears PB. Human ecology: a problem in synthesis. Science 1954;10:959–963.

70. Bradley MD. Decision-making for environmental resource management. J Environ Mgt 1973;1:289–302.

71. Peter G. Peterson in an address to the National Environmental Information Symposium of the Environmental Protection Agency. Reported in *The National Observer* (week ending December 2, 1972).

72. PL 91-190 (1970); 83 STAT. 852; 42 U.S.C. S 4332(2)(B).

73. Brdyaev N. The fate of man in the modern world. Translated from the Russian by Donald A. Lowrie. London: SCM Press, 1935:78.

74. Hardin G, Baden J, eds. Managing the commons. San Francisco: W.H. Freeman, 1977.

75. Baden J, ed. Earth day reconsidered. Washington, DC: Heritage Foundation, 1980.

76. Melsner AJ. Creating a policy analysis profession. In: Nagel SS, ed. Improving policy analysis. Beverly Hills, CA: Sage, 1980:234–244.

77. Cellarius R, Platt J. Councils of urgent studies. Science 1969;166:115–1121.

78. Institutions for effective environmental management: report of the environmental study group to the environmental study board of the national academy of sciences, national academy of engineering. Part 1. Washington DC: National Academy of Sciences, 1970.

79. Dahlberg KA. Beyond the green revolution. New York: Plenum Press, 1979.

80. Kendall HW, Pimentel D. Constraints on the expansion of global food supply. Ambio 1994; 23/3:198–205.

81. Polunin N, Nazim M, eds. Population and global security. Geneva: Foundation for Environmental Education, 1994.

Index